MYOBLAST TRANSFER THERAPY

ADVANCES IN EXPERIMENTAL MEDICINE AND BIOLOGY

MYOBLAST TRANSFER THERAPY

Edited by

Robert C. Griggs

University of Rochester
Rochester, New York

and

George Karpati

McGill University
Montreal, Quebec, Canada

PLENUM PRESS • NEW YORK AND LONDON

Proceedings of a Muscular Dystrophy Association International
Conference on Myoblast Transfer Therapy, held June 10–12, 1989,
in New York, New York

ISBN-13: 978-1-4684-5867-1   e-ISBN-13: 978-1-4684-5865-7
DOI: 10.1007/978-1-4684-5865-7

FOREWORD

I am pleased to introduce this volume on Myoblast Transfer Therapy on behalf of the Muscular Dystrophy Association and all of its Advisory Committees. The international conference which led to this volume brought together leading basic scientists and clinical investigators for the purpose of coordinating the development of this new field in the fight against muscular dystrophy.

The Muscular Dystrophy Association is the nation's most rapidly growing voluntary health agency in terms of its programs of patient care, research, and professional and public education. Success is attributable to its National Chairman, Jerry Lewis, to its effective corporate membership, and to the many physicians and scientists who give their time freely to advise on policies, to review grant applications, and to participate in meetings such as this.

I should like to acknowledge a large number of other individuals to whom we are indebted: the broad segment of the American public which continually and generously supports our spectrum of services. The Muscular Dystrophy Association, next year, should raise in excess of $115,000,000. These contributions are derived from more than 10 million American families. These families are not only pledging their money but expressing their hopes that we will find answers to the tragic problem of neuromuscular disease.

We are confident that the fruits of this meeting will move the frontier of research forward toward that goal.

Leon Charash
Chairman, Medical Advisory Committee
Muscular Dystrophy Association

v

PREFACE

The cloning of the Duchenne muscular dystrophy gene and the identification of its protein product, dystrophin, were momentous discoveries. However, it has become clear that replacing the missing gene product will not be an easy task since dystrophin appears to be a cytoskeletal protein. Nevertheless, by exploiting the unique characteristics of skeletal muscle fibers and their progenitor cells, investigators have been able to transform dystrophin-negative skeletal muscle fibers in a model of Duchenne muscular dystrophy (the mdx mouse) to dystrophin-positive fibers by normal myoblast transfer. This exciting observation has raised hopes that a similar approach may be beneficial in Duchenne muscular dystrophy. This technique has been termed myoblast transfer and holds promise in other genetically determined muscle diseases. The importance of fostering this research and exploring its possible clinical application was the impetus for an International Conference on Myoblast Transfer Therapy held in June, 1989. The goal of the conference was to provide a forum to discuss the many important investigative questions and clinical issues that arise before one contemplates the use of myoblast transfer in patients.

This volume is organized into sections that review topics that are relevant to myoblast transfer: the molecular biology of myogenesis and regeneration; the antigenicity of myoblasts; the practical aspects of myoblast implantation; in situ fusion, defining nuclear domains and the migration of mRNA and protein; the development of cultures of myoblasts; and the best ways of monitoring clinical success and the transformation of the dystrophic phenotype. The discussions which followed the papers have been included since they focus attention on many of the controversies and unresolved questions in myoblast biology and transfer experiments. We have also provided a brief synopsis of the issues considered during a workshop held in conjunction with the International Conference. This workshop addressed clinical issues in the implementation of myoblast transfer and brought together those groups with specific plans to develop clinical strategies for using myoblast transfer in patients.

We wish to thank the Muscular Dystrophy Association for its support of the International Conference and this volume. Even more importantly, the MDA has provided much of the impetus and research support which has lead to the development of this exciting new area of research and brought it to the point of being considered for the treatment of patients with muscular dystrophy.   We would also like to thank the many individuals who were involved with the planning of the scientific program, particularly the session chairmen: Drs. Thomas Caskey, Andrew Engel, Terry Partridge, Donald Fischman, Richard Strohman, Louis Kunkel and Jerry Mendell.

Robert C. Griggs
George Karpati

CONTENTS

5 DEVELOPMENT OF MYOGENIC CELL CULTURES

6 MONITORING CLINICAL SUCCESS: PHENOTYPIC TRANSFORMATION

7   IMPLEMENTATION OF HUMAN TRIALS

SECTION 1

MOLECULAR BIOLOGY OF MYOGENESIS AND REGENERATION

MyoD: A REGULATORY GENE OF SKELETAL MYOGENESIS

Stephen J. Tapscott, Robert L. Davis,
Andrew B. Lassar and Harold Weintraub

Fred Hutchinson Cancer Research Center
1124 Columbia St.
Seattle, WA 98104

MyoD is expressed only in skeletal muscle (1), and could be considered a master regulatory gene of myogenesis. When expressed in non-muscle cells, MyoD converts those cells to myoblasts (2). Specifically, the mouse fibroblast cell line 10T1/2 does not express either MyoD or any of several muscle specific genes. Forced expression of MyoD in 10T1/2 cells converts this cell to a myoblast: In growth medium the cell divides and expresses MyoD and other muscle lineage markers, but does not express myosin, desmin, or other markers of muscle terminal differentiation; in differentiation medium the cell withdraws from the cell cycle and initiates expression of desmin, myosin and other proteins of terminal differentiation (1). In addition to activating muscle structural genes, MyoD also positively regulates its own transcription, as well as myogenin, a related muscle regulatory protein (3). This positive autoregulatory circuit could be a mechanism for amplifying expression of muscle regulatory genes and might serve to maintain expression of these genes once a cell has become committed to the myoblast lineage.

Large populations of fibroblasts, or other cells, can be recruited to the myoblast lineage using a retroviral vector engineered to express both MyoD and a selectable marker (2). In addition to fibroblasts, numerous other cell types can be converted to muscle, including the adipoblast cell line TA1, the hepatoma cell line BNL, the neuroblastoma line B50, the melanoma cell line B16, and the teratocarcinoma cell line P19 (1,2).

MyoD protein is localized to the cell nucleus in interphase cells (4). A mutational analysis of the MyoD protein identifies a region of the protein that is both necessary and sufficient for its biological activity (4). This 68 amino acid region is conserved in a large family of proteins that include myc (5), E12/E47 (6), myogenin (7), myf-5 (8), daughterless (9) and transcripts from the achaete-scute complex (10). Sequence analysis reveals that this region has two stretches of amino acids that are capable of forming

Myoblast Transfer Therapy
Edited by R. Griggs and G. Karpati
Plenum Press, New York, 1990

amphipathic alpha helices separated by a short non-helical region, referred to as a helix-loop-helix motif (6). Immediately amino-terminal to the helix-loop-helix is a region rich in basic amino acids. Site directed mutagenesis together with in vitro gel shift assays and immunoprecipitations demonstrates that the potential helical domains mediate protein-protein interactions among members of the helix-loop-helix family, mediating both homodimer and heterodimer formation, and the basic region mediates sequence specific DNA binding (11,12,13).

How can MyoD be used for diagnosis and treatment in muscular dystrophy? One possibility would be in diagnosis. Cells from a patient, or from amniocentesis, could be transfected with a MyoD expression vehicle and the expression of specific muscle products analyzed, e.g., dystrophin, either by western or immunohistochemical analysis. A second possibility would be in treatment, specifically in myoblast transfer therapy. Large populations of donor fibroblasts could be converted to the muscle lineage by infection with a MyoD expressing retrovirus. Finally, it remains unclear what effect the lack of dystrophin expression in Duchenne's muscular dystrophy might have on the expression and stability of MyoD or related regulatory proteins.

REFERENCES

1. R.D. Davis, H. Weintraub and A.B. Lassar. 1987. Expression of a single transfected cDNA converts fibroblasts to myoblasts (1987). Cell, 51: 987-1000.

2. H. Weintraub et al. 1989. Activation of muscle specific genes in pigment, nerve, fat, liver and fibroblast cell lines by forced expression of MyoD. PNAS, 86: 5434-5438.

3. M.J. Thayer et al. 1989. Positive autoregulation of the myogenic determination gene MyoD1. Cell, 58: 241-248.

4. S.J. Tapscott et al. 1988. MyoD1: a nuclear phosphoprotein requiring a myc homology region to convert fibroblasts to myoblasts. Science, 242: 405-411.

5. F.W. Alt, R.A. DePinho, K. Zimmerman, E. Legouy, K. Hutton, P. Ferrier, A. Tesfaye, G.D. Yoncopoulos and P. Nisen 1986. The human myc-gene family. Cold Spring Harbor Symp. Quant. Biol., 51: 931-941.

6. C. Murre, P.S. McCaw and D. Baltimore. 1989. A new DNA binding and dimerization motif in immunoglobulin enhancer binding, daughterless, MyoD, and myc proteins. Cell, 56: 777-783.

7. W.E. Wright, D.A. Sassoon and V.K. Lin. 1989. Myogenin, a factor regulating myogenesis has a domain homologous to MyoD. Cell, 56: 607-617.

8. T. Braun, G. Buschhausen-Denker, E. Bober and H.H. Arnold. 1989. A novel human muscle factor related to but distinct from MyoD1 induces myogenic conversion in 10T1/2 fibroblasts. EMBO J., 8: 701-709.

9. M. Caudy et al. 1988. Daughterless, a drosophila gene essential for both neurogenesis and sex determination, has sequence similarities to myc and the achaete-scute complex. Cell, 55: 1061-1067.

10. R. Villares and C.V. Cabrera. 1987. The achaete-scute gene complex of D. melanogaster: conserved domains in a subset of genes required for neurogenesis and their homology to myc. Cell, 50: 415-424.

11. A.B. Lassar et al. 1989. MyoD is a sequence specific binding protein requiring a region of myc homology to bind to the muscle creatine kinase enhancer. Cell, 58: 823-831.

12. C. Murre et al. 1989. Interactions between heterologous helix-loop-helix proteins generate complexes that bind specifically to a common DNA sequence. Cell, 58: 537-544.

13. R.L. Davis, P.-F. Cheng, A.B. Lassar and H. Weintraub. 1990. The MyoD DNA binding domain contains a recognition code for muscle-specific gene activation. Cell, 60: 733-746.

Discussion of Dr. Tapscott's paper

Dr. Caskey: We can take comments and questions from the floor.

Dr. Blau: How general a phenomenon is activation, as a regulatory mechanism? Do you see transfected myo-D activation of endogenous myo-D in cells other than 10T1/2?

Dr. Tapscott: No. It is not seen in all cell types.

Dr. Blau: Is it seen in any cell types other than 10T1/2?

Dr. Tapscott: There are mouse fibroblast cell lines that undergo autoactivation, but it is not necessarily a general phenomenon. And I actually don't have data for primary fibroblasts.

Dr. Blau: But for the majority of cell types that you showed, then, in which muscle genes are activated, the endogenous gene is not activated?

Dr. Tapscott: We have not assayed for the endogenous gene in the majority of cell types.

Dr. Strohman: In light of what you just said about the over-expression of genes, could you comment in the context of the use of transformed cells as donor cells in transplant therapy? In other words, is there any evidence that these cells will be regulated in response to in vivo conditions (e.g. in response to nerve)? Will they mature?

Dr. Emerson: In a practical context, I don't think there's enough known yet. But I would say that with transfection of myo-D into other cell types there is a difference between generating a stem cell, or a cell which can proliferate, versus generating a cell which is now differentiated into muscle.

And with adipocyte transfection, it is clear that the phenotype which results is not a myoblast but a sort of mixture in the sense that it is a colony forming cell. If one would take a nerve (although this is not my data), apparently you can activate differentiation but not necessarily the colony formation: the active part of the program. There is a subtle but important distinction in my mind as to whether one is talking about the generation of a cell which is proliferative and has the potential to differentiate in response to growth factors or other signals versus a cell which is already differentiated.

And with the process of transfection, in which myo-D, for instance, can lead to both phenotypes in 10T1/2 cells. But the phenotypes of other cell types may be very complex. For fat cells, there's a mixed proliferative phenotype. Dr. Tapscott may want to comment.

Dr. Tapscott: Following up on your comment, our experience with fat cells is that if you look at an individual cell, even though a colony may have a mixed phenotype, an individual cell rarely expresses the characteristics of both fat cells: as an example, a muscle protein such as myosin heavy chain. And so within a given cell that differentiates, there seems to be a choice between the two differentiation programs. And that is interesting in the sense that fat cells and muscle cells, as mesodermal derivatives, are fairly closely related. On the other hand, in the case of melanoma cells, there seems to be an ability to co-express both programs, where both pigment granules as well as myosin heavy chain can be expressed within the same cell.

Dr. Moxley: A question for Dr. Tapscott. I wonder if you could speculate further on serum factors that might influence myo-D? Are we talking about IGF-1 or other growth factors?

Dr. Tapscott: We probably are. I can only give you a brief piece of data. For example, TGF Beta, while it inhibits myo-D formation, has only modest effects on myo-D protein expression. It's not to say it doesn't affect the modification and activity of myo-D. But I think it's going to be complex, both with such growth factors as well as possibly other factors of the myc-related family that may also be regulated by serum factors.

MYOBLASTS, SATELLITE CELLS, AND MYOBLAST TRANSFER

Frank E. Stockdale, E. Janet Hager, Susan E. Fernyak
and Joseph X. DiMario

Stanford School of Medicine
Stanford, California 94305-5306

The prospects for introducing "foreign" nuclei through the cell fusion process have been considered since it first became clear that skeletal muscle fibers form by fusion of many mononucleated myoblasts rather than the proliferation of skeletal muscle nuclei within fibers (Stockdale and Holtzer, 1961). While cell fusion is cell type specific, there is no restriction in myoblast fusion across species, as skeletal myogenic cells from different species or classes of vertebrates are capable of adding nuclei to forming fibers (Yaffe and Feldman, 1965). The introduction of nuclei by PEG-mediated cell fusion has made it feasible to introduce a limited number of nuclei of non-myogenic cells into fibers in cell culture (Blau, 1983). However, specificity of cell recognition and fusion limits the introduction of new nuclei into existing fibers or into newly forming fibers to those cells that have been committed to a myogenic fate. The prospects of converting non-myogenic cells to myogenic cells by the transfection of commitment or determination genes broadens the repertoire of cells that could be used for contributing nuclei to muscle fibers using normal fusion mechanisms (Konieczny et al., 1986; Lassar et al., 1986; Weintraub et al., 1989).

Cells that are capable of transferring nuclei to already formed or forming fibers in any significant numbers are those committed to a myogenic fate. Thus it becomes very important to consider which myoblasts should be used in transfer and whether there is diversity within the myoblast populations that are available for forming fibers. Recent analyses show that fibers of differing properties are formed at discrete developmental periods during muscle formation and growth (Stockdale, 1989; Stockdale et al., 1989). The so-called primary fibers of the embryonic musculature are those formed at the inception of anatomic muscle formation (Kelly and Zacks, 1969). The myoblasts available for primary fiber formation were shown to differ from myoblasts at other stages of muscle formation and among themselves on a variety of parameters (Stockdale, 1987). Secondary fibers appear throughout fetal development, and in some classes of vertebrates perhaps briefly in the neonatal period of life. Likewise the myoblasts or satellite cells isolated from skeletal muscles capable of repair in the adult also have different properties than those associated with primary and secondary fiber formation. Each of these fiber categories, primary, secondary, and reparative, have their origin in myoblasts that are diverse in their properties.

Particularly important in considerations of the contribution of foreign nuclei or nuclei from different developmental time periods to a forming or existing fiber, is any restraint, or facilitation, that such nuclei may place on the properties of fibers. In the chicken, it has been shown that the fibers formed from a subset of myoblasts found in the embryonic limb bud musculature at the time of primary fiber formation, produce new fibers in cell culture that synthesize different isoforms of myosin heavy chain than do fibers formed from

Myoblast Transfer Therapy
Edited by R. Griggs and G. Karpati
Plenum Press, New York, 1990

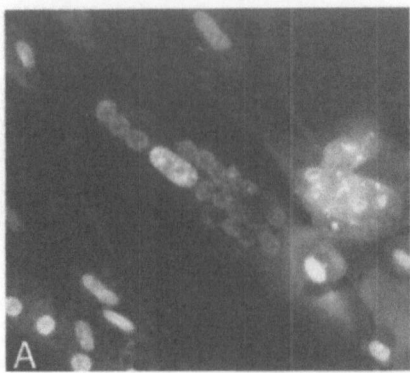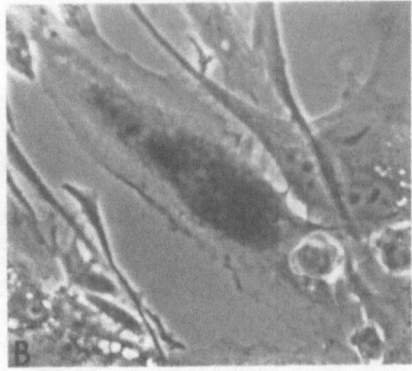

Fig. 1.   Culture of mouse C2 myogenic cells infected with ß-gal-nuclear localizer
sequence mixed with quail myoblasts from 12 day limb muscles of a fetal quail.
A) Stained with the Hoecht stain. The mouse nuclei are apparent because of size
and the intense fluorescent staining. B) Stained with X-gal. The quail nuclei are
stained though they lack the ß-gal fusion gene.

myoblasts (satellite cells) isolated from the same muscles in fetal or adult life (Feldman and
Stockdale, 1988; Miller et al., 1985; Miller and Stockdale, 1986a; Miller and Stockdale,
1986b; Miller and Stockdale, 1989; Schafer et al., 1987; Stockdale and Miller, 1987;
Stockdale et al., 1986). How important is the selection of myoblast type for myoblast
transfer, or the committed state of a nucleus from a myoblast when it is added to a forming
or reparative fiber?

Fibers of either a homotypic or a heterotypic type can be used to answer to this
question and assess fiber formation and gene activation. Either homotypic nucleation,
where all the nuclei in a fiber are of the same myogenic type, or heterotypic nucleation,
where fiber nuclei come from different muscle cell types, different cell types, or different
species, can be used for analysis. The work of Blau and associates *in vitro* (Blau et al.,
1983; Blau et al., 1985; Pavlath and Blau, 1986; Pavlath et al., 1989; Silberstein et al.,
1986), and Mintz and associates *in vivo* (Mintz and Baker, 1967), are examples of
heterotypic and homotypic fiber nucleation. They found that nuclei of various types can be
incorporated either by normal or facilitated fusion mechanisms and that genes specific to the
"foreign" nucleus were expressed. Similarly, the fusion of homotypic myogenic cells,
some of which carry mutations that produce variants of muscle specific proteins, has
shown that the "wild type" and the mutant type nuclei can be active within the same fiber
both in cell culture and *in vivo* (Partridge et al., 1989; Ralston and Hall, 1989).

New approaches in cellular biology utilizing molecular genetic techniques, permit
additional avenues for analysis of the contributions of various myoblast types to fibers
either *in vivo* or *in vitro* by myoblast transfer. Studies of myoblast transfer *in vitro* as well
as *in vivo* have been fostered by the ability to infect myoblasts with retroviruses that carry
reporter genes. Myoblasts marked with the E. coli ß-galactosidase (ß-gal) gene product
demonstrate the feasibility of identifying nuclei or nuclear gene activity within newly
formed or growing fibers (Nicolas and Bonnerot, 1987; Sanes, 1989). An example of this
type of analysis can be seen in the work of Ralston and Hall (1989) and from our
laboratory (Fig. 1). In these studies, C2 mouse myoblasts were infected with a retrovirus
containing the E. coli ß-galactosidase gene fused to a nuclear localizing sequence (NLS).
When this gene was active within C2 nuclei, active nuclei and adjacent inactive nuclei were
detected by staining of the cells with X-gal. C2 cells, infected with the ß-galactosidase
NLS retrovirus, were mixed with quail fetal myoblasts in cell culture (Fig. 1). The mouse
myoblasts with the nuclear marker fused with quail cells to produce heterotypic nuclear
fibers. Presumably, the mRNA from the ß-gal gene entered the cytoplasm, and was
translated into the fusion ß-galactosidase enzyme-NLS protein which has the potential to
return not only to the nucleus of origin (the mouse nucleus) but to the heterotypic quail
nuclei as well. There is a gradient in the distribution of the fusion protein that is

reciprocally related to the distance from the quail nuclei - as the distance increases less ß-galactosidase NLS protein is detectable in quail nuclei (Fig. 1). This has been observed, as well, in homotypic nucleation of myotubes by Ralston and Hall (1989).

Studies of heterotypic nucleation of fibers serve as model systems with which to answer questions important to myoblast transfer. For example, the experiment described in Fig. 1 is a model of how transacting proteins may move within a polynucleated cell such as the muscle fiber. The transacting factors responsible for muscle gene activation during normal development are thought to move within and between nuclei. The distance a putative regulatory protein can traverse becomes important in a fiber containing different types of nuclei. In experiments with mouse C2 and quail myoblasts, the C2 nucleus is the only nucleus within the polykaryon fiber with a gene that encodes a protein that can travel from one nucleus to another through the cytoplasm. Similar analyses and considerations of activation at much shorter distances in heterotypic polykaryons have been performed and discussed (Blau, 1988; Hall and Ralston, 1989) and were foreshadowed by the work of Rao and Johnson (1970) on the initiation of DNA synthesis in polykaryon cells.

The introduction of heterotypic nuclei into fibers poses two questions. Do heterotypic nuclei "entrain" homotypic nuclei by transacting factors in a heterotypically nucleated muscle fiber? What is the cytoplasmic distribution of muscle-specific products relative to the nucleus actively transcribing the mRNA for that product? Work on these questions demonstrates that some muscle-specific products are remarkably localized with regard to the active nucleus (Salviati et al., 1986; Pavlath et al., 1989; Ralston and Hall, 1989). Other analyses show that some products are widely dispersed within the cell even though most nuclei remain inactive with regard to transcribing that gene product. Hall and Ralston (1989) have analyzed the existing data on nuclear domains and provided a framework for an explanation of the demonstrated differences .

Nuclear entrainment and nuclear domains are key considerations in myoblast transfers where nuclei are added to already formed fibers. Fig. 2 illustrates the use of the retroviral marked cell technique to investigate the importance of these concepts in normal muscle formation. Here specific myoblast types (for example embryonic myoblasts of the fast/slow type, or satellite cells) can be cloned and infected so they and their progeny can be identified. These myoblasts can be transferred to developing limb buds. When genetically marked quail myoblasts are introduced into chickens embryos, heterotypic nucleation of

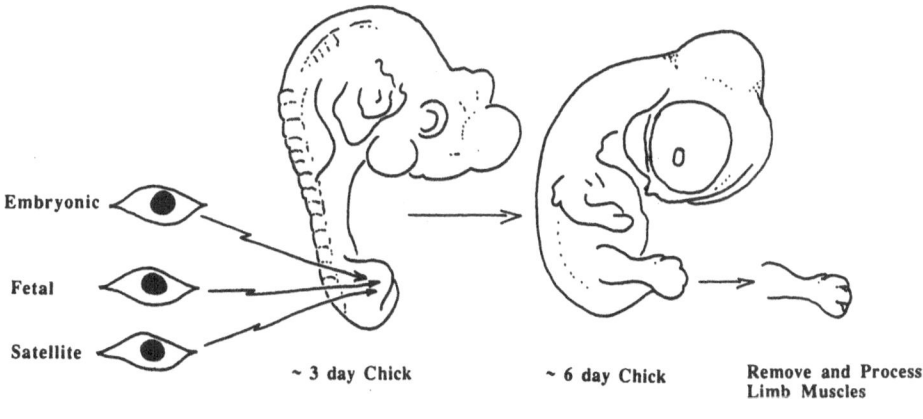

Fig. 2. Diagram of myoblast transfer experiments.. Myoblasts from embryos, fetuses, or adults are infected with ß-gal nuclear localizer sequence fusion gene and cloned. The clonal progeny are injected into the developing limb bud of 3-4 day embryos and the limbs are permitted to form muscle. The injected limbs and control limbs are removed and processed for fiber type and expression of ß-gal.

fibers will occur. By introducing nuclei into the fibers of developing limbs at varying stages of development, it is possible to monitor both nuclear influences on the formation of new fibers, as well as the influence of such nuclei when they enter already existing fibers.

If myoblast transfer is to be used to add additional nuclei to already existing fibers, rather than in the formation of new fibers, a better understanding of nuclear domains is required. For example, where the intent is to introduce a dystrophin gene into a long multinucleated fiber lacking such a gene, nuclear domain questions are germane. Do the heterotypic nuclei express not only the dystrophin gene, but also entrain endogenous nuclei such that other normal myogenic genes that were inactive become activated by the heterotypic nucleus? Differences in the repertoire of active genes between nuclei of different myoblast types become important if some nuclei can entrain or alter the muscle-specific genes of endogenous nuclei. Even if transferred nuclei provide enough transcripts for a product such as dystrophin, some nuclei may be more dominant than others in entrainment of nuclei within a multinucleated fiber thus activating genes for muscle specific functions that provide no functional advantage. For example, there is the possibility that embryonic myoblasts or satellite cells may be a poor source for therapeutic nuclear transfer to correct muscular dystrophy. Although these cells could correct the genetic defect, entrainment by their nuclei could restrict expression of myosin or other muscle specific genes within the resident nuclei of the fiber.

In summary, myoblast transfer as a therapeutic technique requires an understanding of the importance of nuclear contribution to fiber functions beyond correction of just the disease phenotype. Experiments using heterotypic nucleation of fibers will indicate if all nuclei are equivalent in their influence on fiber phenotype, if myoblasts of different developmental origins are equivalent with regard to correction and other fiber functions, and if heterologous nuclei have deleterious effects on the long-term function of a fiber.

References

Blau, H. M., 1983, Cytoplasmic activation of human nuclear genes in stable heterocaryons, Cell, 32:1171.

Blau, H. M., 1988, Hierarchies of regulatory genes may specify mammalian development, Cell, 53:673.

Blau, H. M., Pavlath, G. K., Hardeman, E. C., Chiu, C.-P., Silberstein, L., Webster, S. G., Miller, S. C., and Webster, C., 1985, Plasticity of the differentiated state, Science, 230:758.

Feldman, J. L., and Stockdale, F. E., 1988, Commitment to formation of distinct myotube types in chicken satellite cells, J. Cell. Biochem., 12 C:325.

Hall, Z. W., and Ralston, E., 1989, Nuclear domains in muscle cells, Cell, 59, 771.

Kelly, A. M., and Zacks, S. I., 1969, The histogenesis of rat intercostal muscle, J. Cell Biol., 42:135.

Konieczny, S. F., Baldwin, A. S., and Emerson, C. P., Jr., 1986, Myogenic determination and differentiation of 10T1/2 cell lineages: evidence for a simple genetic regulation system, in: "Molecular Biology of Muscle Development," Alan R. Liss, Inc., New York, 21.

Lassar, A. B., Paterson, B. M., and Weintraub, H., 1986, Transfection of a DNA locus that mediates the conversion of 10T1/2 fibroblasts to myoblasts, Cell, 47:649.

Miller, J. B., and Stockdale, F. E., 1986a, Developmental origins of skeletal muscle fibers: Clonal analysis of myogenic cell lineages based on fast and slow myosin heavy chain expression, Proc. Natl. Acad. Sci. USA, 83:3860.

Miller, J. B., and Stockdale, F. E., 1986b, Developmental regulation of the multiple myogenic cell lineages of the avian embryo, J. Cell Biol., 103:2197.

Miller, J. B., and Stockdale, F. E., 1989, Multiple cellular processes regulate expression of slow myosin heavy chain isoforms during avian myogenesis in vitro, Dev. Biol.,139:393.

Miller, J. B., Crow, M. T., and Stockdale, F. E., 1985, Slow and fast myosin heavy chain content defines three types of myotubes in early muscle cell cultures, J. Cell Biol., 101:1643.

Mintz, B., and Baker, W. W., 1967, Normal mammalian muscle differentiation and gene control of isocitrate dehydrogenase synthesis, Proc. Natl. Acad. Sci. USA, 58:592.

Nicolas, J.-F., and Bonnerot, C., 1987, Recombinant retrovirus, cell lineage and gene expression in the mouse embryo, in: "Cellular Factors in Development and Differentiation - Embryos, Teratocarcinomas and Differentiated Tissues," Alan R. Liss, Inc., New York, 1.

Partridge, T. A., Morgan, J. E., Coulton, G. R., Hoffman, E. P., and Kunkel, L. M., 1989, Conversion of mdx myofibres from dystrophin-negative to -positive by injection of normal myoblasts, Nature, 337:176.

Pavlath, G. K., and Blau, H. M., 1986, Expression of muscle genes in heterokaryons depends on gene dosage, J. Cell Biol., 102:124.

Pavlath, G. K., Rich, K.,.Webster, S. G., and Blau, H. M., 1989, Localization of muscle gene products in nuclear domains, Nature, 337:570.

Ralston, E., and Hall, Z. W., 1989, Transfer of a protein by a single nucleus to nearby nuclei in multinucleated myotubes, Science, 244:1066.

Rao, P.N, and Johnson, R.T., 1970, Mammal cell fusion: Studies on the regulation of DNA synthesis and mitosis, Nature 225:159.

Salviati, G., Biasia, E., and Aloisi, M., 1986, Synthesis of fast myosin induced by fast ectopic innervation of rat soleus muscle is restricted to the ectopic endplate region, Nature, 322:637.

Sanes, J. R., 1989, Analyzing cell lineage with a recombinant retrovirus, Trends Neurosci., 12:21.

Schafer, D. A., Miller, J. B., and Stockdale, F. E., 1987, Cell diversification within the myogenic lineage: In vitro generation of two types of myoblasts from a single myogenic progenitor cell, Cell, 48:659.

Silberstein, L., Webster, M., Travis, M., and Blau, H. M., 1986, Developmental progression of myosin gene expression in cultured muscle cells, Cell, 46:1076.

Stockdale, F. E., and Holtzer, H., 1961, DNA synthesis and myogenesis, Exp. Cell Res., 24:508.

Stockdale, F. E., and Miller, J. B., 1987, The cellular basis of myosin heavy chain isoform expression during development of avian skeletal muscles, Dev. Biol., 123:1.

Stockdale, F. E., 1989, Skeletal muscle fiber specification during development and the myogenic lineage, in: "The Assembly of the Nervous System," L. Landmesser, ed., Alan R. Liss, Inc., New York, 37.

Stockdale, F. E., Miller, J. B., Feldman, J. L., Lamson, G., and Hager, J., 1989, Myogenic cell lineages: Commitment and modulation during differentiation of avian muscle, in: Cellular and Molecular Biology of Muscle Development," L. Kedes and F. E. Stockdale, eds., Alan R. Liss, Inc., New York, 3.

Stockdale, F. E., Miller, J. B., Schafer, D. A., and Crow, M. T., 1986, Myosins, myotubes, and myoblasts. Origins of fast and slow muscle fibers, in: "Molecular Biology of Muscle Development," C. Emerson, D. Fischman, B. Nadal-Ginard, M. A. Q. Siddiqui, eds., Alan R. Liss, Inc., New York, 29:213.

Weintraub, H., Tapscott, S. J., Davis, R. L., Thayer, M. J., Adam, M. A., Lassar, A. B., and Miller, A. D., 1989, Activation of muscle-specific genes in pigment, nerve, fat, liver, and fibroblast cell lines by forced expression of MyoD, Proc. Natl. Acad. Sci. USA, 86:5434.

Yaffe, D., and Feldman, M., 1965, The formation of hybrid multinucleated muscle fibers from myoblasts of different genetic origin, Dev. Biol., 11:300.

REGENERATION OF SKELETAL MUSCLE FIBERS AFTER NECROSIS

Stirling Carpenter

Montreal Neurological Institute
McGill University
Montreal, Quebec

I am going to review the normal process of regeneration after necrosis in skeletal muscle and compare it with what happens in Duchenne dystrophy. Dr. Karpati and I have devised an experimental method for producing a focal lethal injury of skeletal muscle cells in which the sequence of steps can be closely followed[1]. We call it experimental micropuncture injury, because the muscle cells are transfixed by a fine tungsten wire with a 10μm tip. All fibers that are pierced go on to necrosis.

The distinction between necrosis and reaction to nonlethal injury is important. Necrosis means that a cell has passed the point of no return in its efforts to maintain homeostasis, so that it is in the course of becoming debris. Necrosis gives rise to a series of stereotyped reactions, and in skeletal muscle normally it is followed by regeneration mediated by proliferation of mononuclear myoblast-like cells. This is in contrast to nonlethal injury which is followed by various metabolic adjustments in the still viable muscle cells.

In micropuncture injury, necrosis starts focally where the fiber was pierced and then spreads from that spot up and down the elongated muscle fiber. By two hours one can identify necrotic segments. Myofibrils are torn and dense and matted. Z discs disappear. Mitochondria become rounded and show fluffy matrix densities. The plasma membrane is largely absent and T tubules can not be recognized.

Between 2 and 4 hours after puncture, extent of necrosis increases considerably in individual fibers, and one can find long segments in which the Z disc has disappeared but sarcomere order is well preserved. At this time in most fibers no membrane separates the necrotic segment from the surviving stumps on either end. There is however a dense band of contracted myofibrils, in which copious precipitated calcium can be demonstrated, at the junction of a stump and a necrotic segment. This band probably acts as a temporary and not always effective protective barrier for the stump. In most fibers necrotic segments are demarcated by a membrane between 3 and 6 hours after puncture, so that after that time necrosis will not extend further.

In our model neutrophils invaded the muscle around 3 hours. This is not seen in human myopathies. Macrophages appeared at 8 to 10 hours. Progressive phagocytosis of necrotic debris took place. By 72 hours necrotic debris was gone, but macrophages were still present inside the old basal lamina tubes which were now lined by regenerative myoblasts.

Regenerative myoblasts are derived from satellite cells, small mononuclear cells enclosed by the basal lamina of muscle fibers. Schmalbruch has calculated that roughly 4% of the nuclei seen in normal adult muscle are those of satellite cells[2]. They are difficult to identify in cryostat sections by routine staining methods, although they are beautifully outlined by antibodies to N-CAM[3]. Satellite cells serve as a source of nuclei for growing muscle fibers[4]. In immature animals a high percentage of them take up thymidine. They may increase in number in denervated muscle. When there is necrosis of muscle fibers, satellite cells generally survive and give rise to the regenerative myoblasts. They will migrate along muscle fibers towards and into necrotic segments, where they undergo division[5]. From tissue culture evidence, the first division is not before 24 hours[6].

After the regenerative myoblasts have proliferated in the necrotic segments, they fuse longitudinally and transversely with one another and with the surviving stumps. Fusion was in progress at 78 hours in our model, and, by 96 hours, full continuity of muscle fibers had been reestablished across necrotic segments. The regenerated segments were thin but they contained myofibrils.

In Duchenne dystrophy muscle fibers repetitively undergo necrosis and regeneration. Necrosis does not seem to affect regenerated fiber segments until they reach maturity and normal size. Signs of subnecrotic damage are absent with one exception. Reduplicated segments of basal lamina are commonly found around the fibers, and occasional stretches of plasma membrane lack an apposed basal lamina[7].

Dystrophin has been localized to the region of the plasma membrane[8]. From what we know of the configuration of the molecule it is probably restricted to the inner side of the membrane and has no direct attachment to components of the basal lamina. Present theory suggests that it may provide mechanical support to the membrane to help it withstand the stresses induced by contraction.

Another characteristic feature of Duchenne dystrophy is the laying down of collagen between muscle fibers in dense bundles parallel to the fibers[7]. This characteristic feature is probably not merely a reaction to necrosis. Possibly detachment of long stretches of basal lamina is a sufficient stimulus to fibroblasts.

In biopsies from Duchenne dystrophy, regeneration is easy to find, but it does not compensate for necrosis. Schmalbruch has shown, by serial sections of resin embedded biopsies, that many fibers become forked or narrow down to a tiny size or end blindly[9]. These aberrations of regeneration suggest that the satellite cells in Duchenne dystrophy are exhausting their proliferative potential. Lack of sufficient regenerative myoblasts in a necrotic segment leads to incomplete lateral fusion and thus forking, or to failure to reach an adequate girth, or to failure of regeneration. This failure is attested also by the findings on electron microscopy of empty skeins of basal lamina where necrosis took place but regeneration did not reach. The muscle of Duchenne patients is not, however depleted of satellite cells. They are in fact reported as being somewhat increased over normal, but obviously their regenerative capacity is diminished and cannot compensate for the repetitive necrosis[10].

The purpose of this conference is to discuss the possibility introducing new satellite cells into the muscle of Duchenne patients. It should be stressed that this is not in order to provide new energetic satellite cells to regenerate muscle fibers. It is to introduce nuclei with normal dystrophin genes into muscle fibers so that these fibers will be less prone to undergo necrosis.

References

1. S. Carpenter and G. Karpati, Segmental necrosis and its demarcation in experimental micropuncture injury of skeletal muscle fibers, J Neuropathol Exp Neurol, 48:154-170 (1989).
2. H. Schmalbruch, U. Hellhammer, The number of satellite cells in normal human muscle, Anat Rec, 185:279-287 (1976).
3. N.R. Cashman, J. Covault, R.L. Wollman, J.R. Sanes, Neural cell adhesion molecule (N-CAM) in normal, denervated, and myopathic muscle, Ann Neurol, 21:481-489 (1987).
4. P.F. Moss, C.P. Leblond, Satellite cells as the source of nuclei in muscles of grow in rats, Anat Rec, 170:421-435 (1971).
5. F. Schultz, D.L. Jaryszak, C.R. Valliere, Response of Satellite cells to focal skeletal muscle injury, Muscle & Nerve, 8:217-222 (1985).
6. R. Bischoff, Tissue culture studies on the origin of myogenic cells during muscle regeneration in the rat, in "Muscle Regeneration," A. Mauro, ed., Raven Press, NY (1979).
7. S. Carpenter and G. Karpati, "Pathology of Skeletal Muscle," Churchill Livingstone, NY pp.121-129 (1983).
8. E. Zubrzycka-Gaarn, D.E. Bulman, G. Karpati, A.H.M. Burghes, B. Belfall, H.J. Klamut, J. Talbut, R.S. Hodges, P.N. Ray and R.G. Worton, Duchenne muscular dystrophy gene product is localized in the sarcolemma of human skeletal muscle fibers, Nature, 33:466-469 (1988).
9. H. Schmalbruch, Regenerated muscle fibers in Duchenne muscular dystrophy: a serial section study, Neurology, 34:60-65 (1984).
10. Y. Wakayama and D.L. Schotland, Muscle satellite cell population in Duchenne dystrophy, in "Muscle Regeneration," A. Mauro, ed., Raven Press, NY (1979).

DETECTION OF TRUNCATED DYSTROPHIN IN FETAL DMD MYOTUBES

Ieke B. Ginjaar[1,2], Egbert Bakker[1], Johan T. den Dunnen[1], Andy Wessels[2], Marleen M.B. van Paassen[1], Maarten D. Kloosterman[3], Elizabeth E. Zubrzycka-Gaarn[4], Kenneth H. Fischbeck[5], Antoon F.M. Moorman[2] and Gert-Jan B. van Ommen[1]

[1]Dept. of Human Genetics, Sylvius Laboratory 2333 AL Leiden, the Netherlands
[2]Dept. of Anatomy and Embryology, AMC 1105 AZ Amsterdam, the Netherlands
[3]Dept. of Prenatal Diagnostics Hervormd Diaconessenhuis Arnhem, the Netherlands
[4]Hospital for Sick Children, Toronto, Canada
[5]Dept. of Neurology, Hospital of the University of Pennsylvania, Philadelphia, USA

INTRODUCTION

Duchenne muscular dystrophy (DMD) is the most common of all muscular dystrophies (1 in 3000 live male births). It is an X-linked recessive lethal disorder, characterized by a progressive muscle degeneration. The application of DNA technology has made it possible to carry out reliable DNA diagnosis with more than 99% certainty in most familial DMD cases[1,2,3].

Recently the Duchenne gene protein product was detected and dubbed dystrophin[4]. By immunohistochemical studies it was localized to the plasma membrane of normal muscle fibers[5,6,7,8] and possibly the T-tubuli[9]. In muscle biopsies of Duchenne patients no dystrophin could be detected, whereas in BMD patients dystrophin of abnormal size and/or quantity was observed[5,6,11]. The combined results of protein and DNA studies on DMD and BMD patients have led to the hypothesis[10] that the Duchenne phenotype might be caused by an "out of frame" mutation of the gene, producing a truncated and unstable protein. This will probably be recognized as abnormal and degraded by the cell[12]. According to this hypothesis BMD is caused by "in frame" mutations of the gene resulting in a semifunctional protein of altered molecular weight.

We describe here that in a twelve-week old fetus affected with DMD, abnormal dystrophin is initially present indicating that it disappears at a specific later stage. Its detection by NH$_2$-terminal antibodies and not with COOH-terminal antibodies shows that immunohistochemical analysis of DMD and BMD fetuses should be interpreted with great care.

MATERIALS AND METHODS

We have studied the leg muscles of several twelve-week old normal fetuses obtained by aspiration. Intact legs of normal fetuses were fixed in a mixture of methanol, acetone, acetic acid and water (35:35:5:25 by volume) for 1-3 days at room temperature, followed by dehydration in acidified 2.2.dimethoxypropane (Merck) and embedding in paraplast[13] (Monoject). This fixation procedure has been extensively tested in combination with other methods and is found to produce the best immunohistochemical preservation, sensitivity and accessibility of antigens over a wide range of antibodies and tissues. Because the DMD fetus arrived frozen in liquid nitrogen, the tissue was freeze-substituted in the same methanol/acetone/acetic acid/water mixture at -40°C for 7 days, dehydrated and embedded in paraplast. This freeze substitution procedure is routinely used and gives excellent histological preservation and immunohistochemical results virtually indistinguishable from directly fixed tissue. DNA was isolated from whole blood; restriction enzyme digestions and blot hybridizations were as described by Bakker et al.[3] Restriction endonucleases were purchased from Pharmacia. DNA probes pD2 and 754 were used as flanking markers and 87-15 and 87-1 as intragenic probes and 5b-7 as cDNA probe (4.4 - 6.9 kb of the cDNA)[2,3].

For immune-histochemical staining, 7 μm thick sections of the legs of both normal and DMD fetus were deparaffinated and immunostained as described by Moorman et al.[14] All sera were diluted in PBS. The polyclonal antibodies used for incubations were affinity purified rabbit polyclonal antibodies: one directed against a 30 kD subpeptide[3] made of cDNA 3.7 to 4.4 kb of the DMD gene (RA30-2, raised by K.H.F.) and one directed against a peptide corresponding with the last 17 amino acids (11.212 - 11.263 of the cDNA) of dystrophin (P1461, raised by E.E.Z-G.). As a control we also used a sheep polyclonal antibody raised against a 60 kD antigen[3], made of cDNA 1.3 to 2.7 kb (SA60, a gift from E.P. Hoffman).

RESULTS

Figure 1 shows the pedigree of the DMD family (DL 187) for which haplotype carrier detection and prenatal diagnosis had shown 3 carriers and an affected fetus (7). With cDNA 5b-7 an intragenic duplication was detected both with Southern blot analysis and field inversion gel electrophoresis (FIGE) analysis[15]. Given the severity of the disease (Fig. 1), the duplication in the gene probably results in a frameshift producing a truncated dystrophin. To examine this further and to investigate whether the NH$_2$-proximal part of dystrophin might be present in this Duchenne fetus, we performed immunohistochemical studies.

Sections were made of a leg of the twelve-week old Duchenne fetus. The sections were stained with 3 different

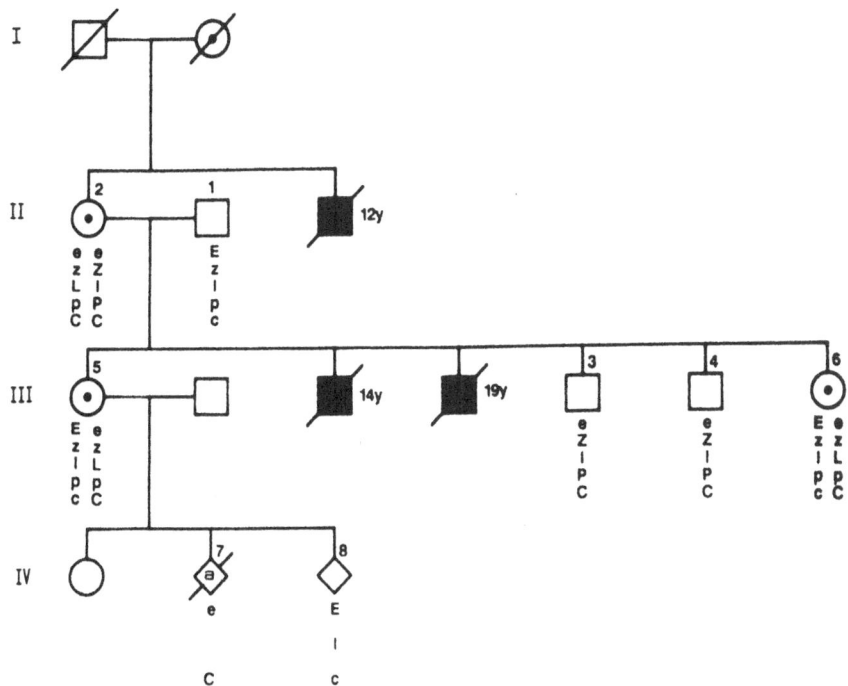

Figure 1. Pedigree of a Dutch family (DL 187) with a familial
case of segregating DMD. For carrier detection in this
family the markers pD2 (E), 44-1 (Z), 87-15 (L), 87-1 (P)
and 754 (C) were used. Daughters 5 and 6 have inherited
the at risk chromosome e-z-L-p-C. Prenatal diagnosis (7)
was possible using the flanking markers pD2 and 754.

polyclonal antibodies, which were directed against a 60 kD
(SA60) and a 30 kD (RA30-2) segment of the NH_2-terminal half of
dystrophin and against the last 17 amino acids of dystrophin
(P1461). The antibody RA30-2 clearly stains the ends of the
myotubes in the leg muscle of both the normal fetus (see
Figure 2A) and the Duchenne fetus (see Figure 2B). A similar
staining was found with the SA60 antibody (not shown). The
signal is located within the sarcolemma, between the distal
nuclei and myotendinous junction, which is typical for this
stage of embryonic development[17]. So NH_2-proximal dystrophin
antigens are present in the myotubes of this Duchenne fetus and
no difference could be made between a normal fetus and the
Duchenne fetus using these antibodies.

In the Duchenne fetus, no signal was detected in the leg
myotubes with the COOH-terminal antibody P1461 (Figure 2D),
while this antibody shows a normal staining pattern in a
healthy fetus (Figure 2C). This indicates that with the COOH-
terminal antibody a clear distinction could be made between the
normal and dystrophic fetus. Besides providing a strong
independent control for the specificity of the reactions, these
results show that early in embryonic development, abnormal
truncated dystrophin is present in the DMD fetus on the same
location as its normal counterpart in a healthy fetus.

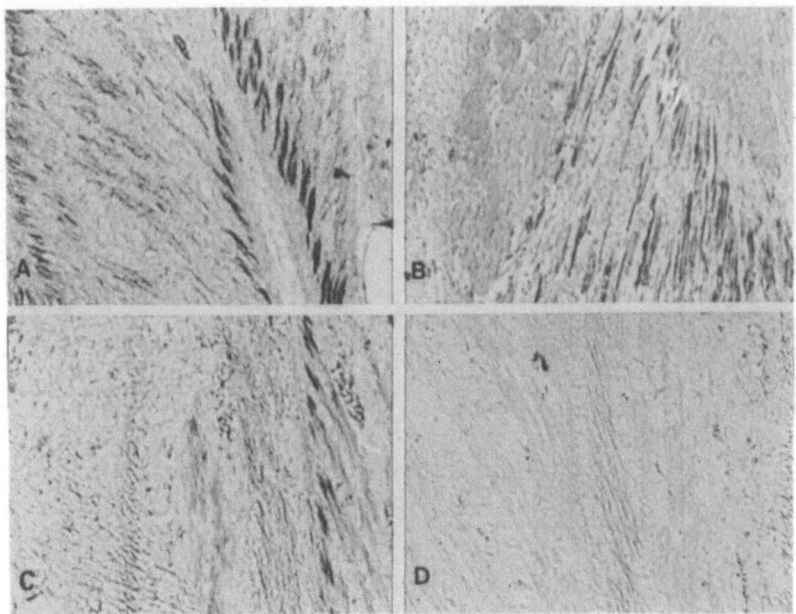

Figure 2. Dystrophin expression in a normal human fetus of about 11 weeks development. Sagittal sections of a fetal leg incubated with affinity purified RA30-2, diluted 1:200 in PBS (A) and with affinity purified P1461, diluted 1:10 in PBS (B). Dystrophin expression in a DMD fetus of about 12 weeks of development. Sagittal sections of a leg incubated with affinity purified RA30-2 and diluted 1:100 in PBS (C) and with affinity purified P1461, diluted 1:10 in PBS (D). All panels X 100.

DISCUSSION

On the basis of DNA analysis of members of this family and the severity of the phenotype, we conclude that the intragenic duplication detected by FIGE and Southern blotting causes a translational frameshift of the Duchenne gene resulting in a truncated protein[11]. This is borne out by the immune histochemical results.

However, our results show that in embryonic development, abnormal dystrophin is initially present in the myotubes of the DMD fetus, on the same location as in a normal fetus. The reported absence of dystrophin in DMD patients later in life indicates that truncated dystrophins are probably degraded in the developing muscle at a later stage during development. It is thought that the COOH-terminal domain of dystrophin is involved in the integration in the plasma membrane, tightly bound to one of several glycoproteins[16]. As this domain is missing in the dystrophin of this and most other fetuses with DMD mutations, it is plausible to propose that dystrophin is degraded when no integration in the sarcolemma of the muscle cell can take place. So the potential to bind to the sarcolemma could be a major point of discrimination between normal and defective dystrophins.

The abnormal size and/or reduced quantity of dystrophin found in muscle cells of BMD patients[10] could be explained as a partial integration of dystrophin in the sarcolemma due to the presence of the last domain of dystrophin, thus protecting it (partially) against breakdown. Our results suggest that BMD/DMD may in some cases be considered as a defect of integration rather than synthesis. In addition, we clearly show that in order to avoid false-positive results in the diagnostic confirmation of DMD in aborted fetuses with immunohistochemistry, both NH_2- and COOH-terminal antibodies should be used.

In the near future optimally, a series of antibodies covering the entire protein should be used to get a better understanding of the determinants of dystrophin stability, which may be critically important for a potential therapeutic intervention in the course of the disease.

SUMMARY

An immunohistochemical study was carried out on a twelve-week old fetus, aborted for high risk of Duchenne muscular dystrophy. Southern and FIGE analysis showed an intragenic duplication in the DMD gene, which had previously resulted in a severe Duchenne phenotype in three relatives. Polyclonal antibodies directed against the NH_2-terminal half of dystrophin showed a positive reaction an a similar distribution of dystrophin in the skeletal myotubes of a twelve-week old normal fetus and the affected fetus. In contrast, a polyclonal antibody directed against the COOH-terminus of dystrophin, i.e., distal to the mutation in this family, did only react with the myotubes of the normal fetus and not with those of the affected fetus. This indicates the presence of a truncated dystrophin in the affected fetus. Apparently at this stage, before binding of dystrophin to the sarcolemma, no distinction is made yet between normal and abnormal dystrophins. This implies that the potential to bind to the sarcolemma could be a major point of discrimination between normal and defective dystrophins. The truncated dystrophin will probably be degraded in a later stage during fetal development.

So it appears that the use of dystrophin immunostaining to confirm high Duchenne risk abortions requires great caution. To prevent false-positive results, the combined use of NH_2- and COOH-terminal antibodies is mandatory.

ACKNOWLEDGMENTS

We thank Jacqueline Vermeulen, Myriam van Miert, Willy Geerts and Sander Kneppers for expert technical assistance, L.M. Kunkel and M. Koenig for probes, Eric Hoffman for kindly providing sheep anti-dystrophin and Cees Hersbach and Lau Blonden for the photography. This work was supported by the Netherlands Foundation for Medical Research MEDIGON and the Dutch Prevention Fund to GJBvO, the Muscular Dystrophy Association (USA) and the Muscular Dystrophy Group of Great Britain to GJBvO and the Princess Beatrix Funds to GJBvO and AFMM.

REFERENCES

1. E. Bakker, M.H. Hofker, N. Goor, J.L. Mandel, K. Wroggemann, K.E. Davies, L.M. Kunkel, H.F. Willard, W.A. Fenton, L. Sanduyl, D. Majoor-Krakauer, A.J. van Essen, M.G.J. Jahoda, E.S. Sachs, G.J.B. van Ommen and P.L. Pearson. Prenatal diagnosis and carrier detection of Duchenne Muscular Dystrophy with closely linked RFLPs. Lancet I: 655-658 (1985).

2. M. Koenig, E.P. Hoffman, C.J. Bertelson, A.P. Monaco, C. Feener and P.L. Pearson. Complete cloning of the Duchenne Muscular Dystrophy (DMD) cDNA and preliminary genomic organization of the DMD gene in normal and affected individuals. Cell 50: 509-517 (1987).

3. E. Bakker, E.J. Bonten, H. Veenema, J.T. den Dunnen, P.M. Grootscholten, G.J.B. van Ommen and P.L. Pearson. Prenatal diagnosis of Duchenne muscular dystrophy: a three year experience in a rapidly evolving field. J. Inher. Metab. Dis. 12 suppl. I: 174-190 (1989).

4. E.P. Hoffman, R.H. Brown and L.M. Kunkel. Dystrophin: the protein product of the Duchenne Muscular Dystrophy locus. Cell 51: 919-928 (1987).

5. K. Arahata, S. Ishiura, T. Ishiguro, T. Tsukuhara, Y. Suhara, C. Eguchi, T. Ishihara, I. Nonaka, E. Ozawa and H. Sugita. Immunostaining of skeletal and cardiac muscle surface membrane with antibody against Duchenne muscular dystrophy peptide. Nature 333: 861-863 (1988).

6. E.E. Zubrzycka-Gaarn, D.E. Bulman, G. Karpati, A.H.M. Burghes, B. Belfall, H. Hajklamut, J. Talbot, R.S. Hodges, P.N. Ray and R.G. Worton. The Duchenne muscular dystrophy gene product is localized in the sarcolemma of human skeletal muscle fibers. Nature 333: 466-469 (1988).

7. S.C. Watkins, E.P. Hoffman, H.S. Slayter and L.M. Kunkel. Immunoelectronmicroscopic localization of dystrophin in myofibers. Nature 333: 863-866 (1988).

8. E. Bonilla, C.E. Samitt, A.F. Miranda, A.P. Hays, G. Salviati, S. DiMauro, L.M. Kunkel, E.P. Hoffman and L.P. Rowland. Duchenne muscular dystrophy: deficiency of dystrophin at the muscle cell surface. Cell 54: 447-452 (1988).

9. E.P. Hoffman, C.M. Knudson, K.P. Campbell and L.M. Kunkel. Subcellular fractionation of dystrophin to the triads of skeletal muscle. Nature 330: 754-758 (1987).

10. E.P. Hoffman, K.H. Fischbeck, R.H. Brown, M. Johnson, R. Medori, J.D. Loike, J.B. Harris, R. Waterston, M. Brooke, L. Specht, W. Kupsky, J. Chamberlain, C.T. Caskey, F. Shapiro and L.M. Kunkel. Dystrophin characterization in muscle biopsies from Duchenne and Becker muscular dystrophy patients. N. Engl. J. Med. 318: 1363-1368 (1988).

11. A.P. Monaco, C.J. Bertolson, S. Liechti-Gallati, H. Moser and L.M. Kunkel. An explanation for the phenotypic difference

between patients bearing partial deletions of the DMD locus. Genomics 2: 90-95 (1988).

12. E.P. Hoffman and L.M. Kunkel. Dystrophin abnormalities in Duchenne/Becker muscular dystrophy. Neuron 2: 1019-1030 (1989).

13. A.F.M. Moorman, M.P.A. Schalekamp, P.A.J. De Boer, W.J.C. Geerts, W.H. Lamers and R. Charles. Immunohistochemical analysis of the distribution of histone H5 and hemoglobin during chicken development. Differentiation 34: 161-167 (1987).

14. A.F.M. Moorman, N.A.J. de Boer, M.Th. Linders and R. Charles. The histone H5 variant in Xenopus laevis. Cell Diff. 14: 113-120 (1984).

15. J.T. den Dunnen, E. Bakker, G.J.B. van Ommen and P.L. Pearson. The DMD gene analyzed by field inversion gel electrophoresis. Brit. Med. Bull. 45: 644-658 (1989).

16. K.P. Campbell and S.D. Kahl. Nature 338: 259-262 (1989).

17. A. Wessels, I.B. Ginjaar, A.F.M. Moorman and G.J.B. van Ommen. Different localization of dystrophin in developing and adult human skeletal muscle. Muscle and Nerve (in press).

GENERAL DISCUSSION

MOLECULAR BIOLOGY OF MYOGENESIS AND REGENERATION

Dr. Strohman: Dr. Carpenter, when you look at the late stages or termination of regeneration, as you put it, in injured muscle, is there any relationship of the satellite cell to repair of necrosis in relationship to the amount of connective tissue?

Dr. Carpenter: I don't think that the connective tissue proliferation has much influence on the course of regeneration. For instance, I don't think it prevents nutrients coming from the vessel.

Dr. Strohman: So it's not a matter of other cell types diluting out a satellite cell population?

Dr. Carpenter: No, I doubt it very much.

Dr. Fischman: I'd like to ask Dr. Frank Stockdale to comment. In line with what Dr. Stockdale presented, Zach Hall's group in the last issue of Science has a nice paper on the targeting of retroviral encoded proteins to either nuclei or cytoplasm. But what's striking about your work and theirs is the large spread of cytoplasmic proteins that develop without a nuclear signal. And I think this cytoplasmic process has relevance to the isoform transitions that we see in muscle fibers. It is clear that when one examines fibers during normal development or during regeneration, you don't see segmental insertion of new isoforms along the myofiber, but they are diffusely inserted along the growing myofibers. And I think, although we are talking about nuclear domains, which are obviously of interest to many of us, it's clear that even large cytoplasmic proteins (in the Hall study, they were examining a protein of about 500,000 molecular weight) diffuse unimpeded for long distances in the cytoplasm.

Dr. Stockdale: I'm not sure that I can say much more than Dr. Fischman said. Obviously the domain over which the particular protein that we're looking at is dispersed, is large relative to the size of the nucleus. So it's clear that a single nucleus can only affect not only many nuclei in a myotube, but can also have an influence in the cytoplasm associated with those nuclei over very large distances, because the cytoplasmic markers don't necessarily relate to any of those nuclei. Clearly the index nucleus is affecting the distance over which you see the distribution of the color. If we actually look at myosins as opposed to ß-galactosidase to determine the domain of an individual, index nucleus, you see different isoforms of myosins appearing in the cytoplasm relative to the index nucleus. And that observation will be interesting with regard to the point you're making.

Dr. Brown: A question for Dr. Stockdale. Did I understand you to say that there are impediments to fusion of embryonic and fetal myoblasts in your cultures?

Dr. Stockdale: I'm not sure that they're impediments. It appears that myoblasts from embryos of different chronological ages don't fuse with equal facility with fetal myoblasts. But I don't know how much they differ.

Dr. Brown: It's striking that, with only a few days separating one population from another, there is a hindrance to fusion. I suppose, given the reports from Helen Blau and others that mouse and human myoblasts can fuse together, perhaps there's a difference in fusion efficiency. But I'm wondering if you could speculate on why there may be limitations in fusion between those two populations.

Dr. Stockdale: Well, you may be making more of this than probably exists. What most people have shown is that cells of differing species can fuse with one another. I think probably most people who do those experiments would agree that if you took the same species and the same developmental age of those myoblasts, they would fuse with greater facility. So it depends what you're trying to emphasize. If you're trying to emphasize whether or not cells are capable of undergoing the same things among species, then the answer is that they clearly do.

Whether they do it with the same ease, I think is less clear. But, in the early development of the limb, the early embryonic myoblast can't be cloned from that limb after about the first week of development, so they actually disappear. So the organism, in a strict sense, doesn't have to worry about the problem of embryonic myoblasts fusing with fetal myoblasts because the embryonic ones don't exist after the first week of development and the entire population is of the fetal type. In the artificial setting, where we can put fetal and embryonic myoblasts together, they do fuse with one another, but probably not with the same ease that either one would itself.

Dr. Hauschka: I have at least a partial answer to an earlier question about whether satellite cells exhibit both myo-D and myogenin. In the cell lines that David Yaffe developed, the C-2 line, and that Helen Blau has worked with, and the cell line that we work with, all are derived from adult satellite cells, and all of those lines first express myo-D and then express myogenin when they differentiate.

Unidentified Speaker: If I could add to that answer, it's also true that chick muscle, when cultured, will express myo-D. Presumably this cultured chick muscle represents a satellite cell population. However, I don't know what the satellite cell population expresses prior to being dispersed or cultured.

Dr. Miller: I want to ask Dr. Carpenter about the regenerating capability of muscle in Duchenne muscular dystrophy.

Does this capability decline linearly with age? Can you give us some idea about the tempo of the impaired regenerating capability?

Dr. Carpenter: I don't think we have enough data to draw a line. However, my guess is that it's not linear. I wouldn't expect it to be. Clearly regeneration is adequate early on. Symptoms don't begin right away but there is cellular necrosis virtually from the time of birth. It probably just gradually becomes more and more manifest.

Dr. Engel: I'd just like to add that if you count satellite cells relative to muscle fibers in Duchenne dystrophy, the number of satellite cells per muscle fiber remains high, even in advanced stages of the disease, but the total number of muscle fibers decreases. So the total number of satellite cells per unit volume of muscle decreases. Secondly, we know from Dr. Blau's study that the regenerative potential of a satellite cell will decline with repeated divisions. Perhaps these two factors account for the observation that regeneration becomes less efficient.

Unidentified Speaker: I have a comment to add to Drs. Fischman's and Dr. Stockdale's comments about index nuclei and how far the products from these index nuclei can progress. We also have evidence on cytoplasmic β-galactosidase and, of course, it does migrate quite far. But the information that might be pertinent is that another protein, the skeletal muscle calcium channel, also appears to migrate far away from the index nucleus. And depending on the time that you allow for differentiation, we see the protein appearing at distances as close as two to three nuclei and perhaps as far away as 10 to 15 nuclei.

Dr. Tapscott: Dr. Carpenter, in your light microscopy studies of N-CAM staining of satellite cells, can you comment on the size of the cell? It appears huge when compared to the size of the fiber. Is this a normal staining pattern? Or is this unique? What is the frequency that you have seen such N-CAM staining?

Dr. Carpenter: I think it's a normal pattern for satellite cells stained for N-CAM in children. So you see such staining in 100 percent of the satellite cells. These are preliminary findings on our part. This preparation was done in Dr. Karpati's lab.

Dr. Stedman: Has anyone yet looked at sarcoma lines with anything resembling an antibody against a myo-D to see whether expression can be detected in such cells. For instance, a cell that is expressing even a low level of myosin heavy chain?

Dr. Tapscott: A few sarcoma lines have been looked at for RNA expression of myo-D. In general I believe every sarcoma line that has been analyzed and that has the capability of muscle differentiation, also expresses high levels of myo-D RNA.

Obviously, it would be of interest if expression of myo-D could overcome the transformed phenotype in these cells. And along those lines, Andrew Lassar has been looking at transformed muscle cells of the rat. In this case, he has found that myo-D expression of these cells can overcome the rat muscle blockade to differentiation with the result that these cells will differentiate. However, when injected into nude mice, the cells are still able to form tumors with breakthrough differentiation. And so, at least in the rat, myo-D lacks the ability to completely overcome the transformed phenotype.

Dr. Moxley: Could Dr. Stockdale speculate about the fact that in Duchenne dystrophy, we believe that there is involvement of heart and smooth muscle, but satellite cells, at least based on my understanding, are missing from these tissues. And I'm curious about the thinking concerning the embryogenesis of different muscle cell types, and whether or not the panel has thoughts about our opportunities for treating these tissues.

Dr. Stockdale: I can answer that with regard to smooth muscle. In the chicken, the smallest muscles are slow muscles. And they probably do arise from the original myoblast population of the organism, so that we have some evidence that suggests that most of the type 1 fibers actually come from that embryonic myoblast population. So I suspect that different muscles have their predominance in terms of total number of fibers in the adult that come from different chronological ages of myoblasts. So most muscle fibers really originate with the fetal cell population, and the embryonic population merely forms a scaffolding for initiation, but not the bulk of the fibers that are found in the adult.

Something like the pectoralis muscle, for example, comes almost exclusively from the fetal population. How these findings relate to the heart may, by analogy, be that there's just one population, the initial population, that gives rise to those fibers. When we culture cells from the heart, it's clear that there are different kinds of myogenic cells in the early heart as well, in terms of the myosins they express. And it's known that the cardiac conduction system, for example, does contain different myosins. So there probably are different kinds of myoblasts, but clearly there aren't the satellite cell types that emerge later in development in the heart.

Unidentified Speaker: Could I also just add a brief response to that question? Myo-D and myogenin are not expressed in cardiac muscle, and yet we're postulating that these genes, either directly or indirectly, act to initiate transcription of the muscle-specific program. In cardiac muscle, many of the same genes are active. It's possible, although obviously not certain at this point, that other genes related to myo-D may be active in the development of cardiac muscle and then go on to activate these same muscle structure genes and cardiac myocytes as well as smooth muscle cells. And the isolation or characterization of the group of genes may be helpful in that regard.

SECTION 2

MYOBLAST/SATELLITE CELL ANTIGENICITY

IMMUNOLOGICAL ASPECTS OF HISTOINCOMPATIBLE MYOBLAST TRANSFER

INTO NON-TOLERANT HOSTS

George Karpati

Neuromuscular Research Group
Montreal Neurological Institute
3801 University Street
Montreal, (Quebec), Canada, H3A 2B4

One of the most pressing issues requiring clarification before considering myoblast transfer in humans is the immunological aspects of the transfer of partially or fully histoincompatible myoblasts into muscles of non-tolerant hosts.

Four possible immunological or inflammatory reactions may develop in non-tolerant hosts in response to histoincompatible myoblast transfer:

(a)   immune rejection by the host
(b)   delayed hypersensitivity reaction
(c)   anaphylaxis
(d)   sterile inflammation

Immune Rejection by the Host

Since rejection of histoincompatible cells by a non-tolerant host is largely dependent on the expression of Class 1 and Class 2 major histocompatibility complex protein products (MHCP) on the cell surface[1], it is important to examine the expression of MHCP in all muscle-type cells that could have any conceivable relevance to myoblast transfer. Table I provides such information (Minor histocompatibility complexes may also play a role but their detection is very difficult).

In the light of the data provided in Table I, it appears that after histoincompatible myoblast transfer into non-tolerant hosts, the following cell types may be targeted for rejection[2]:

(1)   Myoblasts while they are still in the interstitial space
(2)   Myoblasts while they are in the satellite cell position of the host fibers
(3)   Regenerated fiber segments that contain some transferred nuclei with alien MHC loci.

While all of these cells express some Class 1 MHCP, the degree of expression is low compared to most other cell types[2,3,4]; thus, the vigor of immunorejection may be low.

Myoblast Transfer Therapy
Edited by R. Griggs and G. Karpati
Plenum Press, New York, 1990

Table I

Microscopically detectable immunoreactive Class 1 and Class 2
MHC determinants in sarcolemma of human muscle type cells

Cell Type	Class 1	Class 2
Embryonic myoblasts	?	?
Embryonic myotubes	?	?
*Cultured myoblasts	Low	0⁺
Cultured myotubes	Low	0
*Satellite cells	0	0
*Regenerating fibers	Mod	0
*Normal fibers	0	0
Muscle fibers in certain inflammatory myopathies	Strong	0

*Relevant to myoblast transfer
⁺In the presence of interferon-gamma, Class 2 MHCP expression
occurs [Hohlfeld, R., Engel, A.G., 1990. *Am. J. Pathol.* 136:
503-508].

Mature muscle fibers do not express Class 1 MHCP[2,3,4]; thus,
their rejection is unlikely even if they contain transferred
nuclei, unless expression of minor histocompatibility factors
may become operative[1].

Possible methods of mitigating immunorejection include the
following:

(1) Perfect or the best possible HLA matching of donors and
recipients. This would be the ideal situation, but it is time
consuming and appears to be cost-ineffective.

(2) Partial HLA matching by using the father or an unaffected
brother as donor (The mother who is a carrier of Duchenne
muscular dystrophy is not an ideal donor, since about 50% of
the myoblasts cultured from her satellite cells are dystrophin-
incompetent[5]).

(3) The use of pure myoblasts is advantageous over mixed
cultures of myoblast plus fibroblasts, since Class 1 MHCP
expression in fibroblasts is much greater (our unpublished
data).

(4) Immunosuppression of the host. This may be antigen-
specific or antigen-nonspecific[1]. The commonest form of
antigen-nonspecific immunosuppression is drug therapy. The
most important requirements for the optimization of immunosuppressive drug therapy include: high efficacy with wide
safety margin, minimum toxicity, lack of major negative effects
upon the natural history of the dystrophy, convenience of usage
and cost effectiveness. Possible drugs include cyclosporin A,
azathioprine, cyclophosphamide, methotrexate, prednisone, anti-
lymphocytic globulin, as well as X-irradiation. A combination
of more than one agent may be advantageous. At present, the
dosage and the minimum duration of administration of an agent
for obtaining the desired effect after myoblast transfer is
unknown. We have recently determined that azathioprine,

cyclosporin A, cyclophosphamide and large dose prednisone had no appreciable effect on the microscopic features of the natural history of mdx dystrophy after 3 months of administration. We have also determined that the administration of cyclosporin A or cyclophosphamide to mdx mice for 6-8 weeks after allogeneic myoblast transfer, significantly increased the number of dystrophin-positive muscle fiber segments. However, extrapolation from animal experiments (mdx mouse or even dystrophic dogs) to the human situation may be fallacious.

Antigen-specific immunosuppression of the host by induction of specific tolerance is difficult and may be unreliable.

Delayed Hypersensitivity

A certain portion of the injected myoblasts will disintegrate after injection in the interstitial space of the muscle and immunogenic proteins or haptens may trigger a delayed hypersensitivity reaction that might mimic autoimmune myositis. However, muscle tissue or muscle cells are notorious for low immunogenicity, and clinically manifest experimental allergic myositis is very difficult to establish in animals[6]. Furthermore, in Duchenne dystrophy or in mdx, it may be hypothesized that the longstanding active breakdown (necrosis) of muscle fibers acts as a desensitizing factor minimizing the risk of an allergic myositis-like picture even from repeated myoblast transfer. Since myoblasts do not contain acetylcholine receptors, the risk of producing allergic myasthenia gravis is negligible.

Anaphylaxis

If the transferred myoblasts are cross-species, or even of the same species, but during culturing some traces of animal sera adhered to the cell surface, a sensitization for subsequent anaphylaxis can occur, and a full-blown reaction may be provoked by repeat injection. These dangerous complications can be avoided by using only human myoblasts for human hosts and preferably culturing the human myoblasts in AB human serum, at least for some days, prior to transfer.

Sterile Inflammation

Since the transferred myoblasts must be thoroughly dispersed during transfer, multiple injection tracks must be made that could cause muscle fiber tears, necrosis and sterile inflammation with the outpouring of polymorphonuclear leucocytes which can pose serious cytotoxic dangers[7]. Additionally, the protein from the disintegrating myoblasts in the interstitial space may also provoke a nonspecific inflammatory response. While these reactions may also cause swelling and pain for a few days after myoblast transfer, it can be effectively suppressed by low-dose glucocorticoid therapy.

CONCLUSION

While immunological rejections after allogeneic myoblast transfer to non-tolerant hosts requires careful attention and much further study, it is anticipated that most of the major problems are soluble.

REFERENCES

1. I. Roitt, J. Brostoff and D.K. Male. Immunology. Mosby Co., St. Louis, pp 4.1-4.11 (1986).
2. G. Karpati, Y. Pouliot and S. Carpenter. Expression of immunoreactive major histocompatibility complex products in human skeletal muscles, Ann. Neurol. 23: 64-72 (1988).
3. A.M. Emslie-Smith, K. Arahata and A.G. Engel. Major histocompatibility complex Class 1 antigen expression, immunolocalization of interferon subtypes and T cell mediated cytotoxicity in myopathies. Hum. Pathol. 20: 224-231 (1989).
4. R.M. McDougall, M.J. Dunn and V. Dubowitz. Expression of Class I and II MHC antigens in neuromuscular diseases. J. Neurol. Sci. 89: 213-226 (1989).
5. A.F. Miranda, V. Francke, E. Bonilla, G. Martucci, B. Schmidt, G. Salviati and M. Rubin. Dystrophin immunocytochemistry in muscle culture: detection of a carrier of Duchenne muscular dystrophy. Am. J. Med. Genet. 32: 268-273 (1989).
6. D. Manghani, T.A. Partridge and J.C. Sloper. The role of the myofibrillar fraction of skeletal muscle in the production of experimental polymyositis. J. Neurol. Sci. 23: 489-503 (1974).
7. S.J. Weiss. Tissue destruction by neutrophils. New Engl. J. Med. 320: 365-376 (1989).

Discussion of Dr. Karpati's paper

Dr. Partridge: I wonder if you have markers of your cells? Do you know where you put them in? Do you have any details of how far they can move? Do you have any indications as to how far they can move from the injection site?

Dr. Karpati: We mark the nuclear DNA with 3H thymidine during culture. I have seen injected cells moving across a single fascicle, which would be, in a mouse, about the distance of 20 to 30 host fibers. So the injected cells could move about at least 1 mm.

Dr. Partridge: You showed a number of pictures of labelled nuclei which were on the edges of muscle fibers. How can you be assured that those labelled nuclei are genuinely myonuclei rather than adjacent nuclei to the muscle fiber?

Dr. Karpati: These are good points. When you see dystrophin expression in the corresponding fiber segments, you assume that the labelled (donor) nuclei were myonuclei; i.e., situated within the fibers, since otherwise most mdx muscle fibers would show no dystrophin expression. Of course, there is a distinct possibility that some labelled surface nuclei represent donor myoblasts in the satellite position. This could be verified by immunostaining with antihuman N-CAM antibody, which we have been doing. N-CAM is positive only in the satellite cell surface membrane.

Dr. Partridge: You say you have evidence that the fibers into which these cells have been incorporated do not ever degenerate. One of the bits of evidence for that was this peripheral nucleation. And you contrasted that with your results in older animals, where you said the nuclei were central. How do you explain that?

Dr. Karpati: If an mdx fiber segment acquires normogenomic donor nuclei and expresses dystrophin before the necrotic phase commences (i.e., before age 15 days), it will not become centronucleated. However, if an already centronucleated mdx muscle fiber acquires normogenomic donor nuclei, as is often the case after injecting older animals, the already existing centronucleation will not be altered even though dystrophin is present in that fiber segment.

A COMPARISON OF LONG-TERM SURVIVAL OF MUSCLE PRECURSOR CELL

SUSPENSIONS AND MINCED MUSCLE ALLOGRAFTS IN THE NON-TOLERANT

MOUSE

Diana J. Watt

Department of Anatomy
Charing Cross and Westminster Medical School
Fulham Palace Road, London W6 8RF, England

The recent discovery that the myopathic state of muscle fibres manifest in both the Duchenne and Becker forms of muscular dystrophy, is due to a deficiency or defect in the protein dystrophin (Hoffman et al., 1988; Arahata et al., 1988), supports the idea of treating recessively inherited diseases of skeletal muscle. By implanting normal muscle precursor cells, with a normal gene complement, into the multinucleate muscle fibres of affected individuals it should be possible to remedy this biochemical defect. In the experimental mouse model, we have already achieved a partial correction of an inherited biochemical defect of skeletal muscle by the introduction of grafts of muscle precursor cells into myopathic muscle deficient in the enzyme phosphorylase kinase (Morgan et al., 1988). More recently, Partridge et al., (1989), following the introduction of normal precursor cells into muscles of the X-linked Muscular Dystrophic (mdx) mouse which are normally deficient in dystrophin, have raised the levels of this protein by up to 30-40% of normal levels, indicating that a single injection of such precursor cells leads to synthesis of the missing gene product in substantial amounts.

The eventual therapeutic use of allografts of normal muscle precursor cells in recipient human patients in order to alleviate a primary myopathic condition, would require some means of preventing the immune rejection of the allografted tissue. At present little is known about the susceptibility to rejection of either myogenic cells, or of the muscle fibres to which they give rise. It is known that cultured myoblasts express surface detectable Class 1 major histocompatibility gene products (Karpati et al., 1988), although such expression is not evident in normal multinucleate muscle fibres (Ponder et al., 1983; Appleyard et al., 1985).

In the past we have shown that survival of muscle introduced into the leg of a host mouse in the form of a crude minced muscle allograft of the donor strain will only survive for prolonged periods of time if the host animal has first been

Myoblast Transfer Therapy
Edited by R. Griggs and G. Karpati
Plenum Press, New York, 1990

made immunologically tolerant to the donor tissue. Without such immune tolerance induction (Partridge & Sloper, 1977; Partridge et al., 1978; Grounds et al., 1980; Watt, 1982), or the use of immunosuppressive drugs such as Cyclosporin-A (Watt et al., 1981; Gulati et al.,1982; Watt et al.,1984a; Law et al., 1988), rejection of the allografted tissue ensues within 12-15 days after grafting, after which the graft is replaced by fibrofatty connective tissue of host origin. The rejection of such minced grafts occurs in the non-tolerant animal at this early stage after grafting, regardless of whether neonatal or adult muscle is used as a source of minced muscle allograft, and regardless of whether histocompatible or non-histocompatible strains of mice are used as host and donor.

If muscle precursor cells are to be used therapeutically to alleviate a myopathic condition, it is imperative to prepare the cells in such a manner that they are readily injectable into the affected muscles. Again, using the mouse as an experimental model, we have prepared muscle precursor cells, obtained from a neonatal donor source, in the form of a single cell suspension. These cells, prepared by the enzymatic disaggregation of neonatal donor muscle yield a single cell suspension containing a mixture of myogenic and non-myogenic cells, the latter presumably including fibroblasts, endothelial cells, etc. Such mixed cell suspensions have been introduced into the regenerating (Watt et al., 1982) or growing (Watt et al., 1984b) muscles of the host mouse. In the majority of these experiments donor precursor cells were implanted into a regenerating minced muscle autograft of the tibialis anterior muscle of the host, thus ensuring the presence of sufficient host precursor cells to fuse with those of the donor in the formation of mosaic host/donor muscle fibres. Survival of donor tissue was evidenced by the donor form of the isoenzyme variant of Glucose-6-phosphate Isomerase (GPI) characteristic of the donor strain used. Fusion of donor precursor cells with those of the host to form mosaic muscle fibres containing myonuclei of both host and donor origin was evidenced by the detection within grafts of a "hybrid" isoenzyme form of GPI. This "hybrid" isoenzyme form indicates that donor gene products are being expressed within the newly regenerated muscle fibres formed in the graft, a valuable feature if precursor cells are to be used to alleviate a primary myopathic condition.

Grafts were made between strains compatible at the major histocompatibility locus i.e. C57Bl/6j and 129/ReJ strains (both H-2^b), and strains which are not compatible at that locus, i.e. A (H-2^a) and CBA (H-2^k). As hosts were not made tolerant to the donor tissues, we would therefore expect the allografted suspension of donor muscle precursor cells to be rejected some 12-15 days after grafting, as had been the case in the minced muscle grafting situation.

Results from such an experiment surprisingly showed the presence, as late as 105 days after donor precursor cell implantation, of host, donor and hybrid GPI isoenzyme forms, in the majority (26 of 30 performed) of grafts using the histocompatible strain combinations, and in a few of the non-histocompatible ones (3 out of 35). The presence of the "hybrid" form of GPI in these situations was evidence not only of survival of donor cells and their incorporation into newly formed host muscle fibres which were abundant in all the grafts studied, but also the expression of donor genes and dissemination of products within these mosaic muscle fibres.

In similar experiments, we used adult muscle as a source of the suspension of donor muscle precursor cells. The major drawback in using adult muscle however is the decreased yield of stem cells, as compared to neonatal muscle, that can be prepared. To increase this yield, the tibialis anterior muscle from an adult donor mouse was minced up, replaced back into its original site within the leg, left for three days in situ, removed and enzymatically disaggregated to produce a single cell suspension. Regardless of whether grafts were made between histocompatible or non-histocompatible strains, donor cells were rejected in these non-tolerant animals in all cases studied by 16 days following implantation. We interpreted the rejection of precursor cells from such a regenerating adult source to be due to contaminating donor inflammatory cells from the autograft which were included in the preparation of the single cell suspension.

In further attempting to prevent the rejection of muscle precursor cells prepared from adult muscle from the non-tolerant host, we prepared precursor cells from "mature", i.e. non-regenerating, adult muscle. Despite the low yield of cells from this source we were able to perform 16 implantations of these cells. Of these 16, "hybrid" GPI was again identified in 6, of which, 4 were observed between 28 and 130 days after grafting using the non-histocompatible strain combination.

It thus appears that even crude suspensions of cells obtained from simple disaggregation of muscle, when introduced into the non-tolerant host mouse, are less immunogenic than grafts of minced muscle and thus do not raise an immune response from the host. These results also indicate that the presence of inflammatory cells is to be avoided, as it seems to provoke immune rejection of myogenic cells even from donors compatible at the major histocompatibility locus.

Such results may prove valuable in the design of strategies for myoblast transfer therapy. It seems likely that mononuclear muscle precursor cells may well be tolerated by the host in the initial and vulnerable stages of implantation, prior to their incorporation into mosaic host/donor muscle fibres. After this period, their rejection is less likely, for multinucleate muscle fibres have been shown to lack the expression of major histocompatibility complex proteins (Ponder et al., 1983; Appleyard et al., 1985; Karpati et al., 1988).

ACKNOWLEDGEMENT

This work was supported by the Muscular Dystrophy Group of Great Britain and Ireland.

REFERENCES

Appleyard, S.T., Dunn, M.J., Dubowitz, V. and Rose, L.M. 1985. Increased expression of HLA ABC class 1 antigens by muscle fibres in Duchenne Muscular Dystrophy, inflammatory myopathy and other neuromuscular diseases. Lancet I: 361-363.

Arahata, K., Ishiura, S., Ishiguro, T., et al. 1988. Immunostaining of skeletal and cardiac muscle surface membrane with antibody against Duchenne muscular dystrophy peptide. Nature 333: 861-863.

Grounds, M.D., Partridge, T.A. and Sloper, J.C. 1980. The contribution of exogenous cells to regenerating skeletal muscle: An isoenzyme study of muscle allografts in mice. J. Path. 132: 325-341.

Gulati, A.K. and Zalewski, A.A. 1982. Muscle allograft survival after cyclosporin A immunosuppression. Exp. Neurol. 77: 378-385.

Hoffman, E.P. et al. 1988. Characterization of dystrophin in muscle-biopsy specimens from patients with Duchenne's or Becker's muscular dystrophy. New Engl. J. Med. 318: 1363-1368.

Karpati, G., Pouliot, Y. and Carpenter, S. 1988. Expression of immunoreactive major histocompatibility complex products in human skeletal muscles. Ann. Neurol. 23: 63-72.

Law, P.K., Goodwin, T.G. and Li, H-J. 1988. Histoincompatible myoblast injection improves muscle structure and function of dystrophic mice. Trans. Proc. 20: 1114-1119.

Morgan, J.E., Watt, D.J., Sloper, J.C. and Partridge, T.A. 1988. Partial correction of an inherited biochemical defect of skeletal muscle by grafts of normal muscle precursor cells. J. Neurol. Sci. 86: 137-147.

Partridge, T.A. and Sloper, J.C. 1977. A host contribution to the regeneration of muscle grafts. J. Neurol. Sci. 33: 425-435.

Partridge, T.A., Grounds, M.D. and Sloper, J.C. 1978. Evidence of fusion between host and donor myoblasts in skeletal muscle grafts. Nature 273: 306-308.

Partridge, T.A., Morgan, J.E., Coulton, G.R., Hoffman, E.P. and Kunkel, L.M. 1989. Conversion of mdx myofibres from dystrophin-negative to -positive by injection of normal myoblasts. Nature 337: 176-178.

Ponder, B.A.J., Wilkinson, M.M., Wood, M. and Westwood, J.H. 1983. Immunohistochemical demonstration of H2 antigens in mouse tissue sections. J. Histochem. Cytochem. 31: 911-919.

Watt, D.J. 1982. Factors which affect the fusion of allogeneic muscle precursor cells in vivo. Neuropath. Appl. Neurol. 8: 135-147.

Watt, D.J., Partridge, T.A. and Sloper, J.C. 1981. Transplantation 31: 266-271.

Watt, D.J., Lambert, K., Morgan, J.E., Partridge, T.A. and Sloper, J.C. 1982. Incorporation of donor muscle precursor cells into an area of regeneration in the host mouse. J. Neurol. Sci. 57: 319-331.

Watt, D.J., Morgan, J.E. and Partridge, T.A. 1984a. Long term survival of allografted muscle precursor cells following a limited period of treatment with cyclosporin A. Clin. exp. Immunol. 55: 419-426.

Watt, D.J., Morgan, J.E. and Partridge, T.A. 1984b. Use of mononuclear precursor cells to insert allogeneic genes into growing mouse muscles. Muscle & Nerve 7: 741-750.

Discussion of Dr. Watt's paper

Dr. Engel: You showed very nicely how you monitored fusion of the transplanted cells to host cells by the expression of glucose-6-phosphate-dehydrogenase isoenzymes in the host. Did you also do parallel histological studies?

Dr. Watt: I did not show results for lack of time. In all the grafts, where we had implanted a muscle precursor cell in the form of donor cell suspension, the grafts contained a large proportion of new muscle. When I say a large proportion, we are talking about seventy to ninety percent. In the cases where we had to introduce allografts, i.e., minced mixed muscle, we got new muscle. Single allografts were rejected and we just got fibrofatty connective tissue.

Dr. Engel: So the histological findings were in harmony with the other findings?

Dr. Watt: Yes, they were.

Dr. Engel: What steps have you taken to purify either the single cell suspensions or the minced cell preparations before transplantation? A single cell preparation could contain fibroblasts, macrophages, and endothelial cells with different antigenic properties.

Dr. Watt: We have looked at that. Jenny Morgan produced a much more pure myogenic population, and we find that these cells are productive. By using such a situation, we can purify the cells very well. The other thing to do is we can take the cells straight after disaggregating them, and just pop them into a tissue culture dish, and leave them for forty minutes, in which time the myogenic cells adhere, and then take them out.

Dr. Engel: Did you analyze your cell suspensions for MHC expression?

Dr. Watt: We have not as yet done that.

N-CAM IS A TARGET CELL SURFACE ANTIGEN FOR THE PURIFICATION

OF MUSCLE CELLS FOR MYOBLAST TRANSFER THERAPY

Frank S. Walsh

Department of Experimental Pathology
Guy's Hospital, London Bridge
London SE1 9RT, U.K.

Since the introduction of the monoclonal antibody (McAb) method by Kohler and Milstein over ten years ago, a major goal of cell biologists has been to define and characterize lineage and cell type specific antigens that may allow the identification and manipulation of specific cells. While in principle it is possible to prepare McAbs to any subcellular compartment or organelle, a large amount of effort has gone into the identification of cell surface antigens. There are a number of sound reasons for this preference.

Firstly, analyses can be carried out on small numbers of cells including clonal cultures. These analyses are nondestructive and can lead to characterization of positive and negative populations. In addition it is possible to use bulk isolation procedures based on fluorescence activated cell sorting (FACS) to isolate large numbers of cells based on specific cell surface antigen expression. An alternative to this approach is to use complement mediated cytotoxicity methods to deplete cell cultures of unwanted cells if a specific cell surface antigen specifically reactive with that population can be found. A fourth advantage of cell surface antigens is associated with the ability to generate McAbs that react with a restricted number of animal species. This is most powerful when the antigen in question is of human origin as it can lead to the chromosomal assignment of antigens by testing for reactivity in panels of human X rodent cell hybrids that have differing numbers of human chromosomes. A further advantage of species specificity has been exploited in a series of studies in muscle cells by Blau and colleagues (Blau et al., 1985) where human muscle genes identified by McAbs have been shown to be reactivated in heterokaryons when exposed to muscle transacting transcription factors.

My laboratory has been involved in the isolation and characterization of human muscle cell surface antigens defined by McAbs. As one of our goals has been to define antigens of restricted specificity, we have screened panels of McAbs with human muscle cells in culture (Walsh et al., 1984). The three main cell types in tissue cultures of human skeletal muscle are

Myoblast Transfer Therapy
Edited by R. Griggs and G. Karpati
Plenum Press, New York, 1990

myoblasts, myotubes and fibroblasts. We found a large number of McAbs that reacted with all three cell types and most probably represent so-called housekeeping gene products. A smaller number reacted with myoblasts and myotubes only and one reacted with myoblasts only. The two most useful McAbs are called 5.1H11 (Walsh and Ritter, 1981) and 24.1D5 (Gower et al., 1989) and react with myoblasts and myotubes and myoblasts, respectively.

We have now carried out extensive analyses of the reactivities of both McAbs. McAb 24.1D5 appears to be the most specific myoblast antigen yet defined and recent expression cloning strategies have allowed studies on the structure of the antigen (Gower et al., 1989). Northern analysis confirmed the studies using immunofluorescence in that 24.1D5 mRNAs decreased in abundance following myoblast fusion and 24.1D5 appears to be one of the earliest markers of muscle cells yet defined. No detailed analyses have yet been carried out with McAb 24.1D5 with respect to FACS of human myoblasts although preliminary experiments appear promising (Hurko et al., 1987).

McAb 5.1H11 (Walsh and Ritter, 1981) has been the subject of a number of detailed studies to assess its ability to purify human myoblasts. Clearly any McAb that can be used for purification of myoblasts by FACS would be of immense value for myoblast transfer therapy. Two studies have indicated that McAb 5.1H11 may be a useful reagent in achieving this goal. Webster et al. (1988) showed that greater than 99% enrichment for myoblasts could be achieved and that the sorted myoblasts were capable of fusion to myotubes at high frequency. Similar results have been reported by Hurko et al. (1987). One of the problems with McAb 5.1H11 is that it has not been possible to study the reactive antigen as the epitope is sensitive to detergent and no positive results have been reported using standard immunoprecipitation and Western blotting analyses. This has meant that the 5.1H11 homolog in species other than human has not yet been identified, which has clearly restricted the usefulness of the reagent. Recently, we have readdressed this question and can now report that McAb 5.1H11 reacts with neural cell adhesion molecule (N-CAM). A molecular genetic approach was used to come to this conclusion. We have isolated full length cDNA clones encoding N-CAM from human skeletal muscle (Barton et al., 1988; Gower et al., 1988). These cDNAs have been subcloned into an expression vector for gene transfer studies into N-CAM negative fibroblasts to assess the function of individual isoforms.

These studies have now enabled us to show that N-CAM is one of the recognition molecules used in axonal outgrowth (Doherty et al., 1989). However, as the transfected 3T3 fibroblasts synthesized human N-CAM, it offered an opportunity to determine whether it was possible to identify McAbs reactive with this antigen. This seemed important to determine as a large number of McAbs have been generated, mostly to tumor-associated antigens that have been suggested to be reactive with N-CAM (Patel et al., 1989). As part of a screening programme of McAbs, we were able to show definitely that McAb 5.1H11 reacts with human N-CAM (Walsh et al., 1989). An example of a typical experiment is shown in Figure 1. We show that 3T3 fibroblasts do not react with McAb 5.1H11 as assessed by indirect immunofluorescence assay (Fig. 1a).

However, when we stained 3T3 cells that expressed human N-CAM, a clear positive pattern of reactivity was found. This result was found for many independently derived clones and also with different N-CAM isoforms including transmembrane, glycosylphosphatidylinositol linked and secreted (Walsh et al., 1989). This result suggests that the 5.1H11 reactive epitope is in a region that is common to each of these isoforms and probably not a glycosylation site. Biochemical analyses confirmed our previous results with McAb 5.1H11 in that antibody binding was lost when cells were treated with detergent. No positive results were obtained by immunoprecipitation or Western blot analyses also. This is in contrast to other McAbs reactive with human N-CAM that we have identified and which work well in immunofluorescence and biochemical analyses (Patel et al., 1989). Thus, in the future it should be possible to isolate large populations of human myoblasts for transfer therapy using McAb 5.1H11 or other N-CAM reactive McAbs. It should be possible to monitor the purification and differentiation of such populations of cells by biochemical means also since some of the McAbs work in Western blotting and immunoprecipitation systems.

These studies clearly suggest that N-CAM is likely to be a useful target antigen for the purification and subsequent differentiation of human myoblasts in transfer therapy. The restricted cross reaction with species other than human also allows the possibility of following the behavior of human myoblasts when transferred into rodent model systems such as

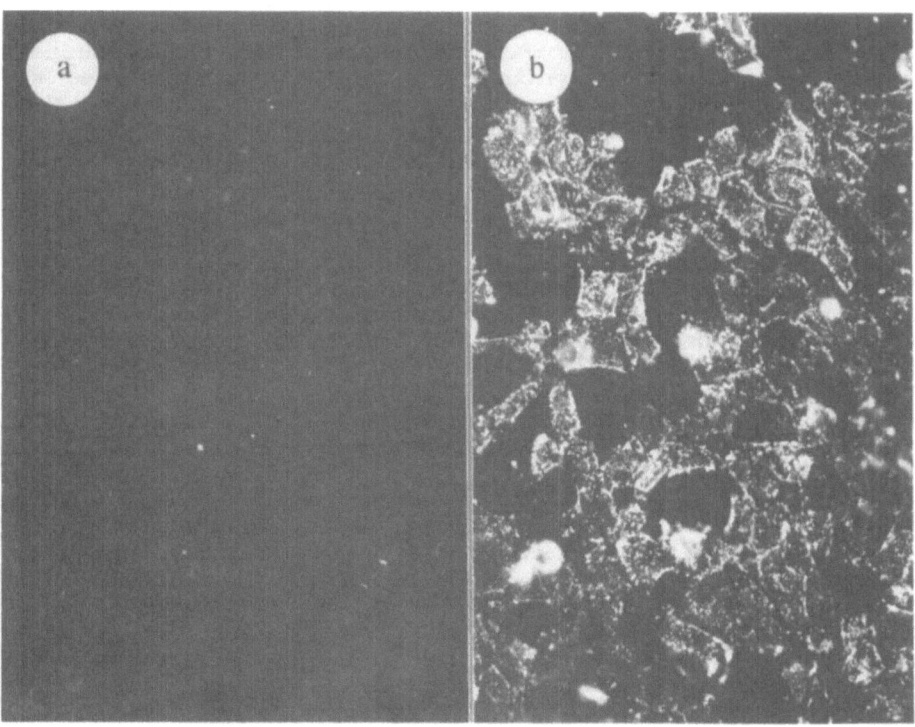

Figure 1.   Indirect immunofluorescence analysis of McAb 5.1H11 with (a) 3T3 fibroblast cells and (b) 3T3 cells following gene transfer with a cDNA encoding human N-CAM.

<u>mdx</u> (Partridge et al., 1989). N-CAM has also been identified and extensively characterized in a number of animal species that are commonly used for studies on normal and pathological processes in muscle including chicken (Cunningham et al., 1987) and rodents (Covault et al., 1986; Moore et al., 1987; Moore and Walsh, 1986). The amino acid sequence of different N-CAM isoforms has now been carried out from chicken, rodent and human sources and it appears highly homologous across species with identical intron-exon boundaries (Walsh and Dickson, 1989). Detailed analyses of patterns of alternative splicing of the N-CAM gene have now been carried out by us in skeletal muscle (Walsh and Dickson, 1989; Thompson et al., 1989). Myoblasts express a predominantly trans-membrane N-CAM isoform of 140 kD while myotubes express mainly a glycosylphosphatidylinositol linked isoform of 125 kD (Moore et al., 1987). As there are differences in usage of exons between these different isoforms, it should be possible in the future to develop new McAbs that react with isoforms characteristic of particular differentiation states. This, in turn, should be of value for the assessment of the proliferative potential and differentiation capacity of transplanted myoblasts.

Finally, for animal species such as the dog which are of increasing interest for studies on inherited mutations in dystrophin expression with consequential pathology in skeletal muscle, it should be possible also to use N-CAM as a myoblast marker for transfer therapy. This could either be using existing McAbs and polyclonal sera or, alternatively, utilizing the high degree of nucleotide sequence homology across species. It should be possible to isolate cDNA clones from skeletal muscle or brain libraries from the dog, once the nucleotide sequence and predicted amino acid sequence has been deduced to prepare antibodies to synthetic peptides. We have used this approach to analyze the expression of muscle specific N-CAM sequences in the human N-CAM gene (Thompson et al., 1989) and the methods have widespread applicability.

REFERENCES

Barton, C.H., Dickson, G., Gower, H.J., Rowett, L.H., Putt, W., Elsom, V., Moore, S.E., Goridis, C. and Walsh, F.S. 1988. Complete sequence and <u>in vitro</u> expression of a tissue specific phosphatidylinositol-linked N-CAM isoform from skeletal muscle. <u>Development</u> 104: 165-173.

Blau, H.M., Pavlath, G.K., Hardeman, E.C., Webster, S.G., Miller, S.C. and Webster, C. 1985. Plasticity of the differentiated state. <u>Science</u> 230: 758-766.

Covault, J., Merlie, J.P., Goridis, C. and Sanes, J.R. 1986. Molecular forms of N-CAM and its RNA in developing and denervated skeletal muscle. <u>J. Cell Biol.</u> 102: 731-739.

Cunningham, B.A., Hemperly, J.J., Murray, B.A., Prediger, E.A., Brackenbury, R. and Edelman, G.M. 1987. Neural cell adhesion molecule: structure immunoglobulin-like domains, cell surface modulation and alternative RNA splicing. <u>Science</u> 236: 799-806.

Doherty, P., Barton, C.H., Dickson, G., Seaton, P., Rowett, L.H., Moore, S.E., Gower, H.J. and Walsh, F.S. 1989. Neuronal process outgrowth of human sensory neurons on monolayers of cells transfected with cDNAs for five human N-CAM isoforms. *J. Cell Biol.* 109: 789-798.

Gower, H.J., Barton, C.H., Elsom, V.L., Thompson, J., Moore, S.E., Dickson, G. and Walsh, F.S. 1988. Alternative splicing generates a secreted form of N-CAM in muscle and brain. *Cell* 55: 955-964.

Gower, H.J., Moore, S.E., Dickson, G., Elsom, V.L., Nayak, R. and Walsh, F.S. 1989. Cloning and characterization of a myoblast cell surface antigen defined by 24.1D5 monoclonal antibody. *Development* 105: 723-732.

Hurko, O., McKee, L., Zuurveld, J. and Swick, H.M. 1987. Comparison of Duchenne and normal myoblasts from a heterozygote. *Neurology* 37: 675-681.

Moore, S.E. and Walsh, F.S. 1986. Nerve dependent regulation of N-CAM expression in skeletal muscle. *Neuroscience* 18: 499-505.

Moore, S.E., Thompson, J., Kirkness, V., Dickson, J.G. and Walsh, F.S. 1987. Skeletal muscle neural adhesion molecule (N-CAM): changes in protein and mRNA species during myogenesis of muscle cell lines. *J. Cell. Biol.* 105: 1377-1386.

Partridge, T.A., Morgan, J.E., Coulton, J.R., Hoffman, E.P. and Kunkel, L.M. 1989. Conversion of MDX myofibers from dystrophin-negative to -positive by injection of normal myoblasts. *Nature* 337: 176-179.

Patel, K., Moore, S.E., Dickson, G., Rossell, R.J., Beverley, P.C., Kemshead, J.T. and Walsh, F.S. 1989. Neural cell adhesion molecule (N-CAM) is the antigen recognized by monoclonal antibodies of similar specificity in small cell lung carcinoma and neuroblastoma. *Int. J. Cancer* (in press).

Thompson, J., Dickson, G., Moore, S.E., Gower, H.J., Putt, W., Kenimer, J.G., Barton, C.H. and Walsh, F.S. 1989. Alternative splicing of the neural cell adhesion molecule generates variant extracellular domain structure in skeletal muscle and brain. *Genes and Development* 3: 348-357.

Walsh, F.S. and Ritter, M.A. 1981. Surface antigen differentiation during human myogenesis in culture. *Nature* 289: 60-64.

Walsh, F.S., Moore, S.E., Woodroofe, M.N., Hurko, O., Nayak, R., Brown, S.M. and Dickson, J.G. 1984. Characterization of human muscle differentiation antigens. *Exp. Biol. and Med.* 9: 50-56.

Walsh, F.S. and Dickson, G. 1989. Generation of multiple N-CAM polypeptides from a single gene. *Bioassays* (in press).

Walsh, F.S., Dickson, G., Moore, S.E. and Barton, C.H. 1989. Unmasking N-CAM. *Nature* 339: 516.

Webster, C., Pavlath, G.K., Parks, D.R., Walsh, F.S. and Blau, H.M. 1988. Isolation of human myoblasts with the fluorescence-activated cell sorter. *Exp. Cell. Res.* 174: 252-265.

Discussion of Dr. Walsh's paper

Unidentified Participant: One short question, Dr. Walsh. Have you tried cross-blocking experiments in trying to see if your 5.1H11 antibody binding is inhibited by LEU 19?

Dr. Walsh: No, we have not; only in one very preliminary experiment have we done anything with LEU 19. We know that LEU 19 reacts with the fibroblasts transfected with human N-CAM and binds to a region on the N-CAM protein (see Schubert et al., 1989, Proc. Natl. Acad. Sci. USA 86: 307-311). However, we have not yet carried out competition experiments with that combination of reagents. The panel of monoclonals I listed on the final slide that are called Cluster 1 have been used in a number of preliminary competition experiments. They have distinct binding sites. There is one epitope that seems very immunodominant on N-CAM and McAb UJ13A and a number of others can cross-compete for this site. However, there are additional sites on the protein and McAb 5.1H11 reacts with one of these. One recently described McAb called ERIC-1 may indeed react with the homophilic binding region. My colleagues George Dickson and Dave Peck have found that addition of Fab fragments of this antibody into a model assay system using C2 muscle cells transfected with human N-CAM will reverse the enhanced process of myoblast fusion that is induced following transfection. This is not the case with McAbs 5.1H11 or UJ13A again suggesting but not providing that there are multiple binding sites for these McAbs on the human N-CAM protein. In addition to these McAbs that likely react with the protein backbone there are a number of McAbs such as HNK1 that react with carbohydrate epitopes.

IMMUNOCHEMICAL ANALYSES OF THE MYOBLAST MEMBRANE AND LINEAGE

Stephen J. Kaufman

Departments of Microbiology and Cell Biology
University of Illinois, Urbana, Illinois 61801

We have been using monoclonal antibodies to analyse cells by immunofluorescence microscopy in an effort to understand the molecules and events on the myoblast membrane that are germane to their differentiation. What follows is a summary of some of our findings.

First, a stage specific remodeling of the cell surface accompanies the differentiation of skeletal myoblasts (1-5). As noted in Figure 1, this remodeling is characterized by quantitative changes in individual membrane molecules, measured by stage specific changes in immunofluorescence on cells reacted with a variety of antibodies. In addition, there are qualitative and topographic changes on the cell surface (1, 4, 5) as well as transient changes in the association of individual molecular species with the cell cytoskeleton (6). We believe this remodeling of the architecture of the membrane is part of the morphologic and biochemical differentiation of skeletal muscle and presumably it reflects changes in the requirements and functional capacities of these cells at distinct stages of development.

We have been focusing on one of these membrane molecules, H36, which is a muscle-specific integral membrane glycoprotein whose expression is developmentally regulated, not once, but at least twice (7). Its pattern of expression during terminal myogenesis is analogous to that depicted in panel B, Figure 1.

Unlike most of the muscle-specific myofibrillar proteins that have come under intensive study, H36 is expressed in replicating myoblasts (8). Using the incorporation of 5-bromo 2-deoxyuridine (BrdUrd) into DNA and anti-BrdUrd antibody to define replicating cells, we have shown that both H36 and the muscle-specific intermediate filament protein desmin are expressed in replicating mammalian myoblasts (Figure 2) (8, 9). Increased expression of these two proteins takes place when myoblasts become postmitotic and is dependent on the further differentiation of these cells and on new transcription (8, unpublished data). Thus one point at which H36 and desmin expression are regulated is during terminal myogenic differentiation.

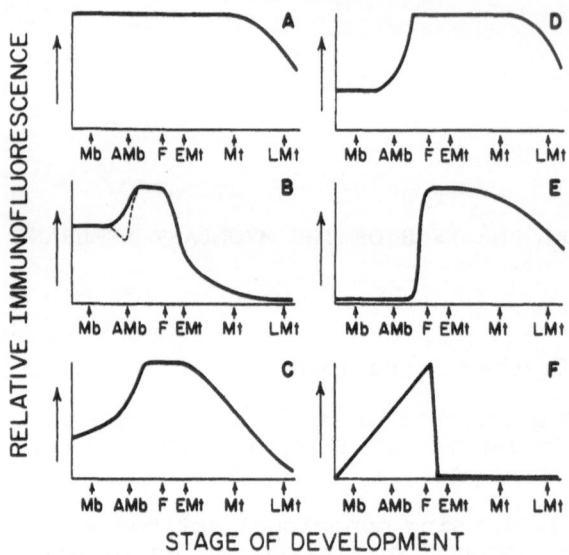

Figure 1. Relative amounts of antigens on developing myoblasts. Forty monoclonal antibodies reactive with surface antigens were used to study changes on the myoblast membrane that accompany differentiation. Indirect immunofluorescence photometry of replicating [Mb], aligned [A] and fusing [F] myoblasts, and early [EMt], mature [Mt] and late myotubes was compared. Six patterns of changes in immunofluorescence were noted and are depicted here.

The second point at which expression of these two proteins is regulated takes place earlier, when H36⁻/Desmin⁻ precursor cells in the rat embryo hindlimb bud first turn on their expression. This takes place between days 12-15 of gestation (9, unpublished data). It is not at present clear whether the initial expression of H36 and desmin are coordinately regulated. Nevertheless this evidence suggests the model of myogenesis depicted in Figure 3 in which distinct cells in a myogenic lineage are defined. One implication from such a model is that multiple regulatory events are involved in the progression of cells through this lineage. Clearly, MyoD1 (10, 11) myogenin (12), Myd (13) and perhaps additional myc related gene products (14, 15) function at at least one regulatory step in this lineage.

The model in Figure 3 suggests two predictions that we have tested: Firstly, that inhibition of terminal myogenic development does not alter the basal level of expression of H36 and desmin. In addition to previously reported results with inhibitors of myogenesis and developmental mutants (7, and Table 1), in recent studies using L8 and C2 myoblasts transfected with adenovirus E1A, provided by Larry Kedes and Keith Webster (16), and ras transfected C2 cells, provided by Eric Olson (17), we have demonstrated that L8 cells continue to express H36 and C2 cells continue to express desmin, even when terminal myogenesis is inhibited by E1A or ras (Table 1).

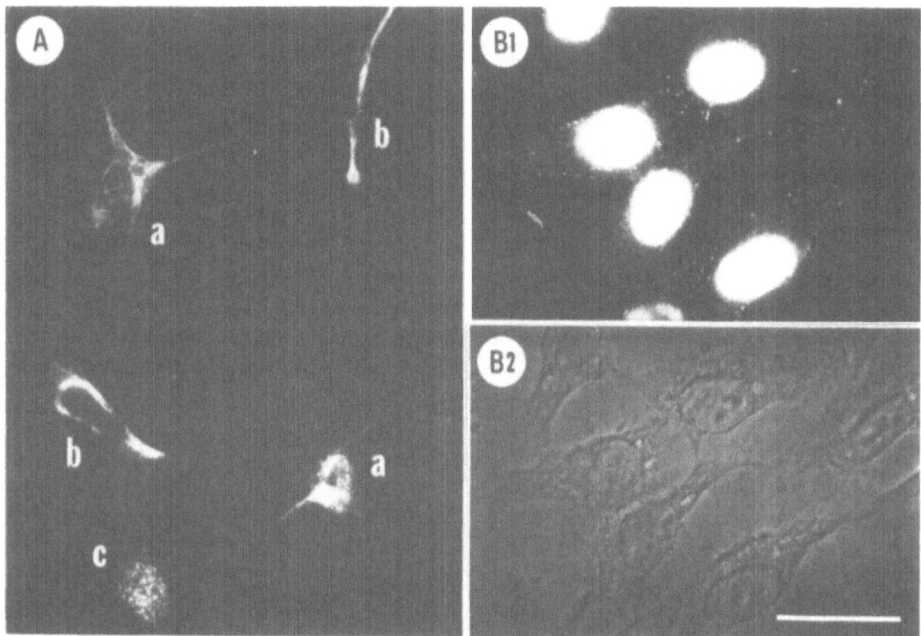

Figure 2. Desmin (A) and H36 (B) in proliferating cells from
newborn rat hindlimb. (A) Cells were grown for 24 hr and
incubated with 40 μM BrdUrd for 90 min, fixed with 95%
ethanol, and stained with anti-BrdUrd and anti-desmin
antibodies. (A) Examples are shown of desmin⁺ BrdUrd⁺ (a)
and desmin⁺ BrdUrd⁻ (b) cells and of a desmin⁻ BrdUrd⁺ cell
(c). (B) A culture grown for 24 hr was enriched for
myoblasts (i.e., H36⁺ cells) by flow cytometry, replated,
and incubated with 4 μM BrdUrd for 15 hr, restained with
anti-H36, and then fixed and stained with anti-BrdUrd
antibody. Most cells (84%) incorporated BrdUrd and
expressed H36 on their surface (B1), indicating that the
myoblast population was replicating. B2, phase-contrast
image of the cells in B1. Bar = 20 μm.

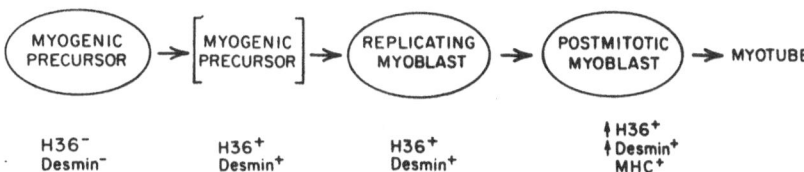

Figure 3. Differentiation within a myogenic lineage. Precursor
cells in the 12 day fetal rat hind limb do not express H36
or desmin but do so upon subsequent growth in vitro.
H36⁺/Desmin⁺ myogenic precursor cells in the 15 day
hindlimb can be distinguished from replicating myoblasts
at later stages of development by their unresponsiveness
to the growth promoting effects of laminin (20).
Regulation of H36 and desmin expression occurs twice,
first, early in the development of the myogenic lineage
and subsequently, upon terminal differentiation into
postmitotic myoblasts and myotubes.

49

Table 1. Desmin and H36 are Expressed in Mutants in which Terminal Myogenesis is Inhibited.

CELLS	% H36+ (IFU)		DIFFERENTIATION	% DESMIN+
L8E63	100	(54)	+	(-)*
L8E1A-Cl1	100	(46)	-	(-)
L8E1A-Cl3	100	(7)	-	(-)
L6	100	(ND)	+	(-)
L6Ama27(-Ama)	100	(ND)	+	(-)
L6Ama27(+Ama)	100	(ND)	-	(-)
C2C12	(-)*		+	100
C2E1A-Cl1	(-)		-	82
C2E1A-Cl5	(-)		-	99
C2	(-)*		+	83
C2RT	(-)		-	57
10T½	(-)*		-	0
23A2	(-)		+	0
14B1	(-)		+	0
20A2	(-)		+	0
P275	(-)		+	8

Desmin and H36 expression were scored by immunofluorescence in wildtype replicating cells (L8E63, L6, C2C12 and C2) and in replicating L6Ama27 cells, which do not differentiate in the presence of α-amanitin (18); in clones 1 and 3 and clones 1 and 5, respectively, of L8 and C2 myoblasts transfected with the adenovirus E1A gene (16); and in C2RT cells, C2 cells in which the ras gene is constitutively expressed (17); 23A2, 14B1, 20A2 are myogenic clones derived by treatment of 10T½ cells with azacytidine; P275 cells are myogenic cells derived by transfection of 10T½ cells with MyoD1 (13). At least 200 cells in each group of 2 day cultures were scored. The mean relative immunofluorescence units (IFU) determined by photometry on 50 cells are given. H36 fluorescence was reduced in L8E1A-Cl3 cells but all cells were positive. ND = not done. * Note: Although primary rat myoblasts do express desmin, L6 and L8 replicating myoblasts do not. C2 mouse myoblasts do not express H36 since the species specificity of anti-H36 antibody is restricted to rat.

Secondly, the model also predicts that if myogenesis is induced distally, that is by invoking a series of events at a stage relatively late in myogenesis, that such myogenic cells may not express desmin nor H36 as replicating myoblasts. To test this possibility we examined three lines of myogenic cells derived from azacytidine treated 10T½ cells that were kindly provided by Charles Emerson and found that as replicating myoblasts these cells do not express desmin. Furthermore, only 8 percent of P275 cells, a clone of MyoD1 transfected 10T½ cells, express desmin.

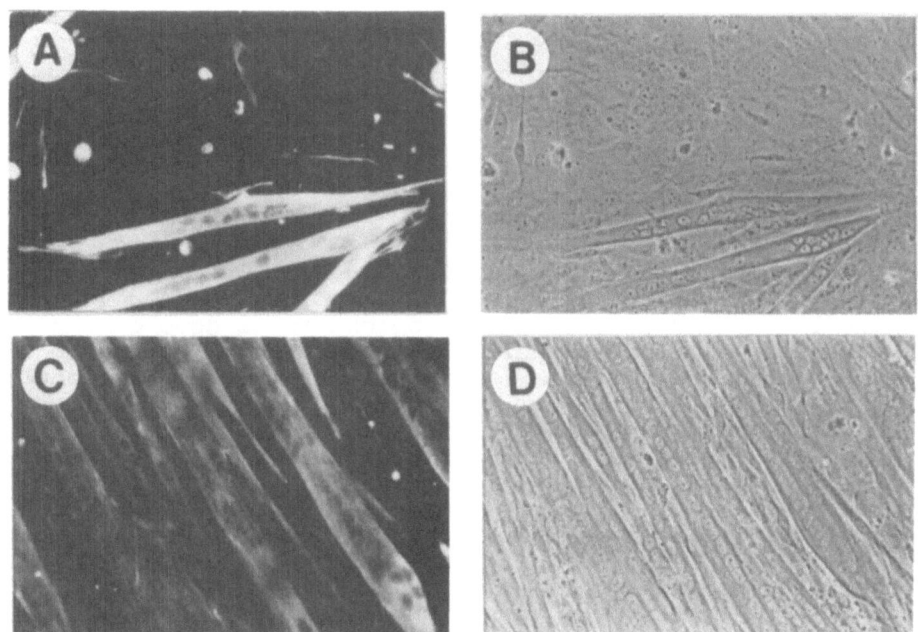

Figure 4. Enhanced myogenic development on a laminin substrate. Cells from newborn rat hindlimb were grown in vitro on collagen (A,B) or polylysine/laminin (C,D) for 5 days. Desmin immunofluorescence (A,C) and phase contrast images indicate the enhanced myotube formation in cultures grown on laminin. This increase in myogenic development is due to the selective maintenance of proliferation of myoblasts grown on laminin (20).

Both of these results support the contention that an earlier regulatory event precedes those that are invoked in terminal myogenesis.

The next point I'd like to make is that the composition of the myoblast membrane can be modulated by fibroblasts. In these experiments, conducted with Jeffery Schweitzer and Marc Dichter, myoblasts were distinguished from fibroblasts in cultures of newborn rat thigh muscle by their expression of H36. Thy-1, a cell surface glycoprotein, was shown not to be expressed on skeletal myoblasts in heterogeneous cultures of myoblasts and fibroblasts. However, when the fibroblasts are removed, the remaining myoblasts do express thy-1. Furthermore, medium conditioned by fibroblasts greatly suppresses this increase in expression of thy-1 by the myoblasts (19). Thus, an interaction between fibroblasts and myoblasts, mediated by a soluble non-dialyzable molecule, modulates expression of thy-1 on the myoblast surface. In these experiments the presence or absence of thy-1 did not affect fusion of these cells. It should also be noted that myoblasts that do express thy-1 exhibit a flatter morphology than the more familiar spindle-shape of thy-1 negative myoblasts and may be confused with fibroblasts. H36 antibody, however, clearly distinguishes both populations as myoblasts.

Myoblast shape and behavior, specifically proliferation, can also be modulated by proteins in the extracellular matrix. As seen in these photomicrographs of cells stained with

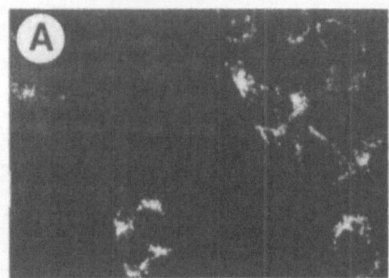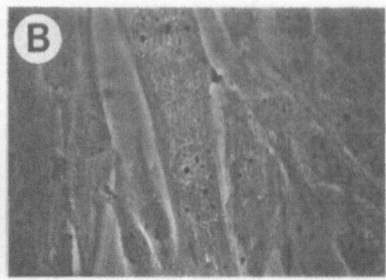

Figure 5. Endocytosis of α2-macroglobulin is developmentally
regulated. A monoclonal antibody against α2-macroglobulin
was used to show the endocytosis of serum α2-macroglobulin
in myoblasts but not in myotubes (21). This cessation of
uptake of macroglobulin occurs subsequent to the cessation
of DNA synthesis during terminal differentiation (22).

antidesmin antibody (Figure 4), many more myotubes develop when
newborn rat hindlimb cells are grown on laminin as compared to
collagen or gelatin.   Using anti-BrdUrd, -desmin and -H36
antibodies, we demonstrated that this was a consequence of the
selective maintenance of proliferation of myoblasts grown on
laminin (20). Furthermore, the capacity of myoblasts taken from
earlier stages of development to respond to laminin in their
extracellular matrix progressively increases (20).   It is
significant to note that this relative responsiveness to
laminin in vitro parallels when laminin is actually localized
in the developing hindlimb and the presence of laminin in vivo
corresponds to the time when rapid muscle growth is occurring.
We suggest that this relative state of responsiveness of
myoblasts to laminin may be useful in further defining stages
in the myogenic pathway.   It is also worth noting that
myoblasts grown on laminin are also much flatter and stretched
out, and thus resemble fibroblasts, rather than the narrow, on
collagen.

        Lastly,   our immunologic evaluations of the myoblast
surface have allowed us to conclude that endocytosis is
developmentally regulated during myogenesis (21, 22). As seen
in Figure 5, the endocytosis of alpha-2 macroglobulin is turned
off in myotubes and we believe this regulation of endocytosis
is a more general phenomena.

        In summary, although immunologic analyses of the myoblast
membrane have not as yet revealed the molecules and mechanisms
by which myoblasts interact and fuse, goals shared by many of
us who have adopted this approach, I have pointed out how in
fact such analyses have resulted in several interesting
discoveries about myogenic development, namely, how muscle-
specific membrane markers have led to the definition of a
myogenic lineage comprised of distinct cells, and how this in
turn expands our view of the molecular regulatory events during
myogenesis.

        In addition,   these analyses of the myoblast membrane
demonstrate   that   myoblast   morphology,   proliferation and
membrane   composition   are   modulated   by   fibroblasts   and
extracellular matrix proteins and, that membrane trafficking
is developmentally regulated.   All of these properties of
myogenic cells may be important to consider as one proceeds to

identify and isolated appropriate cells to be used for therapeutic transplantations.

ACKNOWLEDGEMENTS

We thank Dr. M. Pearson for providing the L6Ama27 cells, Drs. L. Kedes and K. Webster for the L8E1A and C2E1A cells, Dr. E. Olson for the C2RT cells, and Dr. C. Emerson for the 10T½ derived myogenic cells. Rachel F. Foster has played an integral role in the conduct of these experiments. Supported by grant GM28842.

REFERENCES

1. Lee, H. U. and Kaufman, S. J. (1981). Use of monoclonal antibodies in the analysis of myoblast development. Devel. Biol. 81:81-95.

2. Kaufman, S. J. and Foster, R. F. (1984). Antigenic changes on the myoblast membrane accompany development. Exp. Biol. Med. 9:57-62.

3. Lee, H. U., Kaufman, S. J., Coleman, J. R. (1984). Expression of myoblast and myocyte antigens in relation to differentiation and the cell cycle in the rat L8 muscle cell line. Exp. Cell Res. 152:331-347.

4. Kaufman, S. J. and Foster, R. F. (1985). Remodeling of the myoblast membrane accompanies development. Devel. Biol. 109:1-14.

5. Foster, R. F., and Kaufman, S. J. (1985). Cell-surface events during myogenesis: Immunofluorescence analysis using monoclonal antibodies, In: Methodological Surveys in Biochemistry and Analysis, vol. 14, Antibody Combining Sites: Their Investigation and Exploitation in Subcellular Studies, eds. Reid, E. and Moore, D. J., Plenum, 167-176.

6. Lowrey, A. A. and Kaufman, S. J. (1989). Membrane-cytoskeleton associations during myogenesis deviate from traditional definitions. Exp. Cell Res. 183:1-23.

7. Kaufman, S. J., Foster, R. F., Haye, K. R., and Faiman, L. E. (1985). Expression of a developmentally regulated antigen on the surface of skeletal and cardiac muscle cells. J. Cell Biol. 100:1977-1987.

8. Kaufman, S. J. and Foster, R. F. (1988). Replicating myoblasts express a muscle-specific phenotype. Proc. Natl. Acad. Sci. USA 85:9606-9610.

9. Kaufman, S. J. and Foster, R. F. (1989). Preterminal differentiation in the myogenic lineage. In: Cellular and Molecular Biology of Muscle Development, eds. Stockdale, F. and Kedes, L., A. R. Liss, Inc., 47-55.

10. Davis, R. L., Weintraub, H. and Lassar, A. B. (1987). Expression of a single transfected cDNA converts fibroblasts to myoblasts. Cell 51:987-1000.

11.   Tapscott, S. J., Davis, R. L., Thayer, M. J., Cheng, P.-
F., Weintraub, H. and Lassar, A. B. (1988). MyoD1: a nuclear
phosphoprotein requiring a myc homology region to convert
fibroblasts to myoblasts. Science 242:405-411.

12.   Wright, W. E., Sassoon, D. A. and Lin, V. K. (1989).
Myogenin, a factor regulating myogenesis has a domain
homologous to MyoD. Cell 56:607-617.

13.   Pinney, D. F., Pearson-White, S.H., Konieczny, S. F.,
Latham, K. E. and Emerson, C. P., Jr. (1988). Myogenic lineage
determination and differentiation: evidence for a regulatory
gene pathway. Cell 53:781-793.

14.   Edmondson, D. G. and Olson, E. N. (1989).   A gene with
homology to the myc similarity region of MyoD1 is expressed
during myogenesis and is sufficient to activate the muscle
differentiation program, Genes and Devel. 3:628-640.

15.   Braun, T., Buschhausen-Denker, G., Bober, E., Tannich, E.
and Arnold, H. H. (1989). A novel human muscle factor related
to but distinct from MyoD1 induces myogenic conversion of 10T½
fibroblasts. EMBO J. 8:701-709.

16.   Webster, K. A., Muscat, G. E. O. and Kedes, L. (1988).
Adenovirus E1A products suppress myogenic differentiation and
inhibit transcription from muscle-specific promoters. Nature
332:553-557.

17.   Gosset, L. A., Zhang, W. and Olson, E. N. (1988).
Dexamethasone-dependent inhibition of differentiation of C2
myoblasts bearing steroid inducible N-ras oncogenes. J. Cell
Biol. 106:2127-2138.

18.   Pearson, M. L. and Crerar, M. M. (1982). RNA polymerase-
II mutants defective in myogenesis, In: Muscle Development:
Molecular and Cellular Control, eds. Pearson, M. L. and
Epstein, H. F., Cold Spring Harbor Press, 259-267.

19.   Schweitzer, J., Dichter, M. A., and Kaufman, S. J. (1987).
Fibroblasts modulate expression of Thy-1 on the surface of
skeletal myoblasts. Exp. Cell Res. 172:1-20.

20.   Foster, R. F., Thompson, J. M. and Kaufman, S. J. (1987).
A laminin substrate promotes myogenesis in rat skeletal muscle
cultures. Devel. Biol. 122:11-20.

21.   Haye, K. R., Foster, R. F., Goff, J. P., and Kaufman, S.
J. (1986). Endocytosis of a2-macroglobulin is developmentally
regulated during myogenesis. Devel. Biol. 114:470-474.

22.   Kaufman, S. J. and Robert-Nicoud, M. (1985).   DNA
replication in rat myoblasts studied with monoclonal antibodies
against 5-bromodeoxyuridine, actin, and a2-macroglobulin.
Cytometry, 6:570-577.

Discussion of Dr. Kaufman's paper

Dr. Engel: There is an elegant example of fibroblast fusion to myotubes: the rescue of cultured dysgenic muscle cells as recently described by Powell and coworkers. Knowing this, do you think that normal fibroblasts could fuse with dystrophic myotubes?

Dr. Kaufman: In controlled experiments, we've looked for the fusion of primary rat myoblasts and cell lines and never found such fusion of the myoblasts with 10 T1/2 cells. However, there was an abstract at the recent Cell Biology meetings showing that 3T3 myoblasts did fuse into myotubes. That has not been our experience at all.

Unidentified Participant: Laminin will selectively promote and maintain proliferation of myoblasts; Steven Hauschka has shown that as well. It also binds heparin. Heparin binds FGF. So the question is, do you feel that there's a role for either the presentation of growth factors to the myoblast, via their association with extracellular matrix proteins, or the opposite side of that coin is, the sequestration of growth factors via their association with extracellular matrix proteins?

Dr. Kaufman: I also agree; that's an important point. What is important is the way that growth factor is presented to the receptor on a fibroblast, say, compared to the receptor on a satellite cell. The two receptors we've seen here, they're heparin-bound FGF or not, produce a very radical difference in the ultimate growth.

GENERAL DISCUSSION

MYOBLAST/SATELLITE CELL ANTIGENICITY

Dr. Engel: Dr. Burns, would you like to make some comments on the problems inherent in myoblast transplantation from your knowledge of problems after bone marrow transplantation?

Dr. Burns: I'm sitting here trying to remember that these are myoblasts, not myeloblasts that we are talking about and how much easier bone marrow transplantation seems to be from what I've heard compared to the task that you folks have ahead of you. It's much simpler. When we have a stem cell that we can't point to but we know it is there and we know that it is going to behave itself and express itself and go through its differentiation in an appropriate manner, if we take it from one body and put it into the next, but that is not clear at this point with myoblast transfer. What cells are you going to use for your graft? Are they going to be from embryos, from fetuses, single cell suspensions? Are you going to inject them? How will they be delivered? We give bone marrow intravenously. We do not have to worry about getting the cells to the appropriate places to replace the bad cells that are there or that we have gotten rid of. Are you going to inject it into the organ? Will there be any thought, or is it crazy to even think that myoblast put in intravenously, for example, could have some sort of receptor-ligand interaction that would locate them in muscle tissue.

Concerning the immunological aspects, and I do not want to say something that will sound a bit like heresy and I do not mean this to discourage work on the antigenicity of myoblast but whatever you do, since you are not really going to do identical twin kinds of transplants, you are going to have antigenic differences.

Some of those you may control in your selection as well do by looking at the major histocompatibility complex products, Class 1, Class 2 and, of course, we get those primarily from family, but there are now bone marrow registries that you can go to. You may have to do something similar to that. But are you going to get these cells from adults? If so, are you going to have injured muscle tissue in order to get enough of the early cells to be practical?

We know that we can do bone marrow transplants across histoincompatibility, major histocompatibility barriers to a certain degree. One antigen meaning one antigen different either at Class 1 or Class 2 loci do about as well as the complete matches. Of many organ transplants, obviously the HLA is an afterthought. When you have got a heart and a liver you look around for ABO compatibility and the physical size. HLA is usually a second thought in all of that. Here if you are going to be able to culture the myoblast you have the option, since this is not an emergency kind of surgery or doing some matching and trying to get as closely matched as you can at the MHC. But even if you do that, some graft rejection in bone marrow transplantation is not due to MHC differences. They completely match in fact at the gene blocs determining MHC. It is the minor antigens, which we do not even know what they are, that you also have to take into account. We may have some ideas about what some of them are but there are going to be differences and you end up having to approach it from massive global immunosuppression rather than simply trying to match one or two specific antigens.

It might help if you can identify some major MHC determinants on a specific organ. But in all this work the role of the appearance of antigens due to growth factor exposure, to interferon, we have certainly been impressed with the appearance of Class 2 antigens and others due to gamma-interferon and other molecules. Practically any cell you look at is going to express different antigens at different points.

Also, if there is an inflammatory reaction going on, it is going to be exposed to all sorts of growth factors and it would be a massive job to look at this in detail and try to categorize it. Frankly, it may not be worth it in the short term. It may be worthwhile in the long term to do that but in the short term, you are probably going to have to depend on immunosuppression.

Dr. Engel: You have put your finger on many of the vexing problems that will perhaps be addressed further during this meeting. However, graft versus host reaction should not be a problem in myoblast transplantation because lymphocytes are not introduced into the host. The targeting of myoblasts to muscle will be a serious problem. Dr. Brown, could you comment on that from your experiments on intravenous injection of myoblast?

Dr. Brown: Well, we actually have very little hard data that is safe to talk about in this kind of meeting. We have performed initial experiments in conjunction with the laboratory of Dr. Kunkel and, in particular with Dr. Eric Hoffman, looking at direct injection of myoblasts taken from culture. Mixed myoblast/fibroblast cultures into mdx mice and as one arm of that we have begun to do interarterial injections. There is really very little preliminary data worth commenting on. The first thing though one could say is that if one does not do that right, it is a very lethal procedure. Animals injected with any more than a very small number of cells, 1 or 2 or 3 times 10 of the sick cells, die very quickly, perhaps because of pulmonary sludging. But in some instances, with injections of very small numbers, we have had animals survive in one or two preliminary cases; Dr. Hoffman thinks that there is an appearance that looks like successful takes. But I think that is all very preliminary, and really one would not want to say much more at this time.

Dr. Engel: Would you like to make any other comment, Dr. Brown?

Dr. Brown: Well, I have a couple of comments actually for Dr. Karpati. One is that we have also been able to look at Class 1 antigen expression in human muscle in culture and I certainly agree very much with you that we see it expressed on myoblasts and on myotubes. We do not see it expressed in normal human muscle fibers. Just when it is down-regulated, and what the factors are that lead to that we do not know. The other issue that I would raise is the question of whether or not we might learn anything from experience in the heart muscle transplantation field. As you know, rejection of whole organ heart transplants is a major issue in the success of that therapy. It is very clear, as I understand the literature, that normal heart does not express Class 1 or Class 2 antigens but that, in fact, both are expressed. Certain Class 1 and possibly Class 2 during the rejection process and it is interesting and this appears in Dr. Karpati's comments, that the inflammatory process seems to be critical, closely correlated with the expression of both types of antigen or both types of determinants.

The final point from that is simply that the arm of Class 2 expression may be important insofar as if there is activation of helper T cells and lymphokine release. Some of the lymphokines themselves enhance expression of both Class 1 and Class 2 determinants which could positively reinforce the rejection process.

Dr. Engel: Dr. Karpati, could you comment on this?

Dr. Karpati: With regard to the type 2 MHC expression, I do not think that anybody ever saw MHC Class 2 expression on the surface of any type of skeletal muscle cell in vivo even in pathological muscle. However, I cannot be sure that some Class 2 expression may not occur in some of the injected allogeneic myoblasts. We should carefully monitor MHC expression in the transferred cells after injection into the host.

The second thing that emerged from this discussion is that putting in a mixture of cells in the transplantation procedure (i.e., myoblasts plus fibroblasts plus endothelial cells and so on) is probably counter-productive because the nonmyogenic cells, particularly the endothelial cells are MHC "dynamite"; they are richly endowed with Class 1 and 2 MHC. I think that this leads me back to Dr. Engel's earlier question which was this: "How important is it to inject pure myoblasts into these muscles for transfer experiments?" I would say that from the immunological point of view alone, it would make sense to inject purified myoblasts. This is exactly what we have done with our normal human myoblasts prior to injecting them into the mdx mouse. We are in the process of repeating these experiments using various immunosuppressive agents to see if the mosaicization rate could increase.

Dr. Whalen: I want to address a question to Dr. Watt because I am concerned about one aspect of your experimental design. In the experiments you presented, you are actually dealing with two pairs of mice, one tolerant and the other intolerant, which comprises four mice. Correct? But could you not use three mice, instead of four? In other words, always have the same host, and putting in muscle or myoblasts from one of the two strains, one histocompatible and the other not. Because by using four mice, in other words, two pairs, there might be other factors that you cannot know a priori, which might influence the overall results, so if you always use the same host, then to a certain extent, you can overcome that. I would suggest a further model because in light of what we just heard, clearly the clinical problems associated with rejection are going to have a certain importance, at least in the near future and, experimentally, one needs good models to test this. Would it not be advantageous to take a laboratory strain of mice, anyone presumably, chosen judiciously, cross that with the wild stream, then work with the F-1's and subsequent generations in order to do experiments where the polymorphism would be enhanced and do experiments grafting father into son and brother into brother as a means of better mimicking the random polymorphism one generates in human populations in an experimental model where hopefully some of these aspects could be better controlled.

Dr. Watt: I think that the problem here is that with using GPI marker, we cannot just take any strain of mice and cross them over because we have to use one which has passed through which one is slow, so that is the reason we chose these particular strains of mice. I mean, these mice are all inbred strains that we have used anyway.

Dr. Whalen: Between the laboratory strains at least in RLFP's you have very few polymorphisms. That is why I suggest using something totally different, a mouse which is going to introduce a lot more polymorphic antigens. That is the point, of course, of that aspect of the whole thing, and presumably, one could find other genetic techniques or simply looking at other markers, something which would allow you to do the same thing you do with GPI.

Dr. Tapscott: Could you take the patient's own cells, either myoblasts or fibroblasts, expand that cell population and introduce the dystrophin gene or normal X-chromosome, then select the cells that have retained the X-chromosome. Then you could go and put those directly back in; if you used fibroblasts, you could confer the fusion ability of these fibroblasts by expressing one of the myoD related genes in these cells by transferring that gene as well. This would get around the problems of rejection. I think that the initial response might be that it is technically difficult. I think that the panel has pointed out that it is also technically difficult to get around graft rejection as well.

Dr. Fardeau: Dr. Karpati, how do you explain the difference in the expression of Class 1 histocompatibility complex antigens in inflammatory myopathies and in Duchenne dystrophy?

Dr. Karpati: Well, I have to first define what the findings are because there is not an absolute agreement amongst workers. We find that in Duchenne muscle fibers, other than the regenerating ones, there is no Class 1 MHC expression. Whereas there is one group in England which finds MHC expression in most Duchenne muscle fibers. I think that Dr. Engel's findings are more in line with ours but he can speak to that. There is no reason why nonregenerating Duchenne muscle fibers would be stimulated to express their Class 1 MHC loci, whereas, in inflammatory myopathies, there may be reasons. For example, a virus such as HTLV 1 or HIV could stimulate MHC expression in muscle cells.

Dr. Engel: Actually we are not in complete agreement with Dr. Karpati. In Duchenne dystrophy, not only regenerating fibers but also fibers which are focally surrounded by and invaded by T cells are MHC Class 1 - positive.

Dr. Karpati: Yes, but there is a vast difference between this statement and the allegations of the other group which says that 100% of the fiber shows massive Class 1 expression that is comparable to the expression in polymyositis. What you say is right, but it may not involve more than say one percent of the fibers.

Dr. Engel: It depends on the biopsy.

Dr. Partridge: I wish to point out that what Dr. Karpati said about interpreting Dr. Watt's data was not quite correct. What she actually showed was that a very heterogeneous population of cells, extractable from muscle, including endothelial cells no doubt, was not anywhere near as immunogenic as the whole muscle transplant, nor as immunogenic as cells extracted from an inflamed muscle that was regenerating. It is probably the macrophages, I imagine, that are setting off that rejection. So I think that there is some evidence that within certain combinations, even with nonhistocompatible combinations, you can get a tolerance as it were, to inject foreign cells, including cells which are antigen-presenting cells, might be endothelial cells.

The second point I would like to make is that the particular strains used were selected because that is the classical mouse strain which is tolerizable by neonatal injection across major histocompatibility boundaries and so that provides a useful comparative model. If you use other totally histocompatible strains other than the one she particularly used, you can get very prompt rejection. So there are obviously minor histocompatibility antigens about. Probably in those cases, the matching MHC locus would exacerbate that effect because many of the histocompatible antigens are MHC restricted. So a good match in that sense might be disadvantageous.

Dr. Burns: The problem is going to be not so much the antigens you see but the ones that you do not see, and every transplant we do that does not involve an identical twin ends up with tolerance as long as they do not have chronic GBH. They eventually end up with tolerance to lots of minor antigens, hundreds, maybe thousands of minor antigens because of development, we think, suppressor T-cells and in some cases some deletions and that is going to be the case in a practical sense, no matter what one does.

Dr. Partridge: Can I follow up with a further comment which is that in the muscle grafting setup, there is a further complication which does not exist in bone marrow transplants, so far as I know. If we get these mosaic fibers that we are all after, those mosaic fibers will express both the host and the donor histocompatibility antigens and that puts them in a very unusual position. They are mosaic for histocompatible antigens, and that may change their immunological status, as compared with, say, the purely donor type.

Dr. Hauschka: Just a practical extension of Dr. Tapscott's comment. I wish to comment about trying to circumvent the immune response by using the patient's own cells. If we imagine that five or so years down the road, we will know the right way to introduce genes into muscle cells and get them expressed appropriately, I think that people could keep in mind that it might be advisable, in fetuses at risk for dystrophy that are diagnosed and come to term, that it would not be a bad idea to save things like the umbilical cord, perhaps the placenta, i.e., non-invasive sources of the patient's own cells. These could be worked with in the present to get them into myogenic states and we would be ahead of the game when it comes to trying to reintroduce these types of cells back into the patient suffering from this disease.

Dr. Steinman: Dr. Watt's results are encouraging for myoblast transfer. I want to raise a few points about the inducibility of MHC Class 1 and Class 2 and of other antigens. What is very striking is that genotypes of MHC Class 1 and Class 2 are inducible to different degrees which underscores what Dr. Partridge has just said in the mouse studies, H2K strains were far less amenable to transplantation than H2A strains or H2B strains.

Clearly, precise molecular MHC typing is going to be important. A corollary from that is that standard immunohistochemical studies only look at a very restricted number of histocompatibility epitopes. Therefore, one must look at MHC expression in other ways as well.

I also want to comment about two other points. One is passenger antigens. With myoblast transfer we are in a very fortuitous position because we are not transplanting an organ with passenger endothelial cells or lymphocytes. However, the purity of the myoblast will be very important, and I stress that growing myoblast under FDA approved GMP will be extremely important, so as not to transfer passenger viruses or other infectious agents.

The second point is that myoblast transfer will be a very important experiment that immunologists and immunobiologists will watch very carefully. Normally, individuals grow up tolerant to their own dystrophin. The transplant recipients have in some cases never seen dystrophin. What will happen when in a successful transplant in terms of self-tolerance with the appearance of a foreign antigen, even though it may be sequestered and intracellular.

Dr. Engel: Dr. Askanas in the mid-70's has noted that an avian virus is omnipresent in cultured chicken cells. Dr. Askanas, have you made any observations in cultured human cells that might suggest the presence of a virus, under the usual culture conditions?

Dr. Askanas: Not as yet.

Dr. Emerson: Dr. Tapscott and Dr. Hauschka and I commented on using cells from the patient as a vehicle for gene transfer by converting nonmuscle cells to myoblasts by using muscle specific enhancers. An even more radical approach would be to avoid the cells altogether and to develop safe viral vectors, like rhinoviruses that infect muscle cells, to introduce these genes directly into muscle.

Dr. Blau: That may be a very useful approach, especially because the proliferative capacity of myoblast from even the youngest Duchenne patients is extremely limited.

Dr. Karpati: Can one contemplate direct transfection of the dystrophin gene into the dystrophic fibers or the patient's own myoblastic cells in vitro and reinjecting them? This is a two-megabase gene. Is this a realistically possible approach, or will the dystrophin cDNA with a proper promoter suffice?

Dr. Caskey: I was going to make a comment about the possibility of immune response to the dystrophin protein. In the early days of the study of factor 9 deficiency there was the early suggestion that carried deletions and, therefore, were null for factor 9, had a greater frequency of immune response. A subsequent study of that was not held up so the initial that a null gives rise to a greater immune response, even for a soluble protein where the patients have been continually immunized by the administration of factor 9, I do not think it has held up as a significant factor.

In the case of dystrophin where many will be nulls because of the high deletion rate in the disease, the protein is contained intracellularly and I would guess this would be less of a problem. I would like to make a comment to those who do not keep up with the gene transfer literature just to make a couple of quick points about the progress that has been achieved thus far.

This work is about, I would say, four to five years into development. At this point there are safe retroviral vectors, meaning nonlive viruses in the helper cells and they have been engineered now in a variety of redundant safety modalities. Titers of viruses can be obtained at a very high level. One can now get with the use of certain trophic factors in bone marrow transfection, as high as 90 to 95% of all rescued mice carrying the stem cells of mouse and expressing the human gene in periods in excess of six months at levels in excess of the endogenous level of ADA so there has been substantial progress made. Now major pitfalls or major difficulties in moving to the human trials will be: what are the assay systems that can be developed to identify a human stem cell line? So this is the state of the art in the hematopoietic system, and I think we are beginning at a lower level than that in muscle.

Dr. Tapscott: The fact that whole chromosomes can be transferred from one cell suggests that genes of virtually any size could be transferred successfully. In addition, the cDNA of the gene much smaller than the gene and exactly how much of the protein is necessary to complement the defect in the

Duchenne mutation is not yet known. Perhaps one needs just a portion of the protein to replace the function of dystrophin.

As regards to expanding cell populations, neither myoblast nor fibroblast populations can be expanded infinitely but there are approaches that can be taken to temporarily transform or expand these cell populations that would potentially be reversible once they were put back into a host. I also agree with Dr. Emerson that it would be ideal if a virus infecting muscle could also introduce dystrophin into muscle. One potential difficulty of that is maintaining expression in muscle cells, since most viruses will not integrate into the genome.

Dr. Worton: I was just going to make some of the same points. I may also add that one need not put in a two-megabase gene to get in the whole protein. You really only need cDNA perhaps with some processing signals, so it is realistic to think of putting in a gene at some time. Also the vectors are changing over the course of time and with things like yeast artificial chromosomes you have the possibility of mammalian artificial chromosomes and other vehicles for getting genes in. So I think that is possible.

What we have been worried about mostly is the point that Dr. Blau made and that is: where do you get cells with high proliferative capacity because of the problem, of course, in the Duchenne kids with their low proliferative capacity in their myoblasts?

Dr. Ham: I have a partial answer for the question of Dr. Worton just raised. We have cultured muscle cells from two Duchenne patients, aged 8 months and 5 months. The number of clonable cells and the proliferative capacity of the clones were both much higher in these muscle cultures from the young infants. It appears that the model of using up stem cell capacity is correct; therefore, one should obtain cells from young infants.

In those cases where a diagnosis is made very early, it may be possible to take some of the patient's own cells, introduce dystrophin into the cells, and then use these cells for transplants, either immediately or at a later time.

Dr. Cooper: Dr. Burns, from your experience with bone marrow transplantation, could you tell us the range of time in which delayed hypersensitivity reactions occur and, in that context, are you aware of one particular immunosuppressive agent which stands out in terms of a balance between efficacy and safety?

Dr. Brown, in your experiments where you found that myoblasts were lethal to the host, were the injections given as a bolus intravenously? If they were, would a slow infusion over a long period of time with smaller numbers of cells be more effective?

Dr. Burns: The answer to the first questions depends partly on what your immunosuppressive regimen is and whether there is a complication like graft versus host disease. Cyclosporin is the obvious immunosuppressive agent of choice in 1989. We and others are looking at many different new immunosuppressive agents, such as thalidomide, an old drug with cyclosporin-like qualities. Today, cyclosporin would clearly be the drug of choice.

Dr. Brown: I will try to answer the second question. It is certainly worth pursuing the line of either intravenous or interarterial injections hoping that a systemic method of administration will get around the really very primitive nature of direct injection into muscle, which is what is now being tried.

Obviously, questions arise. Do myoblasts sludge in the capillary bed? Do they have motility and the ability to migrate from the vessels into injured muscle? We do not know. Dr. Partridge's work suggests that myoblasts can migrate a considerable distance in minced muscle grafts, but I do not know whether that is pertinent to systematically injected cells.

With respect to the points that Dr. Blau and Dr. Ham raised, I want to ask Dr. Watt if there is any evidence in the successful long-term transplants that residual donor myoblasts were sequestered in a satellite cell-like state and thus were available for subsequent cycles of muscle regeneration.

Dr. Watt: We have noted that transplanted cells can move considerable distances. We have no evidence whether the cells actually have gone through the circulatory system. That is something we have to look at.

Dr. Askanas: Can injected bone marrow cells home in a muscle?

Dr. Watt: We have injured a muscle, and then intravenously introduced donor bone marrow cells to see if they will home in that muscle. We never found them homing into that muscle.

SECTION 3

PRACTICAL ASPECTS OF MYOBLAST IMPLANTATION

THE PRINCIPLES AND PRACTICE OF MYOBLAST TRANSFER

George Karpati

Neuromuscular Research Group
Montreal Neurological Institute
3801 University Street
Montreal, (Quebec), Canada, H3A 2B4

INTRODUCTION

Myoblast transfer is a cell transplantation procedure and it consists
of injecting suspensions of viable myoblasts into muscle(s) of a living host.
Some of the injected myoblasts may ultimately fuse with some of the host
muscle fibers and thereby the transferred myoblast nuclei become functional
myonuclei of certain muscle fibers of the host. Thus, myoblast transfer can
lead to the formation of "mosaic" muscle fibers, in which, in addition to the
native myonuclei, the functional myonuclei of the transferred myoblasts are
present expressing their genome. This type of myoblast transfer has been
used experimentally by several groups of investigators in different rodent
species[1-10] and the formation of "mosaic" muscle fibers in the host
muscle was demonstrated either by the appearance of hybrid forms of
strain-specific isoenzymes or by microscopic techniques using nuclear markers.
The potential usefulness of myoblast transfer in negating gene defects in
some of the host muscle fibers has been raised a few years ago[7-11]. More
recently, transfer of normal allogeneic myoblasts into dystrophin-deficient
muscles produced dystrophin-positive fibers that escaped dystrophic
damage[12,13]. In the dy/dy strain of mice, in which the basic genetic and
biochemical defect is still unknown, normal myoblast transfer was reported to
have negated some of the microscopic pathological features and functional
abnormalities of the injected muscles[14]. On the basis of the successful
normal myoblast transfer experiments in the mdx mice[12,13], investigation
of the feasibility of a similar approach in the treatment of certain genetic
human muscle diseases, such as Duchenne muscular dystrophy appeared to be
justified.

THE RATIONALE FOR MYOBLAST TRANSFER

The following unique characteristics of skeletal muscle fibers and
myoblasts make myoblast transfer feasible.
1)   The skeletal muscle fiber is a long, multinucleated cell in which every
myonucleus expresses a full genome with a characteristic cytoplasmic domain
for a given gene[15]. In such a cell, therefore, there is a possibility of
the coexistence of myonuclei of differing gene expression characteristics
("mosaic" fibers)[16].
2)   Normal myoblasts (satellite cells) are capable of fusing not only with
each other but with established muscle fibers. Such occurrence takes place
during normal growth and development of muscle fibers, when some activated

satellite cells fusing into their parent fibers contribute their nuclei to the myonuclear population of the parent fiber[17].

Thus, after myoblast transfer into a muscle of a young animal, the injected myoblasts may first assume a satellite cell position (in Go stage of the cell cycle) under the basal lamina, and they fuse into the parent host fibers at a predictable rate during growth. Another way by which such transferred myoblasts (satellite cells) could become incorporated into the host parent fibers is during regeneration, after the parent fiber segments had suffered a cycle of necrosis[4]. In mdx, abundant muscle necrosis occurs in young animals, as part of the natural history of the disease[18], providing a convenient natural stimulus for mosaic fiber formation after myoblast transfer.

FACTORS THAT INFLUENCE THE LONG-TERM EFFECTIVENESS OF MYOBLAST TRANSFER IN

DYSTROPHIN-DEFICIENT MUSCLES

There are 2 major categories of factors that have a variable degree of influence on the ultimate aim of normal myoblast transfer in dystrophin-deficient muscles. Within these 2 categories there are several further items of importance.

1. A high degree and widespread mosaicization of the host muscle fibers with donor myoblast nuclei

In other words, the principal aim is to generate many mosaic fibers in which there is a large number of donor (dystrophin-competent) myonuclei. What are the circumstances that favor this outcome?

a) A large number of pure myoblasts should be injected and this requires an efficient culturing system of satellite cells released from biopsied donor muscle and subsequent purification of myoblasts. Optimization of the culture procedure and the possible use of growth factors might be helpful. Purification of myoblasts from a mixed primary culture may be achieved by dilution cloning[19] but if a very large number of cells are required, this approach becomes impractical and runs the risk of senescence of myoblasts by the time of the transfer. A much better alternative of myoblast purification is by automated fluorescent cell sorting using a monoclonal myoblast-specific antibody (5H.1.11) against a cell surface molecule[20] (N-CAM).

b) Good viability and fusion competence of the transferrable myoblasts is essential. Avoidance of senescence has already been emphasized as an important consideration. Myoblasts should also be free of all microbial (bacteria, virus, protozoon) contamination that could impair their viability and also would threaten the host fibers. Myoblasts may also be mechanically traumatized during the injection procedure; thus, the caliber of the injection needle should be chosen carefully according to the cell density in the injectate. If the transferred myoblasts are not totally compatible with the host's class 1 major histocompatibility complex (MHCP), immunological rejection may destroy or damage myoblasts or some of the host fiber segments in which the histoincompatible donor myoblast nuclei got established. Thus, a judicial use of, at least a low dose and temporary immunosuppression of the host may be required.

c) Wide dispersion of the transferred myoblasts in the host muscle is essential. It is clear that it is not only the number of the donor myoblast nuclei that are important, but also their distribution within the mosaicized host fibers, since the longitudinal cytoplasmic domain of dystrophin expression is likely to be limited (probably 100 micrometers or less)[13]. While the injected myoblasts seem to have some mobility in the interstitial space, in our experience, it is quite limited and does not extend beyond 1 or 2 fascicles. Thus, it is preferable that, the injectate is deposited along multiple closely-spaced tracks along the longitudinal axis of the muscle. The needle will produce some tearing, but the resultant segmental necrosis of the host muscle fibers which, as mentioned earlier, could actively enhance the acquisition of transferred myoblast nuclei by the subsequently regenerating segment. A large amount of dense endomysial connective tissuse, which tends to accumulate in Duchenne muscles can also be a major hindrance of appreciable dispersion of the injected myoblasts.

d) The availability of numerous viable myofibers of the host is essential for effective and extensive mosaicization. Thus, in later stages of Duchenne dystrophy, the effectiveness of myoblast transfer is probably seriously compromised by the severe loss of host muscle fibers and the dense endomysial connective tissue.

2. A normal long-term expression of the dystrophin gene by myonuclei derived from the transferred myoblasts in the mosaicized host fibers

The dystrophin gene in the myonculei derived from the transferred myoblasts must normally be expressed in the mosaicized host fibers[12,13] in order the achieve the ultimate goal of myoblast transfer: i.e. negating the mdx or Duchenne gene defect. In certain instances this may not be the case. For example, if myoblasts derived from female heterozygotes are used, 50% of the myonuclei will have an inactive dystrophin gene because of lyonization[12]. Another example is established cell lines, such as L_6 or G_8, in which dystrophin expresion after in vitro formation of myotubes is very low or absent. It has been suggested that transfecting the myoblasts in vitro before transfer with multiple copies of a dystrophin gene-construct might enhance the efficacy of myoblast transfer. However, in light of the probable restricted cytoplasmic domain of dystrophin[13], such difficult manipulations are unlikely to be useful.

A long-term survival and normal dystrophin gene expression by the myonuclei derived from the normal transplanted myoblasts in the mosaic host fiber is an essential requirement for the ultimate aim of myoblast transfer. There is no biological reason to suspect that the donor myonuclei in the host mosaic fibers would be eliminated or inactivated in the mosaic fiber milieu.

There are a couple of manipulations that could experimentally be useful in augmenting the mosaicization rate of the host fibers by the transferred myoblasts. For example, widespread, experimentally created necrosis (induced by crush or bupivacaine injection) can enhance the incorporation of nuclei of the normal transferred myoblasts in the regenerated fiber segments[13]. An even more potent method is a severe depletion of the endogenous satellite cell population by X-irradiation so that the transferred normal myoblasts face a reduced competition from the native satellite cell pool regarding fusion into the host fibers during growth or regeneration.

Unfortunately none of these methods are practical in human myoblast transfer.

THE POSSIBLE APPLICABILITY OF MYOBLAST TRANSFER TO THE TREATMENT OF

DUCHENNE DYSTROPHY

It is clear that, initially, some well-focussed studies are required to establish 3 critical aspects of myoblast transfer in humans:

a)   Is it feasible?
b)   Is it safe?
c)   Is it efficacious?

Only if these issues are satisfactorily resolved can one contemplate the use of myoblast transfer in multiple muscles of Duchenne patients as a treatment.

The following important points must be addressed in the design of initial human pilot experiments on a single muscle in order to obtain unequivocal answers to the previously posed 3 questions:

A. Proper selection of patients.
B. Thorough pretransfer evaluation of patients and donors.
C. Careful selection of the source of myoblasts (father or younger unaffected brother).
D. An effective method of producing abundant and pure myoblasts (fluorescent cell sorting) and rigorous quality control of the injectiable myoblasts. (Microbial contaminants and pyrogens).
E. Optimization of the injection techniques.
F. Immunosuppression (agent(s), dosage, duration).
G. The modalities and timing of the end-point measurements (force generation, dystrophin assessment, pathology of muscle, image of the injected muscle by CT and MRI).
H. Appropriate control, such as a contralateral sham-injected muscle.
I. Ethical and legal considerations.

FUTURE PERSPECTIVES

In the event that successful completion of the initial pilot studies in a single-muscle experiment in humans leads to a conclusion, that myoblast transfer is feasible, safe and efficacious in Duchenne dystrophy, plans can be developed for the injection of multiple, functionally important muscles at several different sessions. Obviously, the injection of certain very important but deeply situated muscles, such as the psoas and particularly the diaphragm could pose difficult but not necessarily insurmountable technical challenges. Unfortunately, the cardiomyopathy of Duchenne dystrophy cannot be treated by myoblast transfer. However, if by myoblast transfer, the longevity of the Duchenne patient can be extended, the cardiomyopathy may be handled by normal cardiac organ transplant.

What can be expected from myoblast transfer in Duchenne dystrophy under the most optimal imaginable circumstances?

If myoblast transfer is instituted at an early age, an appreciable retardation of the dystrophic deterioration or even stabilization at an acceptable functional level of the treated muscles may be expected. A genuine, major improvement of muscle strength is unlikely to occur, except the one that may be ascribed to the effects of natural growth. Myoblast transfer into seriously scarred muscles in advanced stages of the disease is unlikely to be of significant benefit. Myoblast transfer therefore cannot be considered, even in the best of circumstances as a "cure" for Duchenne dystrophy. However, at present, it is probably the most promising approach on the horizon. Nevertheless, efforts should be continuing for the exploration of other possible therapeutic approaches, such as upregulating of putative synergistic molecules of dystrophin that might compensate for dystrophin deficiency[22,23].

REFERENCES

1. B. H. Lipton, and E. Schultz, Developmental state of skeletal muscle satellite cells, Science 205:1292-1294 (1979).
2. P. H. Jones, Implantation of cultured regenerated muscle cells into adult rat muscle, Exp. Neurol. 66:602-610 (1979).

3. T. A. Partridge, M. Grounds, and J. C. Sloper, Evidence of fusion between host and donor myoblasts in skeletal muscle grafts, Nature 273:306-308 (1978).

4. D. J. Watt, K. Lambert, J. E. Morgan, T. A. Partridge, and J. C. Sloper, Incorporation of donor muscle precursor cells into an area of muscle regeneration in the host mouse, J. Neurol. Sci. (1982).

5. D. J. Watt, J. E. Morgan, and T. A. Partridge, Long-term survival of allografted muscle precursor cells following a limited period of treatment with cyclosporin A, Clin. Exp. Immunol. 55:419-426 (1984).

6. D. J. Watt, J. E. Morgan, and T. A. Partridge, Use of mononuclear precursor cells to insert allogeneic genes into growing mouse muscles, Muscle & Nerve 7:741-750 (1984).

7. P.K. Law, and J. L. Yap, New muscle transplant method produces normal twitch tension in dystrophic muscle, Muscle & Nerve 2:236- (1979).

8. S. Champaneria, G. Karpati, D. Brunet, S. Carpenter, and P. Holland, Myoblast transplantation to dystrophic muscles, Muscle & Nerve 9 (Suppl.):232 (1986).

9. G. Karpati, Y. Pouliot, S. Carpenter, and P. Holland, Allogeneic myoblast transplantation into skeletal muscles of mdx mice, J. Cell Biochem. 337:176-179 (1989).

10. G. Karpati, Y. Pouliot, S. Carpenter, and P. Holland, Implantation of nondystrophic allogeneic myoblast into dystrophic muscles of mdx mice produces "mosaic" fibers of normal microscopic phenotype, in: "Cellular and Molecular Biology of Muscle Development", L. Kedes, and F. Stockdale, eds, A. R. Liss Inc., New York (1989).

11. J. E. Morgan, D. J. Watt, J. C. Sloper, and T. A. Partridge, Partial correction of an inherited biochemical defect of skeletal muscle grafts of normal muscle precursor cells, J. Neurol. Sci. 86:137-147 (1988).

12. T. A. Partridge, J. E. Morgan, J. R. Coulton, E. P. Hoffman, and L. M. Kunkel, Conversion of mdx myofibers from dystrophin-negative to -positive by injection of normal myoblasts, Nature 337:1761-179 (1989).

13. G. Karpati, Y. Pouliot, E. E. Zubrzycka-Gaarn, S. Carpenter, P. N. Ray, R. G. Worton, and P. Holland, Dystrophin is expressed in mdx skeletal muscle fibers after normal myoblast implantation, Am. J. Pathol. 134:27-32 (1989).

14. P. Law, T. G. Goodwin, and M. G. Wang, Normal myoblast injections provide genetic treatment for murine dystrophy, Muscle & Nerve 11:525-5353 (1988).

15. E. Ralston, and Z. W. Hall, Transfer of protein encoded by a single nucleus to nearby nuclei in multinucleated myotubes, Science 244:1066-1069 (1989).

16. P. M. Frair, and A. C. Peterson, The nuclear-cytoplasmic relationship in "mosaic" skeletal muscle fibers from mouse chimeras, Exp. Cell Res. 145:167-178 (1983).

17. A. M. Kelly, satellite cells and myofiber growth in the rat soleus and extensor digitorum longus muscles, Dev. Biol. 8:217-222 (1985).

18. G. Karpati, S. Carpenter, and S. Prescott, Small-caliber muscles fibers do not suffer necrosis in mdx mouse dystrophy, Muscle & Nerve 11:795-803 (1988).

19. S. Champaneria, P. Holland, G. Karpati, and C. Guerin, Developmental regulation of cell-surface glycoproteins in clonal cultures of human skeletal muscle satellite cells, Biochem. Cell Biol. 67:128-136 (1989).

20. C. Webster, G. K. Pavlath, D. R. Parks, and H. Blau, Isolation of human myoblasts with the fluorescenc-activating cell sorter, Exp. cell Res. 174:252-265 (1988).

21. S. M. Gartler, and A. D. Riggs, Mammalian X-chromosome inactivation, Annu. Rev. Genet. 17:155-190 (1983).

22. P. Holland, G. Karpati, C. Guerin, and S. Carpenter, Why do dystro-
    phin-deficient regenerated skeletal muscle fibers of mdx mice
    become resistant to further necrosis?, Neurol. 39:153 (1988).
23. D. R. Love, D. F. Hirt, G. Dickson, N. G. Spurr, B. C. Byth, R. F.
    Marsden, F. S. Welsh, Y. H. Edwards, and K. E. Davies, An auto-
    somal transcript in skeletal muscle with homology to dystrophin,
    Nature 339:55-58 (1989).

Discussion of Dr. Karpati's paper

Dr. Partridge: I wonder if you have markers of your
cells? Do you know where you put them in? Do you have any
details of how far they can move? Do you have any indications
as to how far they can move from the injection site?

Dr. Karpati: We mark the nuclear DNA with 3H thymidine
during culture. I have seen injected cells moving across a
single fascicle, which would be, in a mouse, about the distance
of 20 to 30 host fibers. So the injected cells could move
about at least 1 mm.

Dr. Partridge: You showed a number of pictures of
labelled nuclei which were on the edges of muscle fibers. How
can you be assured that those labelled nuclei are genuinely
myonuclei rather than adjacent nuclei to the muscle fiber?

Dr. Karpati: These are good points. When you see
dystrophin expression in the corresponding fiber segments, you
assume that the labelled (donor) nuclei were myonuclei; i.e.,
situated within the fibers, since otherwise most mdx muscle
fibers would show no dystrophin expression. Of course, there
is a distinct possibility that some labelled surface nuclei
represent donor myoblasts in the satellite position. This
could be verified by immunostaining with antihuman N-CAM
antibody, which we have been doing. N-CAM is positive only in
the satellite cell surface membrane.

Dr. Partridge: You say you have evidence that the fibers
into which these cells have been incorporated do not ever
degenerate. One of the bits of evidence for that was this
peripheral nucleation. And you contrasted that with your
results in older animals, where you said the nuclei were
central. How do you explain that?

Dr. Karpati: If an mdx fiber segment acquires
normogenomic donor nuclei and expresses dystrophin before the
necrotic phase commences (i.e., before age 15 days), it will
not become centronucleated. However, if an already
centronucleated mdx muscle fiber acquires normogenomic donor
nuclei, as is often the case after injecting older animals, the
already existing centronucleation will not be altered even
though dystrophin is present in that fiber segment.

MYOBLAST TRANSFER IMPROVES MUSCLE GENETICS/STRUCTURE/FUNCTION

AND NORMALIZES THE BEHAVIOR AND LIFE-SPAN OF DYSTROPHIC MICE

Peter K. Law, Tena G. Goodwin, H.-J. Li,
Ghaith Ajamoughli and Ming Chen

Departments of Neurology and Physiology/Biophysics
University of Tennessee, Memphis, TN

INTRODUCTION

Dystrophic cells degenerate because they lack the normal genome. In our laboratory, transplant techniques have been developed to facilitate cell survival and cell fusion, such that the normal genome can be incorporated back into the dystrophic muscle[1-6].

Our goal is to develop a universal treatment against muscle weakness in hereditary myopathies. The concept of myoblast transfer technology was first published in 1978, with circumstantial evidence[1]. Throughout its development the C57BL/6J-$dy^{2J}dy^{2J}$ dystrophic mice[7] were used as an animal model of hereditary muscle degeneration.

We hypothesize that the incorporation of cultured normal myoblasts into dystrophic muscles can significantly improve muscle genetics, structure, and function, thus alleviating muscle weakness in hereditary myopathies.

Our approach has been to produce mosaic muscles that contained normal fibers, dystrophic fibers, and fibers mosaic with normal and dystrophic nuclei.

The following video provides evidence of behavioral, functional, structural, and genetic improvements as a result of myoblast transfer therapy in dystrophic mice.

DR. LAW IN VIDEO

Muscular dystrophy in the mouse is characterized by progressive degeneration of skeletal muscles in the hind-limbs and in the chest wall. Dystrophic symptoms first appear at twenty to thirty days after birth and consist of sporadic flexion and flaccid extension of the hind-limbs. Occasionally, the dystrophic mouse walks with duck feet[7].

METHODS

Nineteen mice so identified received normal myoblast injection into the hind-limb and intercostal muscles on both sides of the hosts. About 10^6 myoblasts were injected into each of the following muscle groups: extensors, flexors, quadriceps, hamstrings, and external intercostal muscles. Donor myoblasts were histoincompatible myoblast clones cultured from limb-buds of Swiss Webster mouse embryos. Host immuno-suppression was induced by daily subcutaneous injection of cyclosporine-A (50mg/kg body weight) which was withdrawn in six months.

BEHAVIORAL IMPROVEMENT

From two to eighteen months after myoblast injection, eleven dystrophic mice showed such behavioral improvement that their locomotory patterns were indistinguishable from those of the unoperated normal mice. Sporadic flexion and flaccid extension of the hind-limbs were not seen. Occasionally, they still walked on duck feet. The mouse could now run. Muscle power was increased in both legs. They were able to use their hind-limbs and toes. Their hind-limb muscles were strong enough to support them and to allow them to balance themselves on the glass rod (Fig. 1).

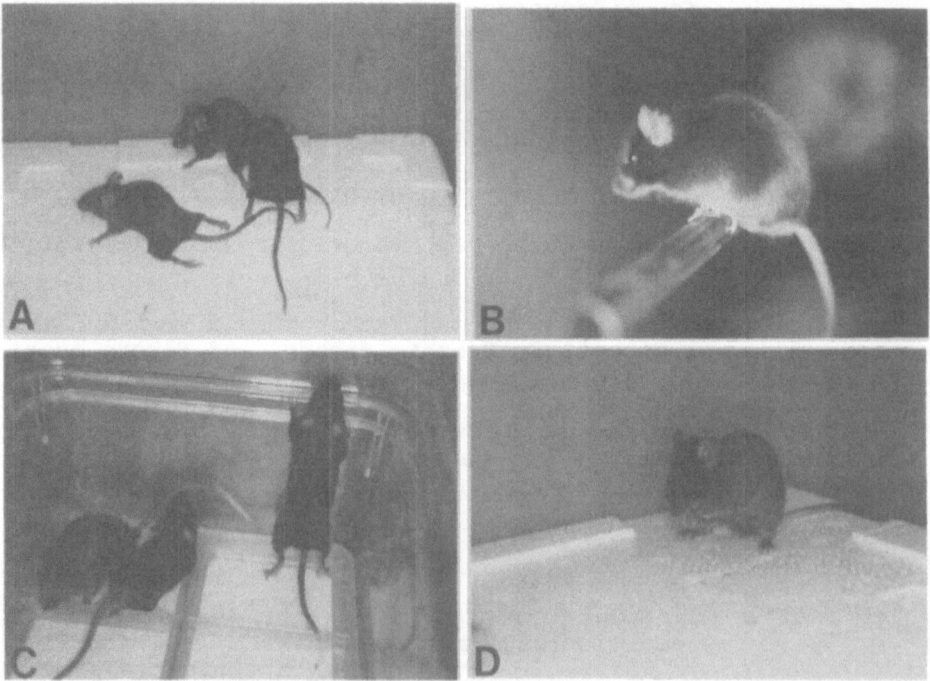

Fig. 1. (A) Various postures of control dystrophic mice; (B) Dystrophic mice responding favorably to myoblast injection and CsA treatment could use their toes to grasp the glass rod; (C) to climb up the cage; and (D) to stand on their hind-limbs.

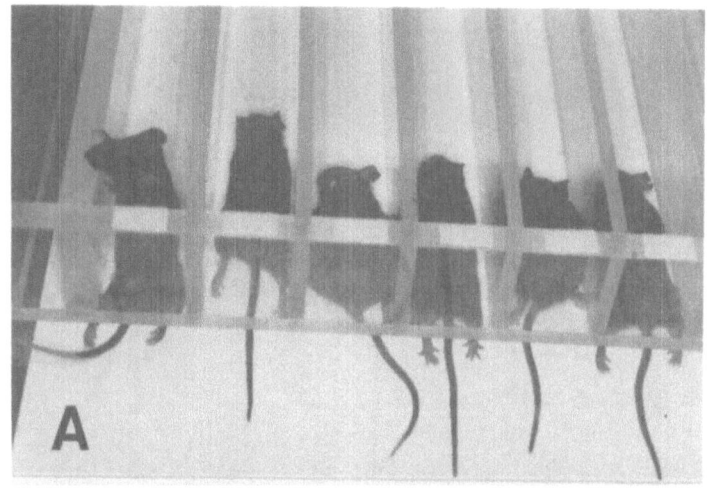

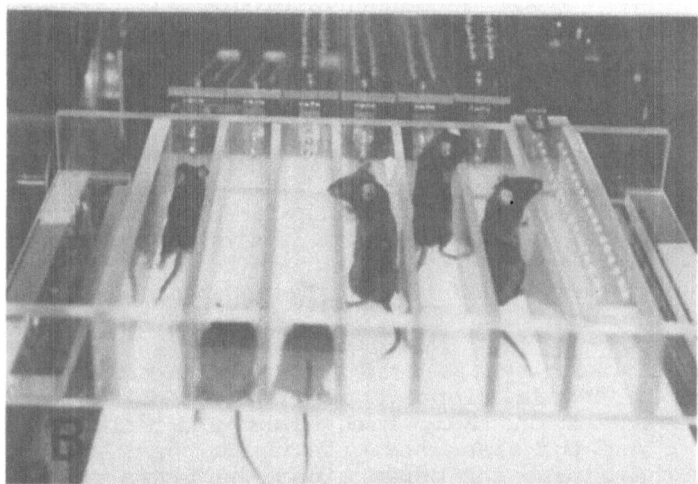

Fig. 2. (A) Before myoblast injection, 4-month old dystrophic mice could not run the treadmill at 2 meters/minute. (B) After myoblast injection, they could run at 2.8 meters/minute. (C) The one that showed most improvement actually outran that speed and could jump out of the treadmill.

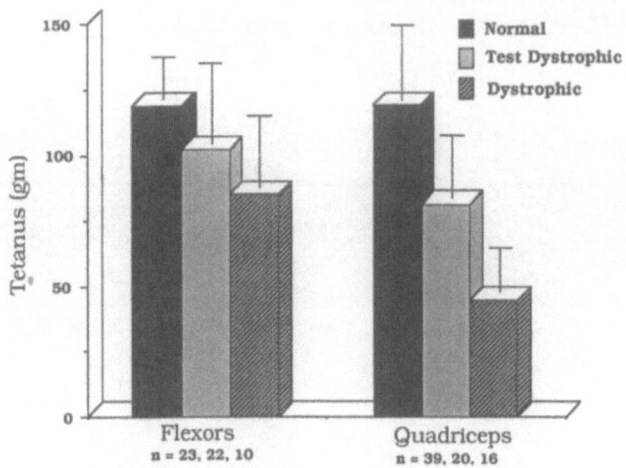

Fig. 3. In vivo mechanophysiology at 37°C showing significantly increased tetanus tensions for myoblast injected dystrophic flexors and quadriceps as compared to dystrophic controls. n is the number of muscle groups studied.

Five mice did not respond at all. Three other mice showed intermediate recovery 2 months after myoblast injection indicating that there was some functional improvement of the muscles. However, when they were tested on the glass rod, their hind-limbs were not strong enough to hold onto the glass rod. Sporadic flexion and flaccid extension of the hind-limbs could still be demonstrated.

One year after myoblast injection, the mice that showed significant behavioral improvement continued to be able to support themselves on their hind-limbs and walked normally. They could grasp the glass rod with their toes, and were able to balance themselves and moved along the glass rod.

The mice that showed intermediate improvement could now walk on the glass rod without showing sporadic flexion and flaccid extension of the hind-limbs. They continued to walk with duck feet.

Fig. 4. (A) The myoblast-injected dystrophic soleus had a larger cross-sectional area, more myofibers, and better fiber-type differentiation than the unoperated dystrophic control or the fibroblast-injected dystrophic control. It closely resembled normal control soleus. Its electrophoretogram showed GPI-1CC, GPI-1BB, and GPI-1BC. Myosin ATPase at pH 9.4. Sections were of the same magnification. (B) Control dystrophic characteristics such as (1) fiber splitting, (2) central nucleation, (3) phagocytic necrosis, (4) variation in fiber shape and size, and (5) infiltration of connective tissues were present in much lower frequency in the dystrophic soleus receiving myoblasts. Myoblast-injected dystrophic soleus closely resembled normal control, although it still showed some abnormalities and its fibers had smaller diameters. Fibroblast injection caused an increase in connective tissues in dystrophic solei. Modified Gomeri Trichrome stain. Sections were of the same magnification. (From Ref. 4)

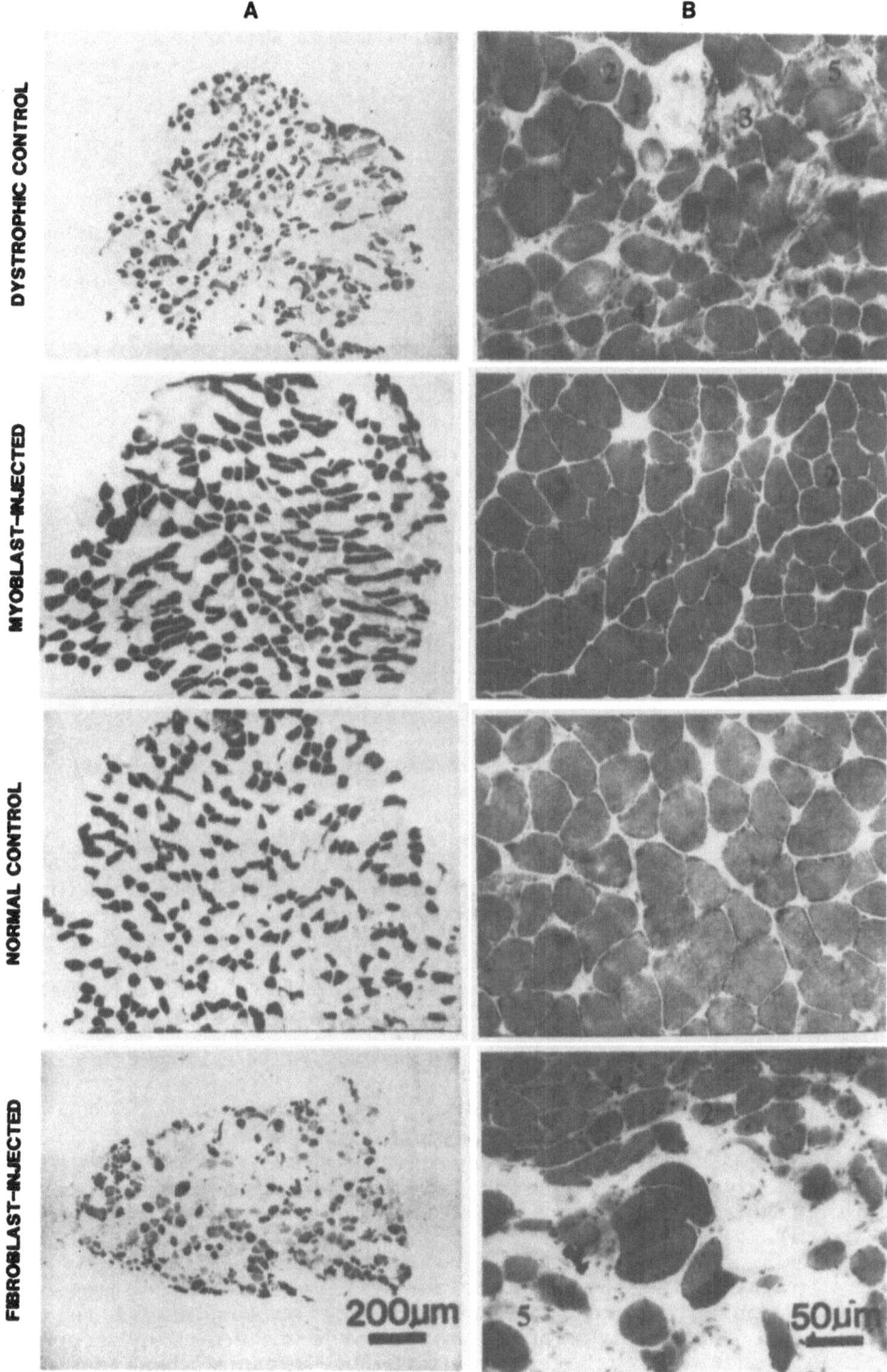

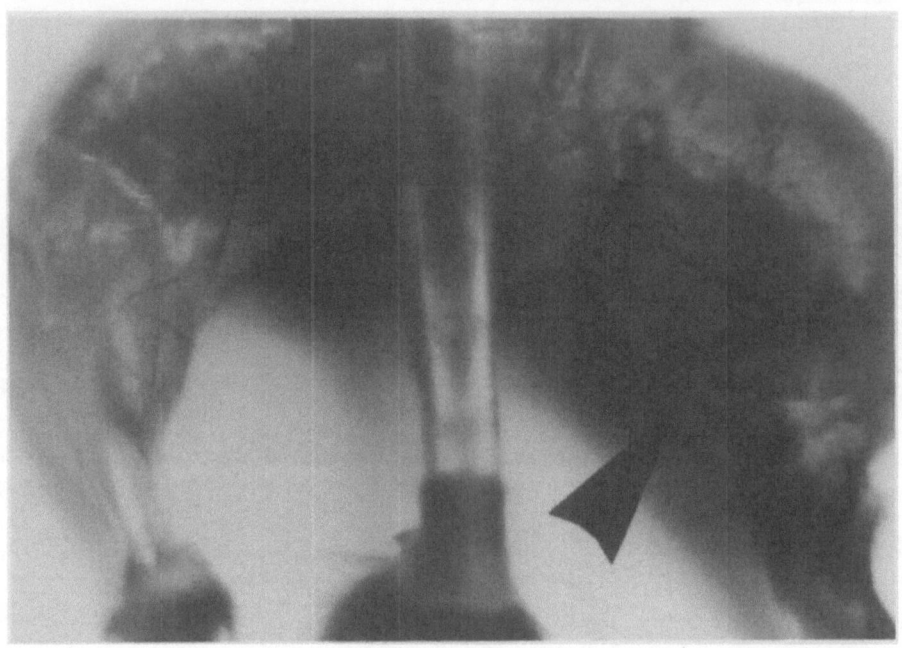

Fig. 5. Cyclosporine-treated dystrophic mouse showed muscle
enlargement in the right leg, which received myoblast
injections 2 months earlier. The contralateral leg was not
injected. (From Ref. 5)

Among the five injected dystrophic mice that did not
respond behaviorally, only one survived to a year after
injection. It could not balance itself on the glass rod.
Nevertheless, its life span was extended.

Before myoblast injection, four-month old dystrophic mice
could not run the treadmill at 2 meters/minute (Fig. 2A).
After myoblast injection, they could run at 2.8 meters/minute
(Fig. 2B). The one that showed most improvement actually
outran that speed and could jump out of the treadmill (Fig.
2C). Young dystrophic mice injected at about one month of age
could run at a speed of 7.1 meters/minute and could easily
climb out of the treadmill, if the speed was slowed down to 2.8
meters/minute.

FUNCTIONAL IMPROVEMENT

Supramaximal in vivo nerve stimulation of the myoblast-
injected dystrophic muscles showed significantly improved
tetanus tension, as compared to the control dystrophic muscles
(Fig. 3).

In another series of experiments, in which histocompatible
normal myoblasts were injected into the soleus muscle in the
right leg of the dystrophic mouse, the injected dystrophic
soleus generated greater twitch and tetanus tensions as
compared to the contralateral uninjected dystrophic control
muscle, indicating that there was significant functional
improvement[4].

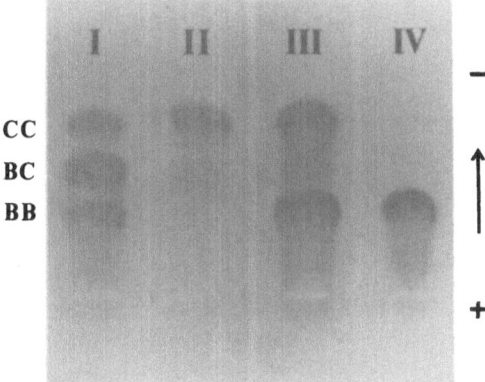

Fig. 6. Agarose gel electrophoresis of muscle GPI of two test solei (lanes I,III) were analyzed six months after injection. The presence of parental and hybrid isozymes in lane 1 indicated a mosaic muscle consisting of normal, dystrophic, and mosaic fibers. Lane III indicated a mosaic muscle of normal and dystrophic fibers. Normal donor cells exhibited GPI-1CC (lane II) and dystrophic host muscles exhibited GPI-1BB (lane IV). (From Ref. 4)

STRUCTURAL IMPROVEMENT

Histologically, the myoblast-injected dystrophic soleus had a larger cross-sectional area, more muscle fiber and better fiber type differentiation, as compared to the contralateral dystrophic control soleus[4]. Now it highly resembled the normal control soleus.

NORMAL

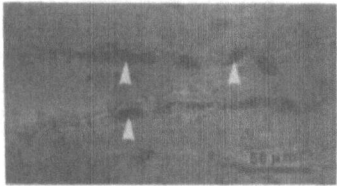

DYSTROPHIC

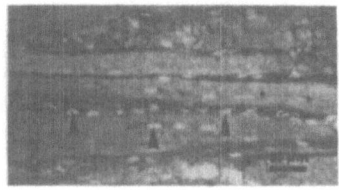

MOSAIC

Fig. 7. Immunocytochemical localization of donor (stained) and host (unstained) nuclei (arrows) in longitudinal section of myoblast-injected dystrophic muscle. Normal and Dystrophic are controls. M indicates a normal-appearing mosaic fiber 18 months after myoblast injection.

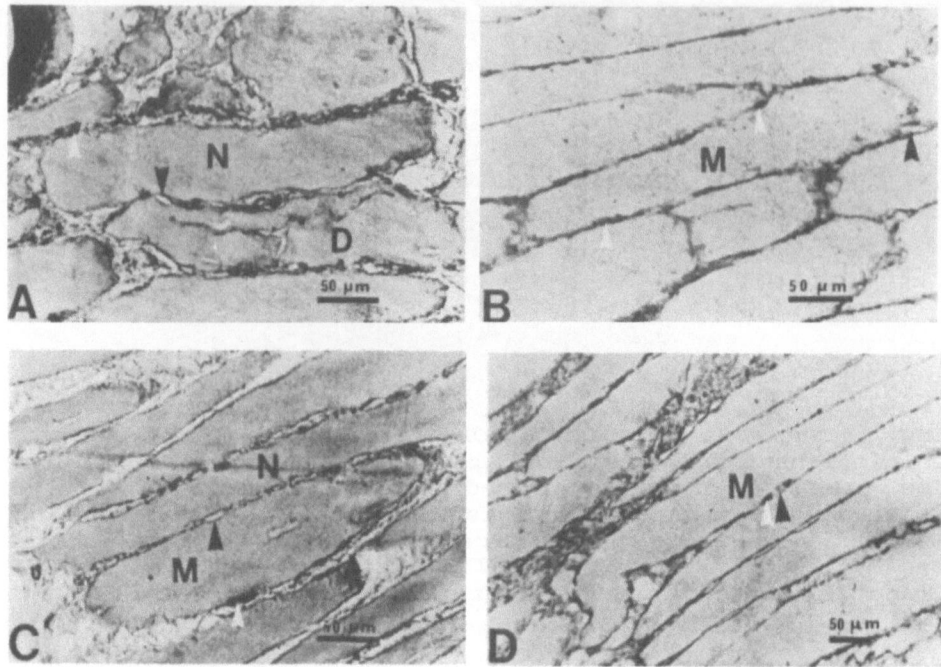

Fig. 8. Immunocytochemical localization of donor (stained) and host (unstained) nuclei (arrows) in oblique sections of myoblast-injected dystrophic muscles (A-D) 18 months after myoblast injection. N indicates normal-appearing fiber. D indicates dystrophic fiber. M indicates mosaic fiber.

At higher magnification, dystrophic characteristics such as muscle fiber splitting, central nucleation, phagocytic necrosis, variation in fiber shape and size, and infiltration of connective tissues were less apparent in the myoblast injected dystrophic soleus (Fig. 4). Some dystrophic characteristics could still be observed, and angular fibers that were the result of denervation were present. Denervation is inherent in this mouse model[8-10]. Injection of fibroblasts in place of myoblasts caused detrimental effects.

The survival and development of donor myoblasts in the host muscles were shown in the injected major muscle groups in the right leg, which were enlarged (Fig. 5). The uninjected left leg muscles served as control.

GENETIC IMPROVEMENT

Two mechanisms were responsible for these phenotype improvements. First, donor myoblasts developed into normal fibers, replenishing lost cells and replacing the myopathic tissue. Second, normal myoblasts fused with dystrophic satellite cells, to form mosaic fibers of normal phenotype. These two mechanisms of genetic complementation were demonstrated using the glucose phosphate isomerases as genotype markers. Host cells produced GPI-1BB. Donor cells produced GPI-1CC. The presence of both isozymes in fifty percent of the myoblast injected muscles indicated the survival and development of donor cells in the host muscles. Furthermore, the presence of the hybrid isozymes, GPI-1BC indicated the fusion between host and donor cells (Fig. 6). GPI-1BC was found in fifty percent of the myoblast-injected muscles.

For immunocytochemical localization of host and donor nuclei within the injected muscles, we have produced polyclonal antibodies that is highly specific for GPI-1CC but not for GPI-1BB. These antibodies stained the nuclei of normal control muscles but not the dystrophic control muscles. Normal-appearing mosaic fibers containing host and donor nuclei could be observed eighteen months after myoblast injection (Fig. 7). Normal nuclei of donor origin were stained. Dystrophic nuclei of the host were not stained.

The mosaic muscles also contained normal and dystrophic muscle fibers (Fig. 8). Analysis of sections of the mosaic fibers will allow the determination of how much genetically normal nuclei is necessary to convert the dystrophic muscle cell back to normal phenotype.

CONCLUSION

Normal myoblast injections provide long-term genetic complementation treatment to muscle degeneration in murine dystrophy. The procedure repairs degenerating cells and replenishes degenerated cells, thus restoring normal structure and function.

Since the treatment design is based on development processes universal to all mammals, it has potential for clinical application to treat all types of muscle diseases. This study lays the scientific groundwork for the development of a clinical treatment, with the promise of bringing restoration of function and quality of life to patients who suffer muscular dystrophy. Such treatment may prevent or remedy muscle weaknesses in locomotive and respiratory muscles, thereby increasing muscle strength and life span of the patients.

ACKNOWLEDGMENTS

The authors gratefully acknowledge Dr. Alan C. Peterson for the supply of GPI-1CC mice, Brian Schafer for constructing the mouse treadmill, and Ms. Claudine O'Steen and Ms. Helen Ham for typing. This research was supported by USPHS Grant NS-20251 and NS-26185 from the National Institute of Neurological and Communicative Disorders and Stroke and by a grant from the Muscular Dystrophy Association, Inc.

REFERENCES

1. P.K. Law. Reduced regenerative capability of dystrophic mouse muscle. Exp. Neurol. 60:231 (1978).

2. P.K. Law and J.L. Yap. New muscle transplant method produces normal twitch tension in dystrophic muscle. Muscle & Nerve 2:356 (1979).

3. P.K. Law. Beneficial effects of transplanting normal limb-bud mesenchyme into dystrophic mouse muscle. Muscle & Nerve 5:619 (1982).

4. P.K. Law, T.G. Goodwin, and M.G. Wang. Normal myoblast injections provide genetic treatment for murine dystrophy. Muscle & Nerve 11:525 (1988).

5. P.K. Law, T.G. Goodwin, and H.J. Li. Histoincompatible myoblast injection improves muscle structure and function of dystrophic mice. Transplant Proc. 20:1114 (1988).

6. P.K. Law, H-J Li, T.G. Goodwin, G. Ajamoughli, X.Y. Zhang and M. Chen. Pathogenesis and treatment of hereditary muscular dystrophy. In press.

7. H. Meier, and J.L. Southard. Muscular dystrophy in the mouse caused by an allele at the dy-locus. Life Sci. 9:137 (1970).

8. P.K. Law and H.L. Atwood. Non-equivalence of surgical and natural denervation in dystrophic mouse muscles. Exp. Neurol. 34:200 (1972).

9. P.K. Law, H.L. Atwood, and A.J. McComas. Functional denervation in dystrophic mouse muscles. Exp. Neurol. 51:434 (1976).

10. P.K. Law, E. Cosmos, J. Butler and A.J. McComas. The absence of dystrophic characteristics in normal muscles successfully cross-reinnervated by nerves of dystrophic genotypes: Physiological and cytochemical study of crossed solei of normal and dystrophic parabiotic mice. Exp. Neurol. 51:1 (1976).

Discussion of Dr. Law's paper

Dr. Partridge: This piece of work is striking. Perhaps I've got to ask you a series of questions which would bring it out to some extent. The first thing that springs to mind is that this myopathy is not related to Duchenne myopathy, which is the one that most people are concerned with. How confident are you that you can apply knowledge taken from an old model of one disease to another disease? For instance, you said yesterday that you could estimate the proportion of the donor genome you'd need to get a certain degree of improvement. And you can do it for that model. But how is that likely to be applicable to, say, Duchenne dystrophy?

Dr. Law: Yes, I think that it is very likely that it would be applicable to myopathies more than Duchenne muscular dystrophy alone. By using the DY2JE mice as animal model, we are using the animal model for hereditary degeneration and weakening. We are not targeting the dystrophic mouse model as just for Duchenne muscular dystrophy alone. We take a model that shows hereditary muscle degeneration. And if we can fulfill our duty of restoring this normal structure and function, then maybe we can help out the various types of diseases.

Dr. Partridge: For instance, do you think that if it takes so much of gene product A to restore function in your mice, that it is necessarily the same proportion of gene product B in another disease would also be sufficient to restore function in that disease?

Dr. Law: Right. It all depends on whether the disease is dominantly inherited or recessively inherited. If it is recessively inherited, I would tend to think that it would be. It would take a very similar amount of normal genetic material for restoration, if not exactly the same. But in terms of autosomal dominant inheritance, then one will have to resort to transplantation of myotubes, that we can obtain from the cell culture.

Dr. Partridge: I was going to ask a question about the life span of these mice. How much longer do the treated animals live? And what's the variation in the life span?

Dr. Law: We have practically doubled the life span of these animals. Normally, they live around nine months and definitely very rarely do you see them living over twelve months. Now we have animals that were transplanted at about one to two months of age and lived 18 months afterwards. And cyclosporin was kicking off six months after myoblast implantation, and still the animals are surviving and doing well.

Dr. Partridge: Right. Now the other thing that you've mentioned is that you've put in myoblasts from a variety of sources. And some of those were nonhistocompatible sources, which was a point you made when suppressing with cyclosporin A. Can you give us an indication of how noncompatible they were? Incompatible at the major locus, or minor locus?

Dr. Law: It belongs to a totally different strain of animal. The donor cells were from Swiss Webster, and then the host animals were a C57 black.

Dr. Partridge: You found that cyclosporin treatment was sufficient to keep these donor cells in the animal for how long?

Dr. Law: We were able to keep the donor cells surviving in their parents 18 months, after myoblast implantation, or 12 months after cyclosporin withdrawal. And we had only about six months of cyclosporin treatment.

Dr. Partridge: So you get long-term survival? Would you consider using cyclosporin for treating DMD boys in the same way?

Dr. Law: Yes. This probably is our first choice of immunosuppression right now.

Dr. Partridge: Now the other thing that you were talking about is the various mechanisms by which normal myonuclei can get into these animals. You suggested that they can form fibers of their own type.

Dr. Law: That is correct.

Dr. Partridge: How can you prove it?

Dr. Law: We have the donor myoblasts in between themselves to form genetically normal fiber. And then we have the donor myoblast fusing with the host cells.

Dr. Partridge: Well, what is your evidence for the former?

Dr. Law: Yes, we did show a slide towards the end of the film that actually contained only the stained nuclei, but not the host nuclei. And also by our GPI electrophoresis.

Dr. Partridge: Among the mosaic muscle fibers that you were describing with the hybrid isoenzyme, you were suggesting that they were forming by coregeneration of the host muscle cells together with the cells you'd put in. Do you have any evidence that these cells in any way repair the segmental necrotic gaps that, for instance, occur in these muscles?

Dr. Law: No, not at all. We have not looked at segmental necrosis whatsoever. Because I think that is too minor to be dealt with. We are looking at all of the muscle cells in the dystrophic muscle. And if one can repair the degenerating fibers, and if we can let the ones that one cannot rescue to keep on degenerating and get the normal donor fibers to keep on regenerating, I think that we have achieved what we can achieve.

Dr. Partridge: Concerning the ages at which you've injected the mice, are you injecting mice when degeneration first begins? Or are you injecting them when there is already significant fibrosis in the muscle pathology?

Dr. Law: Our initial experiments were performed on animals that just showed the dystrophic symptoms, probably around 20 days or about a month old. But eventually, we have extended our animal research into four-month-old hosts, when they receive their transplant. And now we have done some of those in six-month-old animals when they show considerable necrosis and then we inject the myoblasts. And some of them also showed improvement.

Dr. Partridge: Can you reverse the fibrosis that occurs in these later stages? Does that go away?

Dr. Law: It is very difficult. In order to answer your question, one needs to take a necrotic fiber, and then show that muscle fiber to... turning to normal histology, which is not possible to do.

Dr. Partridge: Could you, for instance, have biopsied the mice at a certain age and compared them with injected animals?

Dr. Law: Yes. I think that what one needs to do is to look at our histology of the whole muscle and look at the overall normality of the muscle fibers of the whole muscle, rather than looking at one or two muscle fibers alone. I think that because of that is what gives you the function of the whole muscle, not one or two muscle fibers.

Dr. Partridge: The last question has to do with the sort of dose of cells you think you'd need. I mean, can you get any estimates of this sort of work, as to how many cells do you think you might need to reverse or to significantly improve a myopathy of this sort?

Dr. Law: We are still working on this problem. But our initial benchmark thing right now is half of a million cells per one cubic centimeter of tissue. But it all depends how much cell fusion occurs.

Dr. Partridge: Can we defer general questions until the end of this session?

Unidentified Participant: I have one very specific question.

Dr. Partridge: Okay.

Unidentified Participant: What happens if you treat animals just for six months with cyclosporin alone without giving them myoblasts?

Dr. Law: We have actually done such experiments. And what happened was that the animal became very hyperactive. The animal lost some hair, and that's about it. The animal continued to undergo degeneration. No significant structural, behavioral or functional improvement occurs.

PRACTICAL ASPECTS OF MYOBLAST IMPLANTATION:

INVESTIGATIONS ON TWO INHERITED MYOPATHIES IN ANIMALS

Jennifer E. Morgan

Department of Histopathology
Charing Cross and Westminster Medical School
St. Dunstan's Road, London W6 8RF, U.K.

For some years, our group has been studying the idea of implanting normal muscle precursor cells (mpc) into the skeletal muscles of individuals suffering from inherited myopathies, in an attempt to alleviate the disease. The idea behind this is simple: multinucleated muscle fibres are formed during muscle regeneration by the fusion of mononucleated mpc (Stockdale and Holtzer, 1961); during growth, mpc become incorporated into the muscle fibres (Moss and Leblond, 1971). It follows that if we can implant normal mpc into growing or regenerating myopathic muscle, the implanted mpc would become incorporated into the host muscle fibres (Watt et al. 1982, 1984). The presence of myonuclei of normal origin, carrying the normal genome, within the myopathic fibre should help to alleviate the myopathy.

The previous speaker, Peter Law, concentrated on clinical and structural muscle defects, of unknown aetiology and pathogenesis, in the dy^{2J} mouse, for assessment of his cell implantation experiments (Law et al., 1988). We have adopted a quite different approach, studying two mouse models in which the primary genetic defect is known to be manifested as a biochemical deficiency in the muscle. Both of our animal models are X-linked primary myopathies: the phosphorylase kinase deficient ICR/IAn mouse, whose muscles lack phosphorylase kinase (Cohen et al.,1976) and the dystrophin-deficient mdx mouse (Bulfield et al., 1984), which has the same biochemical and genetic defect as human Duchenne muscular dystrophy (DMD) (Hoffman et al.,1987). Neither of these animal models exhibits a severe functional deficit, but the fact that both involve defects in identified genes, gives them an advantage over the dy^{2J} mouse in terms of the quality and quantity of information they yield. We have determined first, whether we could obtain formation of the correct gene product in a muscle which inherently lacks it and second, how much of this gene product can be formed within the treated myopathic muscle. In addition, in the mdx mouse we have been able to confirm that the introduced dystrophin, of the correct size, is correctly positioned within the muscle fibre and is widely distributed within the treated muscles.

Myoblast Transfer Therapy
Edited by R. Griggs and G. Karpati
Plenum Press, New York, 1990

Mpc were prepared for implantation by enzymatically disaggregating newborn mouse muscle (Morgan et al., 1987). Mpc of Gpi-1s^b allotype were implanted into host mice of Gpi-1s^a allotype: the finding of the BB GPI isoenzyme in the treated muscle shows that the implanted cells had survived, but these cells need not necessarily be muscle. The finding of the heterodimeric AB GPI isoenzyme shows that the implanted cells were myogenic, as they had fused with host mpc to form mosaic muscle fibres (Morgan et al., 1987).

To minimize immunological rejection in animals in our experiments, the phosphorylase kinase gene was bred onto the nude mouse background (Morgan et al., 1988); nu/nu mice have a deficient immune system and will accept grafts from other strains. We also bred the mdx gene onto the nude mouse background (Partridge et al., 1989), but in some of our earlier experiments, the mdx gene was on an immuno-competent strain which had to be made tolerant (Watt et al., 1984) prior to mpc implantation.

IMPLANTATION OF NORMAL MPC INTO PHOSPHORYLASE KINASE DEFICIENT

MUSCLE (Morgan et al., 1988)

Normal mpc were implanted into the growing tibialis anterior muscles of the right legs of 1-8 day old mice, by means of a 5μl Hamilton syringe. We also implanted normal mpc into regenerating muscles in phosphorylase kinase-deficient mice: as the muscles of these mice do not spontaneously degenerate and regenerate, we forced them to do so by autografting the extensor digitorum longus (EDL) muscle of the right leg. Into these autografts, we implanted our mpc.

Following implantation of normal mpc into growing phosphorylase kinase-deficient muscles, our GPI isoenzyme analyses showed that mosaic myofibres, containing myonuclei of donor origin, were found in only 11 out of 192 muscles examined from 64 mice. Of these muscles, 5 contained very low levels of phosphorylase kinase activity.

Myonuclei of donor origin were found in host muscle fibres in 8 out of 9 of the autografts treated with normal mpc, but phosphorylase kinase activity was found only in the three grafts into which very large numbers (10^6 or more) of donor cells had been implanted. Control autografts, into which we had injected either phosphorylase kinase-deficient mpc of Gpi 1-s^b allotype, or no cells, contained no phosphorylase kinase activity.

These results show that regenerating phosphorylase kinase-deficient mouse muscle incorporates mpc more efficiently than does growing muscle. In neither case, however, did the incorporation elicit formation of anything approaching normal levels of the missing gene product. The low level of phosphorylase kinase activity in muscles injected during postnatal growth was probably due to the small numbers of normal mpc incorporated into the muscle. In the regenerated autografts, it was most likely due to the fact that after regeneration, skeletal muscle does not fully re-express its phosphorylase kinase gene (Morgan, 1988).

IMPLANTATION OF NORMAL MPC INTO MDX MUSCLE
(Partridge et al., 1989)

In our first series of experiments, we implanted normal
mpc either into growing (5-7 day old) mdx muscle, or into the
muscle when it was undergoing its first round of degeneration
and regeneration, at around 3 weeks of age. Again, GPI
isoenzyme analysis of the treated muscles showed that the
injected cells had become incorporated far more frequently into
the regenerating than into the growing mdx muscles. We
sometimes found the heterodimeric GPI isoenzyme in muscles
adjacent to the target muscle; this may indicate inaccurate
injection of the cells, or migration of mpc between muscles.

Muscles containing high proportions of donor and
heterodimeric GPI were assessed for their dystrophin content,
by immunofluorescent staining and by immunoblotting. In these
muscles, dystrophin of normal size was present at 30-40% of
normal levels, located at its expected subsarcolemmal position.
The dystrophin-positive fibres were distributed diffusely in
some muscles and formed small foci in others.

DOES THE INTRODUCED DYSTROPHIN DO ANY GOOD?

Because mdx mouse muscle regenerates so well, eventually
becoming bigger and stronger than before (Coulton et al. 1988
b), it is difficult to tell whether the presence of dystrophin
within the host muscle fibres is saving them from necrosis.
In an attempt to make the mdx myopathy more similar to DMD, we
have inhibited the regeneration of the muscle by applying local
doses of X-irradiation (Wakeford et al., submitted for
publication). When the leg of an mdx mouse is X-irradiated
with 16-20 Gy at 16 days of age, before the visible onset of
the degeneration and regeneration of the muscle, the muscles
atrophy. Myonecrosis occurs in these muscles, but there is no
regeneration of new fibres. The few surviving muscle fibres
become rounded and separated from each other by connective
tissue. Almost all of the muscle fibres in these X-irradiated
muscles have peripherally-placed nuclei, indicating that they
have been present from birth and have never undergone
degeneration and regeneration.

This X-irradiated mdx muscle provides a good model to test
any putative therapy: we can see if we can prevent the massive
loss of muscle fibres in the irradiated mdx muscles by the
implantation of normal mpc. We implanted normal mpc into mdx
nude muscles 3-5 days after X-irradiation, i.e at 19-21 days
of age, at which time we expect the host muscle to be beginning
to undergo massive necrosis (Coulton et al. 1988 a).
Preliminary analysis of these injected muscles shows that, in
the majority, the donor cells had survived and been
incorporated into the host muscle fibres. The mechanism by
which the donor cells become incorporated into host muscle
fibres is unclear. In mdx muscles subject to this dose of X-
irradiation, there is no regeneration (Wakeford et al.,
submitted for publication), so mosaic fibres cannot have been
formed by fusion of the implanted cells with host mpc. There
is little growth in muscles of this age (Cardasis and Cooper,
1975), so it is unlikely that the implanted mpc are becoming
incorporated into growing muscle fibres. It seems likely,

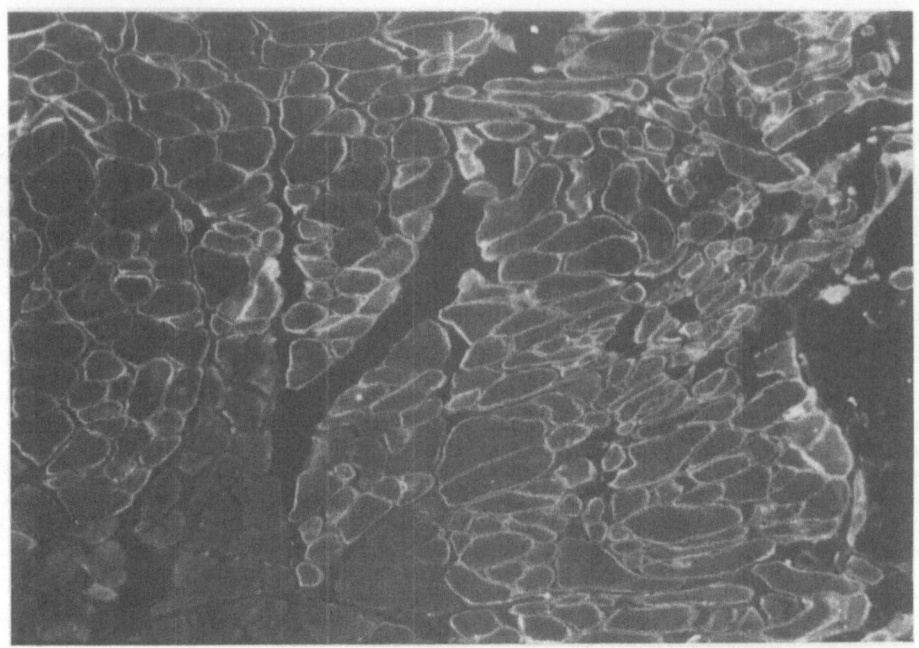

Figure 1. Dystrophin immunostain of a cryostat section of a TA muscle from the leg of an mdx nu/nu mice which had been X-irradiated with 18 Gy at 16 days of age. 5 x 10^5 C57Bl/10 mpc had been implanted into this muscle 3 days after X-irradiation. The muscle was examined 49 days after cell implantation and contained donor and herterodimeric GPI isoenzymes; 69% of the muscle fibres were dystrophin-positive. (x 200)

therefore, that the normal mpc are fusing with necrosing host muscle fibres and repairing them.

Dystrophin immunostaining of irradiated and injected mdx muscles containing mosaic muscle fibres showed large numbers of dystrophin-positive fibres (up to 80% of all of the fibres in a single muscle); the dystrophin-positive fibres are usually found in large blocks (Figure 1).

In the few muscles where we found only host and donor, but no heterodimeric GPI isoenzyme, there were also large blocks of dystrophin-positive fibres. This implies that the dystrophin-positive fibres in these muscles must be derived solely from the implanted normal mpc. This is unequivocal evidence that fibres derived only from the donor cells are formed within host muscle. For successful incorporation of donor mpc into a muscle, therefore, there is no essential requirement for the presence of proliferating host mpc to fuse with the donor mpc. Thus, there is good reason to believe that myopathies where there is poor regeneration of skeletal muscle from host myogenic cells, such as DMD, may be amenable to therapy by myoblast transplantation.

These X-irradiated muscles which had incorporated the injected normal mpc, giving rise to dystrophin-positive fibres, were larger and contained more muscle fibres than irradiated muscles into which only medium had been injected. In addition to this prevention of muscle loss, injection of normal mpc produced a remarkable histological improvement; the treated muscles contained very few central nuclei, the muscle being almost indistinguishable from normal.

In controls, where we injected mdx mpc into pre-irradiated mdx nude muscles, there was, as we might expect, preservation of muscle bulk and fibre number. This shows that the implanted mdx mpc had good regenerative capacity. However, histopathological features of untreated mdx muscle, such as a high level of central-nucleation, and areas of both necrosis and regeneration, were evident, particularly in older muscles.

We occasionally found isolated, dystrophin-positive fibres in our control irradiated mdx muscles which had been repopulated with mdx mpc. Small numbers of dystrophin-positive fibres are also found in untreated mdx muscles and are thought to arise as the result of somatic mutation (Hoffman et al. in preparation). This emphasizes the need for good controls when analyzing the effects of myoblast implantation in human experiments.

Using two different mouse X-linked myopathies, we have demonstrated that some diseases may be more amenable to treatment by myoblast implantation than others. In treated phosphorylase-kinase deficient muscles, myoblast implantation only elicited production of a few percent of normal levels of the missing gene product, whereas in treated mdx muscles, high levels of dystrophin were produced. This difference in outcome argues against the notion put forward by Peter Law, that results obtained from experiments on a myopathy in one species can be applied to other biochemically and genetically unrelated myopathies in a different species.

The fact that the implanted cells repaired necrosing muscle fibres in the irradiated mdx muscles, indicates that DMD may be amenable to myoblast transfer therapy. It is at this basic, cell-biological level that further studies of animal models are likely to yield information which will be of value to human studies. We need to know: how far the implanted mpc can move and have a biochemical and functional effect, whether the dystrophin-positive fibres are protected from undergoing further rounds of degeneration and regeneration and whether there is any physiological improvement in treated, irradiated mdx muscles. We also need to know why mdx mouse and DMD muscles differ so much, although they both lack the same protein, dystrophin. In both cases, the muscle undergoes necrosis, but in the mdx mouse, the regeneration more than compensates for the degenerative events, whereas in DMD, it does not. The elucidation of the basis of this difference may suggest alternative methods for treating DMD.

In fact, many of these basic questions about the cell biology of myogenesis in dystrophin-deficient muscles can only be addressed at present in the mouse. When we have a clearer idea of how myogenic cells behave in vivo in the mouse, it may be possible to construct critical experiments to test whether they behave similarly in the dystrophic dog and ultimately, in man.

ACKNOWLEDGEMENTS

This work was supported by the Muscular Dystrophy Group of Great Britain and Northern Ireland. I would like to thank Dr. Terry Partridge for his advice and criticism, Mr. Ron Barnett for photography and Dr. Eric Hoffman for the anti-dystrophin antibodies.

REFERENCES

Bulfield, G. Siller, W.G., Wight P.A.L. and Moore, K.J. (1984). X-chromosome-linked muscular dystrophy (mdx) in the mouse. Proc. Natl. Acad. Sci. U.S.A. 81, 1189-1192.

Cardasis, C.A. and Cooper, G.W. (1975). An analysis of nuclear numbers in individual muscle fibres during differentiation and growth: a satellite cell-muscle fibre growth unit. J. Exp. Zool. 191, 333-346.

Cohen, P.T.W., Burchell, A. and Cohen, P. (1976). The molecular basis of skeletal muscle phosphorylase kinase deficiency. Eur. J. Biochem. 66, 347-356.

Coulton, G.R., Morgan, J.E., Partridge, T.A. and Sloper, J.C. (1988 a). The mdx mouse skeletal muscle myopathy: I. A histological, morphometric and biochemical investigation. Neuropathol. appl. Neurobiol. 14, 53-70.

Coulton, G.R., Curtin, N.A., Morgan, J.E., and Partridge, T.A. (1988 b). The mdx mouse skeletal myopathy: II. Contractile properties. Neuropathol. appl. Neurobiol. 14, 299-314.

Hoffman, E.P., Brown, R.H. and Kunkel, L.M. (1987). Dystrophin: the protein product of the Duchenne Muscular Dystrophy locus. Cell 51, 919-928.

Law, P.K., Goodwin, T.G. and Wang, M.G. (1988). Normal myoblast injections provide genetic treatment for murine dystrophy. Muscle & Nerve 11, 525-533.

Moss, F.P. and Leblond, C.P. (1971). Satellite cells as the source of nuclei in muscles of growing rats. Anat. Rec. 170, 421-436.

Morgan, J.E., Coulton, G.R. and Partridge, T.A. (1987). Muscle precursor cells invade and repopulate freeze-killed grafts. J. Muscle Res. Cell Motil. 8, 386-396.

Morgan, J.E. (1988). Phosphorylase kinase activities in damaged mouse muscle. J. Neurol. Sci. 86, 149-158.

Morgan, J.E., Watt, D.J., Sloper, J.C. and Partridge, T.A. (1988). Partial correction of an inherited biochemical defect of skeletal muscle by grafts of normal muscle precursor cells. J. Neurol. Sci. 86, 137-147.

Partridge, T.A., Morgan, J.E., Coulton, G.R., Hoffman, E.P. and Kunkel, L.M. (1989). Conversion of mdx myofibres from dystrophin-negative to -positive by injection of normal myoblasts. Nature 337, 176-179.

Stockdale, F.E. and Holtzer, H. (1961). DNA synthesis and myogenesis. Exp. Cell Res. 24, 508-520.

Wakeford, S., Watt, D.J. and Partridge, T.A. X-irradiation improves mdx mouse muscle as a model of DMD. (in press)

Watt, D.J., Lambert, K., Morgan, J.E., Partridge, T.A. and Sloper, J.C. (1982). Incorporation of donor muscle precursor cells into an area of regeneration in the host mouse. J. Neurol. Sci. 57, 319-331.

Watt, D.J., Morgan, J.E. and Partridge, T.A. (1984). Use of mononuclear precursor cells to insert allogeneic genes into growing mouse muscles. Muscle & Nerve 7, 741-750.

Discussion of Dr. Morgan's paper

Dr. Partridge: You've looked at two different models there. And from that do you think it's possible to apply generally the knowledge of any single model to all myopathies?

Dr. Morgan: No. I think they all have to be looked at separately because in the phosphorylase kinase deficient mouse, we didn't cause much improvement at all, whereas in the mdx mouse we have.

Dr. Partridge: Can you account for that by a low level of expression of the phosphorylase kinase?

Dr. Morgan: In the phosphorylase kinase deficient mouse, we showed that in autografted normal muscle there are very low levels of phosphorylase kinase activity. That is why we did not achieve much success in the autografts. I think when you implant into growing muscle, so few cells become incorporated, so it did not have much effect.

Dr. Partridge: You've used the mdx mouse as a model but it obviously has had limitations. What are the principal limitations, in your view, of this animal model?

Dr. Morgan: The mdx mouse regenerates so well that it is not really a good model for Duchenne muscular dystrophy.

Dr. Partridge: But you can remove that effect by X-irradiation. Are there other problems? Does that entirely get rid of the problem?

Dr. Morgan: The problem with that is we are only X-irradiating one leg, so we can't look at any systemic factors or fibrosis.

Dr. Partridge: We can't look at something that Dr. Law is looking at?

Dr. Morgan: No.

Dr. Partridge: In the animals you avoided rejection problems in various ways. Those are ways in which you can use the mice. How applicable are they to the human situation?

Dr. Morgan: In most of the experiments I described, we used nude mice. So we completely avoided any problems of rejection. In our very early work with phosphorylase kinase deficient strain of mouse, all our cells that we injected were rejected. That's why we went on to use nude mice. And also from what Diana Watt says, I think there is a big problem with rejection of implanted cells.

Dr. Partridge: Now you stated that about 70% of your fibers were dystrophin-positive. And you have evidence as to what's actually happening. That mostly comes from your GPI analysis. Do you have any evidence as to what extent they become incorporated as mosaic fibers, and are there also fibers of purely donor origin?

Dr. Morgan: They obviously become incorporated as mosaic fibers, to a great extent, because there's quite a bit of the hybrid GPI isoenzyme. In some of the muscles in which I only found host and donor GPI isoenzyme, there were also many dystrophin-positive fibers. So those fibers must purely have been of donor origin.

Dr. Partridge: What about the repair of segmental necrosis? There's a lot of segmental necrosis in this animal. Do you have evidence that you can actually repair the segments with injections of normal cells?

Dr. Morgan: Yes, in the pre-irradiated mdx model, we've completely knocked out the regenerative capacity of the host; we've completely eliminated that. So when we found mosaic muscle fibers, these must have arisen by the transplanted cells fusing with the regenerating host muscle fibers.

Dr. Partridge: Do we have any other evidence of any beneficial effects of this dystrophin in the fibers? Do you think this is a consequence of the dystrophin being there?

Dr. Morgan: Yes, because in the controls when we injected mdx muscle precursor cells, there was an increasing level of central nucleation with time. So I think it is having a beneficial effect histologically.

Dr. Partridge: Do you think that happens via a direct effect of dystrophin where the nuclei are, or as an effect of stopping degeneration in those muscles?

Dr. Morgan: I think it's probably more likely to be due to a direct effect of dystrophin.

Dr. Partridge: What about those positive fibers? Do you find them very frequently? I mean, the positive fibers that one finds in myoblast-injected mdx muscles. Are they a fairly frequent finding?

Dr. Morgan: Yes, they did seem to be fairly frequent, particularly in the older ones. And we also find them in just mdx muscles from older mice.

Dr. Partridge: Has their incidence increased in these animals that you've injected or not?

Dr. Morgan: Not that we've been able to tell yet.

PURIFICATION AND PROLIFERATION OF HUMAN MYOBLASTS

ISOLATED WITH FLUORESCENCE ACTIVATED CELL SORTING

Helen M. Blau, Cecelia Webster and Grace K. Pavlath

Department of Pharmacology
Stanford University School of Medicine
Stanford, CA 94305

Myoblast therapy, or the transfer of normal myoblasts into dystrophic muscle, requires both a large number and high purity of human myoblasts. Solutions to both of these problems are presented below.

Early evidence that human muscle cells could be successfully grown in culture as small clones was provided by Hauschka.[1,2] In 1981, we extended his elegant work, and developed methods for cloning human myoblasts in large quantity that could be stored frozen for growth at later times in culture. These cells retained their original proliferative and differentiative capacity, permitting ample biochemical and molecular analyses of cell differentiation.[3-16] These methods have now been widely employed by others.[17-21]

More recent experiments in our laboratory demonstrated that the yield of muscle cells from postnatal muscle tissue is significant. From a 0.1 gm biopsy, 10^4 cells can be obtained. To purify the myoblasts from this cell mixture, we developed methods using the fluorescence activated cell sorter (FACS)[15] and the monoclonal antibody, 5.1 H11, now known to recognize human N-CAM (neural cell adhesion molecule).[22] Fibroblasts and myoblasts could be clearly distinguished using two different fluorescent labeling procedures (Figure 1). One procedure employs Texas red, which is a three-step procedure and requires a dual-laser cell sorter, whereas the other procedure uses fluoroscein in a two-step procedure with the single laser FACS, a more readily available instrument (Figure 1).

Table I shows that the FACS can be used to provide relatively pure cultures of myoblasts, circumventing problems caused by contaminating cells which appear to induce an immune response, as shown by Dr. Partridge in this volume. A number of samples were analyzed over time. Three different fetal samples (17-week gestation) were analyzed. Even when the initial composition of cells was 24% myoblasts, and the remaining 76% were fibroblasts, immediately after the sort 99% of the cells recovered were myoblasts. After growth for a further 17 days in culture, or nearly three weeks, a reanalysis revealed that they were 100% pure. Thus, the purity of the sorted cells is maintained over time. Cells isolated from a post-natal biopsy that exhibited an initial composition of 79% myoblasts, could also be enriched using the cell sorter to a purity of 99%. After three weeks of culture they remained 98% pure.

The finding that myoblasts can be purified from postnatal tissue which retain the ability to genetrate large numbers of cells upon proliferation in culture (see below) is of particular significance. First, it contrasts with many kinds of cell types that are used for cell therapy, in that fetal cells are generally required to obtain a sufficient number of cell doublings. Second, not only are postnatal myoblasts more readily available than fetal myoblasts, but this finding also makes it possible to use myoblasts from parents to treat patients with DMD.

Key experiments carried out in the course of this work revealed that the proliferative capacity of normal myoblasts is remarkably high. These experiments involved isolating individual myoblasts from controls, carriers and Duchenne muscular dystrophy patients and

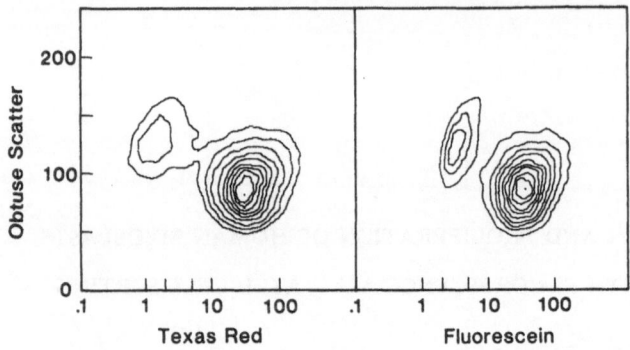

Figure 1. Efficient separation of myoblasts from fibroblasts can be achieved by labeling with either Texas red or fluorescein. The same mixed-cell population is shown labeled with Texas red (A) or fluorescein (B) in two-dimensional contour plots of red or green immunofluorescence (abscissa) and intrinsic obtuse angle light scatter (ordinate), where 10% of the cells lie between each contour line. In this case, 87% of the cells were myoblasts (right-hand plots). *Reprinted by permission from reference 15.*

following them as clones. Serial passaging and cell counts revealed the number of total population doublings of which the cells were capable prior to senescence. As many as 60 doublings, a yield equivalent to 10^{18} cells, could be obtained from an individual myoblast derived from the muscle of a child.[16] This represents a potential yield of kilograms of cells per cell. Moreover, since a 0.1 gm biopsy contains 10^4 myoblasts each with such a proliferative capacity, it is apparent that cell therapy will not be limited by the number of myoblasts needed.

In Duchenne muscular dystrophy, this proliferative capacity is greatly reduced. Even in the youngest patients, who are typically diagnosed at age two years, the reduction in total proliferative capacity, or the number of doublings of which a myoblast is capable, is pronounced.[6] Several studies[6,12,16] suggest that the impaired growth of DMD myoblasts is likely to be secondary to degeneration and regeneration. In conjunction with the pronounced

TABLE 1

Percentage of myoblasts in cell cultures before and after cell sorting

Sample (age)	Initial	Immediately after sort	Following growth in culture (days)
A. Fetal			
I (17 weeks)	80	99.8	100.0 (17) 100.0 (27)
II (17 weeks)	24	99.7	100.0 (17) 100.0 (20) 100.0 (27)
III (17 weeks)	56	99.2	99.4 (14) 99.9 (21)
B. Postnatal			
I (3 years)	81	99.7	97.5 (17)
II (14 years)	79	99.4	97.8 (18)

Reprinted by permission from reference 15

deficit in the number of satellite cells that can be isolated per gram DMD muscle tissue at two years of age,[6] the muscle even at this early stage of the disease, has the potential to generate less than 2% of the muscle precursor cells, or myoblasts, necessary for growth.[16] By age 10, the problem is exacerbated and the number of growing myoblasts that can be isolated is negligible.

What are the implications of our findings for myoblast therapy? The results suggest that a therapeutic approach to DMD directed at isolating a Duchenne child's myoblasts, genetically engineering them by introducing the missing dystrophin gene and injecting them back into muscle, will fail unless cells from very young patients are used. In addition, the use of therapeutic agents designed to stimulate the residual satellite cell population to proliferate is unlikely to succeed because these cells in DMD muscle are capable of very few cell divisions. By contrast, the tremendous proliferative capacity of the normal myoblast suggests that it can be obtained in sufficiently large numbers from a small piece of adult tissue to generate ample material for transplantation. In conjunction with the methods that we have developed for purifying these myoblasts, a strategy based on using normal myoblasts to deliver the missing gene product seems promising. Although other factors remain to be solved, cell number and purity do not appear to be limitations of myoblast therapy.

References

1. Hauschka, S.D., Clonal analysis of vertebrate myogenesis II. Environmental influences upon human muscle differentiation, Dev. Biol., 37:329 (1974).
2. Hauschka, S.D., Clonal analysis of vertebrate myogenesis. 3. Developmental changes in the muscle-colony-forming cells of the human fetal limb, Dev. Biol., 37:345 (1974).
3. Blau, H.M. and C. Webster, Isolation and characterization of human muscle cells, Proc. Natl. Acad. Sci. USA 78:5623 (1981).
4. Blau, H.M., I. Kaplan, T.-W. Tao and J.P. Kriss, Thyroglobulin-independent, cell-mediated cytotoxicity of human eye muscle cells in tissue culture by lymphocytes of a patient with Graves' ophthalmopathy, Life Sci., 32:45 (1982).
5. Blau, H.M., C. Webster, C.-P. Chiu, S. Guttman and F. Chandler, Differentiation properties of pure populations of human dystrophic muscle cells, Exp. Cell Res., 744:495 (1983).
6. Blau, H.M., C. Webster and G.K. Pavlath, Defective myoblasts identified in Duchenne muscular dystrophy, Proc. Natl. Acad. Sci. USA, 80:4856 (1983).
7. Blau, H.M., G.K. Pavlath, E.C. Hardeman, C.-P. Chiu, L. Silberstein, S.G. Webster, S.C. Miller and C. Webster, Plasticity of the differentiated state, Science, 230:758 (1985).
8. Costa, E.M., H.M. Blau and D. Feldman, 1,25 dihydroxyvitamin D_3 receptors and hormonal responses in cloned human skeletal muscle cells, Endocrin., 119:2214 (1986).
9. Hardeman, E., C.-P. Chiu, A. Minty and H.M. Blau, The pattern of actin expression in human fibroblast X mouse muscle heterokaryons suggests that human muscle regulatory factors are produced, Cell, 47:123 (1986).
10. Kaplan, I.K. and H.M. Blau, Metabolic properties of human acetylcholine receptors can be characterized on cultured human muscle, Exp. Cell Res., 166:379 (1986).
11. Shimizu, M., C. Webster, D.O. Morgan, H.M. Blau and R.A. Roth, Insulin and insulin-like growth factor receptors and responses in cultured human muscle cells, Amer. J. Physiol., 251:E611 (1986).
12. Webster, C., G. Filippi, A. Rinaldi, C. Mastopaolo, M. Tondi, M. Siniscalco, and H.M. Blau, The myoblast defect identified in Duchenne muscular dystrophy is not a primary expression of the DMD mutation: Clonal analysis of myoblasts from double heterozygotes for two X-linked loci: DMD and G6PD, Human Genetics, 74:74 (1986).
13. Gunning, P., E. Hardeman, R. Wade, P. Ponte, W. Bains, H.M. Blau and L. Kedes, Differential patterns of transcript accumulation during human myogenesis, Mol. Cell Biol., 7:4100 (1987).
14. Miller, S.C., H.Ito, H.M. Blau and F.M. Torti, Tumor necrosis factor inhibits human myogenesis in vitro, Mol. Cell Biol., 8:2295 (1988).
15. Webster, C., G.K. Pavlath, F.S. Walsh and H.M. Blau, Isolation of human myoblasts with the fluorescence-activated cell sorter, Exp. Cell Res., 174:25 (1988).
16. Webster, C. and H.M. Blau, Extensive proliferative capacity of normal postnatal human myoblasts, 1989 submitted.
17. Yasin, R., F.S. Walsh, D.N. Landon and E.J. Thompson, New approaches to the study of human dystrophic muscle cells in culture, J. Neurol. Sci., 58:315 (1983).
18. Miranda, A.F., T. Mongini, and S. DiMauro, Human myopathies in muscle culture: Morphological, cytochemical, and biochemical studies, Adv. Cell Culture., 4:1 (1985).
19. Martinuzzi, A., V. Askanas, T. Kobayashi, W.K. Engel and S. DiMauro, Expression of muscle-gene-specific isozymes of phosphorylase and creatine kinase in innervated cultured human muscle, J. Cell Biol., 103:1423 (1986).

20. Ham, R.G., J.A. St.Clair, C. Webster and H.M. Blau, Improved media for normal human muscle satellite cells: serum-free clonal growth with low serum, In Vitro Cell. Dev. Biol., 24:833 (1988).
21. Hurko, O., E.P. Hoffman, L. McKee, D.R. Johns and L.M. Kunkel, Dystrophin analysis in clonal myoblasts derived from a Duchenne muscular dystrophy carrier, Amer. J. Hum. Genet., 44:820 (1989).
22. Walsh, F.S., G. Dickson, S. Moore and C.H. Barton, Unmasking N-CAM, Nature, 339:516 (1989).

Discussion of Dr. Blau's paper

Dr. Partridge: How large are those clusters of cells?

Dr. Blau: Depending on the stage of development, when one actually does the injection, the size of the cluster can vary. In older animals it's usually about two fibers, but it can be as great as 10.

Dr. Partridge: And how old are older animals?

Dr. Blau: Older animals are injected about day 18 or so. But if one injects around day 9, or a little bit earlier, one sees clusters as large as 10, and sometimes even 20.

Dr. Partridge: How long is it between injection and looking at clusters? What sorts of time period are you using?

Dr. Blau: We usually look about two weeks after the injection.

Dr. Partridge: What is the time course of the size of those clusters at post-injection?

Dr. Blau: We haven't done detailed time courses at this point. We don't know the exact distances of migration, or the rate of developmental multiple clusters yet. But it's definitely dependent on the stage of development and how much growth is going on at that time because it's correlated with cell division.

Dr. Partridge: A much harder technical question. Do you have any idea about the longitudinal extent of movement from this sort of work?

Dr. Blau: That's a very important point. We want to determine what the longitudinal extent of migration is, and we're doing some experiments now to address that directly.

THE PROLIFERATION AND FUSION OF MYOBLASTS *IN VIVO*

Miranda D. Grounds

Department of Pathology, University of Western Australia
Queen Elizabeth II Medical Centre, Nedlands, 6009, Western Australia

In this paper the term myoblasts is a general term for all muscle precursor cells. The aim of myoblast therapy is to introduce normal myoblasts into dystrophic (dy) hosts, and to fuse the maximum number of normal donor myoblasts with dy host muscle cells. Some of the questions that need to be discussed are: How many host muscle precursors or areas of regenerating muscle fibres are normally available at any one time to fuse with the implanted myoblasts, and therefore what proportion of the dy muscle fibres will actually incorporate donor muscle nuclei after a single injection of myoblasts? We also need to know whether the introduced myoblasts should be encouraged to proliferate locally or should they just fuse after implantation? For how long can these implanted myoblasts persist without fusing? (This may have immunological implications). Finally, what are the conditions *in vivo* that regulate or enhance the proliferation or the fusion of myoblasts?

At all times we should be aware that we will be dealing with dy host myoblasts which will be derived from postnatal muscle of juveniles (presumably from satellite cells), we are not likely to be talking about embryonic host muscle precursors. However, the potential source of implanted donor myoblasts may be postnatal adult or juvenile muscle, or embryonic tissue, or perhaps cell lines. There may be significant differences between postnatal myoblasts and those from embryos or cell lines.

I shall briefly discuss 3 topics.

1. Proliferation of host (dy) myoblasts.
2. Factors controlling the proliferation and fusion of myoblasts *in vivo*.
   Models of myogenesis
   Growth factors
3. Genetic influences in regeneration

Proliferation of host (dy) myoblasts

It has been shown that introduced normal myoblasts will fuse with regenerating dy host muscle cells (Partridge *et al.*, 1989); but it is not known whether the donor myoblasts fuse directly with new membranes formed around damaged areas of the dy muscle fibres, or with host muscle precursors to form myotubes. Are enough dy muscle fibres regenerating at any one time, to take full advantage of the myoblast transfer? If not, then should the number of host muscle cells available for fusion be increased? One way to achieve this it to induce regeneration in dy host muscles. This would result in large numbers of host (dy-) myoblasts being available for fusion with implanted (dy+) myoblasts and the formation of more mosaic muscle fibres. If we are prepared to induce regeneration, then a local anaesthetic like Marcaine which is known to be highly myotoxic (Foster and Carlson, 1980) might be the agent of choice in dy human subjects.

Most of the results that I shall refer to are from autoradiographic studies on muscle regeneration *in vivo* carried out in collaboration with Dr. John McGeachie from the University of W.A. Essentially what we do in these experiments is to injure or transplant muscles in mice, inject ^{3}H-Tdr at various times after injury to label cells synthesising DNA at that time, leave the animals for 10 days until the labelled muscle precursors have fused to form myotubes, and then retrospectively determine the pattern of muscle precursor replication (McGeachie and Grounds, 1989). Autoradiographic results after muscle injury in BALB/c mice demonstrate the onset, peak and duration of muscle precursor cell replication (McGeachie and Grounds, 1987), and show that there is essentially no DNA synthesis in muscle precursors before 30 hours and little after 5 days. The peak of precursor replication is around 2 to 3 days, depending on the type of injury. We know that many of these precursors fuse after only 2 cell divisions (Grounds and McGeachie, 1987) and this is confirmed by the appearance of myotubes from about 3 days.

Similarly, in dy muscles injured with Marcaine, we would expect that many host myoblasts would be proliferating after 2 to 3 days. So if we injured dy host muscles with Marcaine and then around 3 days later injected the donor myoblasts, maximum fusion should be possible between host and donor muscle cells. Although it seems radical, such a procedure might be recommended in humans to maximise the benefit of injecting myoblasts.

Factors controlling the proliferation and fusion of myoblasts

Factors controlling muscle precursor cell behaviour *in vivo* are largely of relevance to satellite cells, but also to the implanted myoblasts whatever their actual source.

Two different models of myogenesis have emerged from tissue culture studies. The cell lineage model says that stem cells give rise to committed cells and that these must undergo a series of obligate cell divisions before they are terminally differentiated and their only option is to fuse. This relatively inflexible model has important repercussions *in vivo* as committed precursors cannot be made to fuse prematurely, and they cannot be made to proliferate beyond the quantal cell cycle. The alternative permissive or opportunistic model says that there is no strict lineage and that, depending on the particular set of conditions, the muscle precursors can continue to proliferate or alternatively can undergo fusion. We have tested the cell lineage model of myogenesis as defined by Quinn and her colleagues (Quinn *et al.*, 1985) in regenerating mature muscle of mice and chickens *in vivo*, and our autoradiographic results do not support its predictions (Grounds and McGeachie, 1987; 1989a). This issue needs to be resolved, as these models have relevance to the potential manipulation of muscle precursor behaviour *in vivo*.

Tissue culture studies also show that various hormones (Florini, 1987; Cossu *et al.*, 1989), growth factors (Florini and Magri, 1989) and components of extracellular matrix (Sanes, 1986; Grounds, 1989) affect the proliferation and fusion of muscle precursors: some of these studies have been carried out on satellite cells from mature muscle (Allen and Boxhorn, 1989; Cossu *et al.*, 1989).

The main effects of some hormones and growth factors (GF) are shown below: their action may depend on the dose used and the combination with other growth factors (Florini and Magri, 1989). It will be very interesting to find out how these substances operate *in vivo*, and whether they can be used therapeutically to improve muscle regeneration, or to specifically enhance proliferation or fusion of the host or donor myoblasts in the proposed therapy.

FACTORS AFFECTING MYOBLAST BEHAVIOUR *IN VITRO*

	Proliferation (mitogen)	Fusion (differentiation)
Fibroblast GF	↑	↓
Insulin GF	↑	↑
Muscle GF	↑	–
Dexamethasone	↑	–
Adrenocorticotropin	↑	–
Prostaglandin E	–	↑
Transforming GFβ	↓	↓
Interferon	–	↓

Genetic factors

In myoblast therapy in humans, donor myoblasts will be fused with host muscle cells of a different genotype. The genetic differences between donor and host may be quite pronounced. There clearly may be problems of immunological rejection. In addition, it has already been recognised that there are problems associated with efficiently fusing together myoblasts from combinations of different genetic mouse strains, and that this also depends on the immunological status of the host (Watt, 1982; Partridge, 1982; Partridge *et al.*, 1989).

Two genetic mouse models of muscle regeneration that we have described may enable us to identify some of the important factors controlling regeneration (Grounds and McGeachie, 1989b). In Swiss mice muscle regeneration is excellent, whereas in BALBc mice it is poor. At 10 days after severe crush injury large numbers of myotubes are present and there is very little fibrosis in lesions of adult Swiss mice, however, in BALBc mice myotube formation is poor and large areas of fibrous and cellular connective tissue are present in the centre of the lesions. It would be useful if conditions in BALBc mice could be manipulated to improve their muscle regeneration. Why is muscle regeneration so different in the two strains? What are the genes involved? It is desirable to know to what extent the general host environment is important in controlling the effectiveness of muscle regeneration, e.g. the capacity to revascularise, macrophage function, the extracellular matrix and circulating factors: as compared with the importance of genetic factors inherent in the muscle itself, such as the capacity to form new membranes to seal the ends of the damaged fibres, the production of specific growth factors (Bischoff, 1986), production of growth factor receptors and responsiveness to growth factors.

An understanding of the genetic factors controlling different aspects of muscle regeneration has direct relevance to myoblast transfer therapy, as the success of this procedure may be determined by the genotype of the host environment and host myoblasts, in combination with the genotype of donor myoblasts.

ACKNOWLEDGEMENTS

Original research was supported by grants from the National Health and Medical Research Council of Australia.

REFERENCES

Allen, R.E., and Boxhorn, L.K., 1989, Regulation of skeletal muscle satellite cell proliferation and differentiation by transforming growth factor-beta, insulin-like growth factor 1, and fibroblast growth factor, J. Cell Physiol. 138:311.

Bischoff, R., 1986, A satellite cell mitogen from crushed adult muscle, Dev. Biol. 115:140.

Cossu, G., Cusella-De Angelis, M.G., Senni, M.I., De-Angelis, L., Vivarelli, E., Vella, S., Bouche, M., Boitani, C., and Molinari, M., 1989, Adrenocorticotropin is a specific mitogen for mammalian myogenic cells, Dev. Biol. 131:331.

Florini, J.R., 1987, Hormonal control of muscle growth, Muscle Nerve 10:577.

Florini, J.R., and Magri, K.A., 1989, Effects of growth factors on myogenic differentiation, Am. J. Physiol. 256: C701.

Foster, A.H., and Carlson, B.M., 1980, Myotoxicity of local anaesthetics and regeneration of the damaged muscle fibres, Anesth. Analg. 58 : 727.

Grounds, M.D., 1989, Factors controlling skeletal muscle regeneration in vivo, in: "New Concepts of Pathogenesis and Therapeutic Strategies for Duchenne and Becker Muscular Dystrophy", B.A. Kakulas and F.L. Mastaglia, eds., Raven Press, New York, In press.

Grounds, M.D., and McGeachie, J.K., 1987, A model of myogenesis in vivo, derived from detailed autoradiographic studies of regenerating skeletal muscle, challenges the concept of quantal mitosis, Cell Tiss. Res. 250: 563.

Grounds, M.D., and McGeachie, J.K., 1989a, Myogenic cells of regenerating adult chicken muscle can fuse into myotubes after a single cell division in vivo, Exp. Cell Res. 280:429.

Grounds, M.D., and McGeachie, J.K., 1989b, A comparison of muscle precursor replication in crush injured skeletal muscle of Swiss and BALBc mice, Cell Tiss. Res. 255:385.

McGeachie, J.K., and Grounds, M.D., 1987, Initiation and duration of muscle precursor replication after mild and severe injury to skeletal muscle, Cell Tiss. Res. 248, 125.

McGeachie, J.K., and Grounds, M.D., 1989, Applications of an autoradiographic model of myogenesis in vivo, in: "New Concepts of Pathogenesis and Therapeutic Strategies for Duchenne and Becker Muscular Dystrophy", B.A. Kakulas and F.L. Mastaglia, eds., Raven Press, New York, In press.

Partridge, T.A., 1982, Cellular interactions in the development and maintenance of skeletal muscle, p555, in: "Cell Behaviour", R. Bellairs, A. Curtis, G. Dunn, eds., Cambridge Uni. Press, G.B.

Partridge, T.A., Morgan, J.E., Coulton, G.R., Hoffman, E.P., and Kunkel, L.M., 1989, Conversion of mdx myofibres from dystrophin-negative to -positive by injection of normal myoblasts, Nature 337:176.

Quinn, L.S., Holtzer, H., and Nameroff, M., 1985, Generation of chick skeletal muscle cells in groups of 16 from stem cells, Nature 313:692.

Sanes, J.R., 1986, The extracellular matrix, p155, in: "Myology", 1: A.G. Engel, B.Q. Banker, eds., McGraw-Hill Book Co, N.Y. 155.

Watt, D.J., 1982, Factors which affect the fusion of allogeneic muscle precursor cells in vivo, Neuropathol. Appl. Neurobiol, 8:135.

DISCUSSION

Partridge: First of all could you talk about the idea of inducing a crisis as a means of enhancing fusion. You're suggesting that one might put Marcaine into dystrophic muscle: could you expand upon the timing if you were led to do so grave a thing?

Grounds: One really needs to know to what extent the donor myoblasts want to fuse when they're introduced, or to what extent they're prepared to stick around and proliferate. But if one assumes that when you implant them they want to fuse, then I think that you have to injure the host muscle prior to implanting the donor myoblasts such that there are large numbers of host muscle precursors available and ready to fuse with the donor myoblasts in order to produce the mosaic fibers.

There is really very little information on the number of host muscle precursors in dystrophic muscle that are available for fusion at any one time. There has been only one study that I know of which provides some information on this subject. Anderson and colleagues (1988) injected tritiated thymidine into mdx mice aged four, or 32 weeks of age, took the samples out one hour later and looked at (premitotic) subsarcolemmal nuclei, which are considered to be satellite cells. They found that two to three percent of the subsarcolemonuclei were labelled. This represents a 200 fold increase over satellite cell turnover (0.014%) in normal adult mouse muscle (McGeachie and Grounds, 1989). Tritiated thymidine is only available for about an hour after injection, so only cells which are actually synthesizing DNA at that time are labelled. The DNA synthesis part of the cell cycle is perhaps roughly a third of the entire cell cycle, so the two to three percent might represent as much as nine percent of satellite cells which were moving through the cell cycle and proliferating at this time: not necessarily synthesizing DNA, but in a proliferative phase. Now, nine percent is a very high proportion because satellite cells are normally considered to represent only about one to five percent of all muscle nuclei in adult host muscle fibres (Allbrook *et al.,* 1971). There may be very large populations of satellite cells available in dystrophic muscle, and I think that it is important to obtain more precise information on this subject.

Another option that exists if you do have quite a large population of activated proliferating dystrophic host muscle precursors, is to directly enhance the proliferation of these cells (not by causing degeneration and regeneration with Marcaine) by putting in a mitogenic growth factor like fibroblast growth factor, which will increase the proliferative capacity of the host muscle precursors, before you put in the donor myoblasts.

I think that if you're going to go through the trauma of injecting the donor myoblasts, then you want to ensure that you get the maximum efficiency out of the procedure. Although it is slightly radical to consider interfering with the dystrophic subject in this way I think that this approach should be seriously evaluated.

Partridge: As Marcaine is a local anaesthetic, at least it would be painless!

Grounds: It is certainly preferable to the use of myotoxic snake venoms.

Partridge: You put forward two extreme versions of the developmental basis of myogenesis. If one of these were to be confirmed to the exclusion of the other in mature muscle *in vivo*, then what sort of changes in strategy would that lead you to adopt with respect to myoblast implantation therapy?

Grounds: Well there are two aspects to this:
One is that the donor cell which you are going to implant will be relatively synchronized, as the muscle precursor cells will have been cultured for a particular length of time and will presumably have passed through several cell cycles. If you assume that the cell lineage model applies - with its obligate cell divisions and a terminal cell cycle - then your donor myoblasts might be fairly close to the stage of terminal differentiation. Whereas your dystrophic host myoblasts, if you haven't modulated them in any way, would presumably be a very heterogeneous population of cells - some would be stem cells, some early committed precursors, and some might be almost terminally differentiated. So when you put these two groups of muscle precursor cells together, there might be difficulty in actually fusing the donor myoblasts (many of which may be essentially terminally differentiated), with the host muscle precursors (which are at various stages of differentiation, many not able to undergo fusion). This simplistic scenario results from the strict cell lineage model, which clearly has implications for the efficiency of fusion between the donor and host muscle precursors. In contrast, the opportunistic idea says that muscle precursors if they find themselves in the right environment, will quite readily fuse together: there is no barrier to fusion.

The second aspect is relevant if you want to entertain the idea of playing around with enhancing the proliferation of either the host or the donor myoblasts, or perhaps delaying of enhancing the fusion between them. This is much more difficult if the cell lineage model is the situation that applies *in vivo*, because you just don't have the flexibility. In contrast the opportunistic model says that if you change the environment around the muscle precursors, then they will respond accordingly.

Partridge: What about growth factors? Could you use them systemically, or would you have to apply them locally?

Grounds: Well, again, not many people have worked with, or explored these avenues *in vivo*. The situation depends upon what questions you're asking.

You may want to use a mitogen like fibroblast growth factor (FGF) or perhaps the Bischoff muscle growth factor. If it is a competence factor like FGF, then it probably only has to be available for a very short period of time, in order to move the cells from G0 into G1. So if the aim is to promote host muscle precursor cell proliferation by putting in a competence factor, the factor only has to be available transiently and could be injected. A local intramuscular injection would seem to be much more efficient than systemic injection which would consume a large amount of expensive growth factor. In dystrophic humans an intramuscular injection like this is feasible.

Another method of administering growth factors is the incorporation into substances like Agarose gel which are biodegradable, would disperse throughout the tissue, and would have a slow release effect (Hayek *et al.,* 1987). This approach is more important when the growth factor needs to be available for a longer period of time, such as with progression factors, or with growth inhibitors like transforming growth factor beta.

A third option which lends itself more to experimental work in mice, is to incorporate the growth factors into plastics such as Hydron (hydroxy methyl methacrylate : Langer and Folkman, 1976) or L-vax 40 (ethylene vinyl acetate copolymer : Rhine *et al.,* 1980). These plastics can be made up into little buttons, pellets or sheets, and implanted under the skin where they are non-inflammatory and slowly release the growth factor. You can devise scenarios whereby pellets are implanted containing combinations of several growth factors which would enhance proliferation but restrict differentiation and fusion (Allen and Boxhorn, 1989): the pellet could later be removed and the large numbers of muscle precursor cells that had been generated could fuse to form myotubes. The use of pellets in this way is attractive for experimental purposes in animals, but is less applicable to humans.

REFERENCES

Allbrook, D.B., Han, M.F., and Hellmuth, A.E., 1971, Population of muscle satellite cells in relation to age and mitotic cycle, Pathol. 3:233.

Anderson, J.E., Bressler, B.H., and Ovalle, W.K., 1988, Functional regeneration in the hindlimb skeletal muscle of the mdx mouse, J. Musc. Res. Cell Motil. 9:499.

Hayek, A., Culler, F.L., Beattie, G.M., Lopez, A.D., Cuevas, P., and Baird, A., 1987, An *in vivo* model for study of the angiogenic effects of basic fibroblast growth factor, Biochem. Biophys. Res. Comm. 147: 876.

Langer, R., and Folkman, J., 1976, Polymers for the sustained release of proteins and other macromolecules, Nature 263:797.

McGeachie, J.K., and Grounds, M.D., 1989, The onset of myogenesis in denervated mouse skeletal muscle regenerating after injury, Neurosc. 28:509.

Rhine, W.D., Hsieh, D.S.T., and Langer, R., 1980, Polymer for sustained macromolecular release : procedures to fabricate reproducible delivery systems and control release kinetics, J. Pharmaceut. Sci. 69: 265.

THE DMD GENE PROMOTER:

A POTENTIAL ROLE IN GENE THERAPY

Peter N. Ray, Henry J. Klamut and Ronald G. Worton

Genetics Department and Research Institute,
Hospital for Sick Children, and
Department of Medical Genetics, University of Toronto
555 University Avenue, Toronto, Canada   M5G 1X8

Introduction

I have been asked to discuss the work our group in Toronto has been doing in examining the DMD gene promoter region at the molecular level so that we might better understand the factors involved in regulating the transcription of this gene. This topic differs significantly from the previous talks we have just heard on the technical aspects of myoblast transplantation but while it may not be essential to fully understand the in vivo regulation of the DMD gene for the immediate experiments being proposed here, a knowledge of DMD gene regulation may well be necessary if long term therapy is to be successful.

It is clear from what we have just heard that there are many unanswered questions regarding myoblast transplantation. While it seems likely that it will be possible, though difficult and expensive, to obtain large numbers of donor myoblasts, we do not yet know how far donor myoblasts will migrate from the site of injection, how well they will fuse to existing myofibers, or how many donor nuclei need to be present to provide enough dystrophin to restore function. If the answers to these questions do not live up to our hopes then myoblast therapy may not provide sufficient benefit, or may inflict a level of physical and psychological trauma on the affected boy that is not acceptable.

May I add my voice here to those who have advocated caution in proceeding with human trials. Since so many questions regarding myoblast transfer remain unanswered there can be no expectation, but only a hope for real success. Most of these parameters could be systematically tested in either of the available animal models while they cannot be adequately addressed by human experimentation. There is the risk that human trials will lead to unfulfilled expectations in the general public resulting in distrust of and lack of support for the scientific community.

One possible way of overcoming some of the technical limitations of myoblast transfer would be to augment the synthesis of dystrophin in the donor nuclei, thereby reducing the number of donor cells required. For the longer term it may be more reasonable to develop true gene therapy approaches where a

Myoblast Transfer Therapy
Edited by R. Griggs and G. Karpati
Plenum Press, New York, 1990

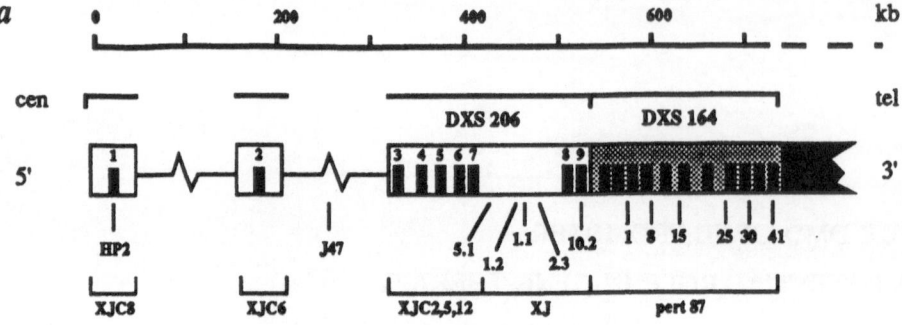

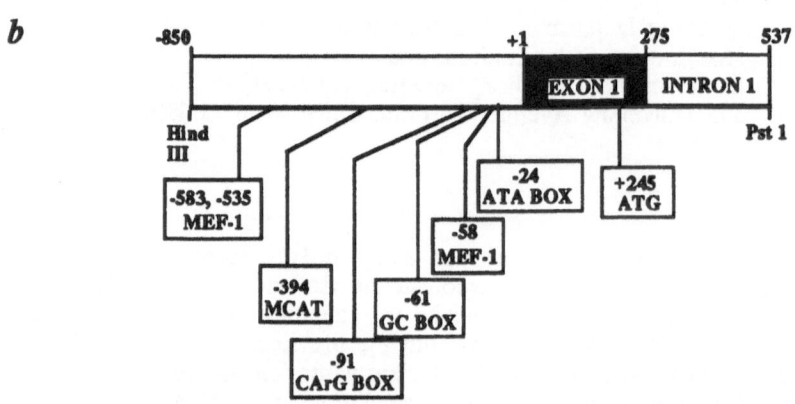

Fig 1. Physical map of the 5' end of the DMD gene and sequence position of promoter elements. (a) Five cosmid clones XJC8, 6,2,5,12 containing the first seven exons of the DMD gene were cloned from an XXXXY human genomic library. The position of the exons within these cosmids and the position of the cosmids relative to the XJ (DXS206) and pERT87 (DXS164) regions is shown. (b) a 1.4 kb HindIII - PstI fragment (designated HP2), containing exon 1 was subcloned from cosmid XJC8 and sequenced. This clone contained 275 bp corresponding to exon 1, 262 bp of intron 1, and 850 bp of upstream sequence. The position of the promoter elements described in the text is indicated.

functional DMD gene is introduced either directly into the affected boys muscle or into explanted myoblasts derived from the affected boy which can then be reintroduced by myoblast injection. If the gene construct used in this therapy included the authentic DMD promoter region, then one might expect that expression of the transgene would be limited to muscle fibers thereby increasing the possibility of success while decreasing the likelihood of adverse complications due to uncontrolled gene expression. While gene therapy is not without its own difficulties, this approach would reduce the potential immunological problems inherent in injecting "foreign" myoblasts. Of course, both of these approaches, transcriptional enhancement and gene therapy, requires a complete understanding of the cis- and trans-acting factors controlling expression of the DMD gene.

Promoter Cloning and Sequencing

In order to examine the regulatory domains governing DMD gene expression we have cloned the 5' region of the DMD gene from a human genomic cosmid library using cDNA probes. Fig 1a shows several of the isolated cosmid clones which contain the first seven exons of the DMD gene and their physical relationship to the DXS206 and DXS164 regions. The position of exon one within cosmid XJC8 was determined and a 1.4 kb Hind III-Pst I fragment containing all of exon one, part of intron one, and sequences upstream of exon one was isolated and sequenced.

The CAP site, or transcriptional start site, was accurately mapped by primer extension and RNase protection (Klamut et al 1989) and was found to be located 37 base pairs upstream from the reported 5' end of the DMD cDNA (Koenig et al 1987). A number of transcriptional consensus sequences were found immediately upstream of this CAP site (Fig. 1b). Two of these are well known core promoter elements, an ATA box (Breathnach and Chambon, 1981) at position -24 and a GC box (Briggs et al 1986) at position -61. In addition, three potential muscle specific promoter elements were also found in this region. A CArG box (Minty and Kedes, 1986) with strong homology (79-100%) to other reported CArG box domains was found at position -91. This region also displayed 75% homology to the 16 bp consensus sequence of the related CBAR domain (Carroll et al 1988, Schwartz et al 1989). The position of this element 91 base pairs upstream of the CAP site is in good agreement with that of other reported CArG box elements. Also present in this region are three potential MEF-1 factor binding sites (Buskin and Hauschka 1989). While these showed 79-86% homology to the published MEF-1 concensus sequence their positions at -58, -538, and -583 do not correspond well to the position of the MEF-1 domain (-1130) within the muscle creatine kinase promoter (Horlick and Benfield 1989). The significance of this variation in position is not yet clear but in general position has very little effect on true enhancer activity. A third potential muscle specific promoter element was found at position -394. This MCAT consensus sequence (Mar and Ordahl, 1988) lies further upstream from the position where this element has been found in a number of other muscle-specific promoter regions.

Promoter Function

In order to investigate the functional significance of these potential regulatory elements the 1.4 kb Hind III- Pst I fragment, designated HP2, was cloned into a eukaryotic expression vector to determine if this region would drive the transcription of the chloramphenicol acetyltransferase (CAT) gene in a muscle specic manner. This construct was transfected into a number of different myoblasts cultures and CAT activity was measured before and after the cells had differentiated to form multinucleated myotubes. As can be seen in Table 1, the human DMD promoter showed considerable variation in activity depending upon the cell type transfected. In clonal human muscle cell cultures, promoter activity was very high once the cells had differentiated into myotubes (hMT) but was a low basal level in myoblasts before fusion and in primary human fibroblasts (data not shown). Thus the 850 base pairs region upstream of the CAP site contained the elements necessary to provide muscle specific, differentiation specific transcription. In studies to compare the DMD promoter activity to other well characterized promoters it was found that the DMD promoter displayed a level of activity, in differentiated myotubes, significantly greater than that of the SV40 promoter and only marginally lower than that of the highly

Table 1. Comparison of pHP2CAT activity in various cell lines.

cell line	n [a]	CAT activity [b]
hMT	22	29.8 ± 19.3
mMT	3	9.35 ± 4.82
C2 MT	6	0.95 ± 0.12
L6 MT	3	0.50 ± 0.19
H9C2 MT	6	33.2 ± 12.2

[a] number of observations

[b] CAT activity (units CAT/mg protein/unit hHG ($\times 10^{-3}$))
corrected for transfection efficiency in various cell lines.

active RSV LTR. This level of activity was unexpectedly high since DMD transcripts have been reported to be present at relatively low levels in muscle tissue (Chelly et al 1988). It is possible that other regulatory sequences situated outside the HP2 region are functioning in vivo to down regulate or repress transcription of the DMD gene.

Another interesting observation in these studies was that the DMD promoter activity was not induced upon differentiation of either the C2 mouse or the L6 rat myoblast lines. This was not evidence of species specificity since promoter activity was induced upon differentiation of primary mouse myoblasts (mMT) and the rat H9C2 muscle cell line. Since both C2 and L6 lines are known to synthesize other late muscle proteins the fact that the DMD promoter is not induced in these cell lines would indicate that the trans-acting transcriptional factors required by this promoter are different from those required by other late muscle promoters.

Summary

The studies I've outlined here are obviously at a preliminary stage but do offer some insight into the complexity of DMD gene regulation and do suggest that an understanding of this regulation may have some potential benefit in gene therapy for this disease. The high promoter activity found was unexpected and suggests that in vivo the activity of the endogenous gene may be repressed by elements not present within this region. Of course, other interpretations are possible. The transcripts in vivo may turn over very quickly, or the very large size of the DMD gene may in itself limit the rate of transcription. Alternatively, the gene may be actively transcribed only during the early stages of differentiation. A more detailed analysis of developmental expression and of DNA sequences surrounding exon one is required to address these alteratives, but the possibility for augmenting dystrophin synthesis during myoblast therapy clearly exists. The high level of activity and the tissue and developmental specificity exhibited by the HP2 construct suggest this may be the promoter of choice in future gene therapy experiments. The high degree of specificity shown by this promoter would reduce the need to target gene constructs to muscle cells and would reduce the potential complications of uncontrolled gene expression.

Of course, before any of these benefits could be realized much more work must be done both in analysing DMD gene expression and in testing potential gene therapy constructs both in culture and in animal models of this disease.

References

1. Breathnach, R., and P. Chambon. 1981. Organization and expression of eukaryotic split genes coding for protein. Annu. Rev. Biochem. **50**: 349-383.

2. Briggs, M.R., J.T. Kadonaga, S.P. Bell, and R. Tijan. 1986. Purification and biochemical characterization of the promoter-specific transcription factor, Sp1. Science **234**: 47-52.

3. Buskin, J.N., and S.D. Hauschka. 1989. Identification of a myocyte nuclear factor that binds to the muscle-specific enhancer of the mouse muscle creatine kinase gene. Mol. Cell. Biol. **9**: 2627-2640.

4. Carroll, S.L., D.J. Bergsma, and R.J. Schwartz. 1988. A 29-nucleotide DNA segment containing an evolutionarily conserved motif is required in cis for cell-type-restricted repression of the chicken alpha-smooth muscle actin gene core promoter. Mol. Cell. Biol. **8**: 241-250.

5. Chelly, J., J.C. Kaplan, P. Maire, S. Gautron, and A. Kahn. 1988. Transcription of the dystrophin gene in human muscle and non-muscle tissues. Nature **333**: 858-860.

6. Horlick, R.A., and P.A. Benfield. 1989. The upstream muscle-specific enhancer of the rat muscle creatine kinase gene is composed of multiple elements. Mol. Cell. Biol. **9**: 2396-2413.

9. Klamut, H.J., Gangopadhyay, S.B., Worton, R.G., and Ray. P.N. 1990. Molecular and functional analysis of the muscle specific promoter region of the Duchenne muscular dystrophy gene. Molec. Cell Biol. **10**:193-205.

8. Mar, J.H., and C.P. Ordahl. 1988. A conserved CATTCCT motif is required for skeletal muscle-specific activity of the cardiac troponin T gene promoter. Proc. Natl. Acad. Sci. USA **85**: 6404-6408.

9. Minty, A., and L. Kedes. 1986. Upstream regions of the human cardiac actin gene that modulate its transcription in muscle cells: Presence of an evolutionarily conserved repeated motif. Mol. Cell. Biol. **6**: 2125-2136.

10. Schwartz, R.J., J.M. Grichnik, K.L. Chow, F. DeMayo, and B.A. French. 1989. Identification of cis-acting regulatory elements of the chicken skeletal alpha-actin gene promoter, p. 653-667. In L.H. Kedes and F.E. Stockdale (ed.), Cellular and Molecular Biology of Muscle Development, Alan R. Liss, Inc., N.Y.

Discussion of Dr. Ray's paper

Dr. Partridge: Is there any basis for speculating on how long it takes to actually begin to apply this sort of knowledge to driving the gene in cells?

Dr. Ray: Well, that's very difficult to answer, of course. Since we have a promoter it is reasonable now to try to construct a minigene, and then we can start doing experiments introducing such a minigene into myoblasts, so those experiments will go along. They are clearly long-term experiments, and it is difficult to predict an exact time frame.

Dr. Partridge: You were suggesting that you've put them into the myoblasts and then perhaps transfect them. Is the idea to use the patient's own myoblasts?

Dr. Ray: Well, there are a number of ways you can go with this. One, of course, is if it turned out to be desirable to use the patient's own myoblast, you could take them out and try to augment synthesis in those by either introducing constructs into those. Alternatively, by understanding what factors control this regulation, you may be able to use some alternative form of gene therapy in inducing or augmenting expression of the endogenous gene.

There are a number of Duchenne patients, or Becker patients, that make some functional protein but not enough to prevent the expression of the disease. But by augmenting the endogenous gene, you may alleviate their symptoms, which might be an alternative form of therapy.

Dr. Partridge: Is that a very large proportion of patients to which this applies?

Dr. Ray: Well, it's not clear. One assumes that nearly all Becker patients have some gene function. There are probably Duchenne patients that have a functional gene which is expressed at such a low level that they have essentially very little functional protein. So those patients you could help, but that might represent 10 to 15% at the most. There may be a number of patients who have actually Becker Muscular Dystrophy, which is not recognized as such, and is included in other myopathies, or is just not found. Those are other possible candidates.

Dr. Partridge: Would this type of thing be applicable to other myopathies as well? Could you do equivalent things?

Dr. Ray: It depends on how much of a generalist you want to be. Certainly understanding the functioning of the Duchenne gene is going to have its most immediate application to treating and providing therapy to Duchenne patients. Understanding of the Duchenne promoter may not help us to understand other promoters but understanding promoter function will help us to understand promoter function in other genes, but that would be an independent investigation certainly.

112

GENERAL DISCUSSION

PRACTICAL ASPECTS OF MYOBLAST IMPLANTATION

Dr. Law: I have some questions for the laboratory of Dr. Partridge, that includes Dr. Watt, Dr. Morgan and Dr. Grounds.

It is my understanding that your initial work was practically all done in normal to normal animal transplant, and it was not until your recent paper in 1988, that you actually have done a transplant from normal myoblasts into mdx mouse, which is not a concrete dystrophic animal model. Is that correct? You show significant improvement after myoblast implantation. Were you able to show using dystrophin antibodies that there was a substantial dystrophin staining?

Dr. Morgan: Yes. There are about 70%.

Dr. Law: I was amazed at the fact that in the other preparations that you had in which you showed more than 75% dystrophin staining that there were a lot of abnormality in that preparation. There was tremendous variation in fiber sizes and there were a lot of central nuclei.

Dr. Morgan: No, there was not.

Dr. Law: In fact, it was...

Dr. Partridge: Could I, just cover that quickly? The preparation in which was 75% or in excess of 70% of dystrophin-positive fibers, had almost no central nuclei. It was almost entirely periphery nucleated and looked normal. There were a few minor abnormalities. The one that had a high proportion of central nuclei and abnormalities was one that was mdx.

Dr. Roses: I do not have any problem looking at dystrophic mouse that gets weak and saying that it gets injected and it gets better. I do have a problem with the mdx mouse concerning some of the results presented here. How do the "mdx people" explain the fact that the molecule is expressed late? I also understand that in some experiments where people have done saline injections, expression of dystrophin was also seen. Is it a function of the particular antibody or is it seen with all the antibodies?

Dr. Partridge: Dr. Karpati, do you want to answer that first?

Dr. Karpati: Are you referring to those extremely rare native dystrophin-positive fibers in untouched mdx mice?

113

Dr. Roses: No, I am referring to the rather common ones that are present when you do saline injections as controls.

Dr. Karpati: I can say that most mdx investigators find that in some animals in certain colonies, you find isolated, or sometimes small clusters of dystrophin-positive fibers in intact animals. This amounts to about 1 dystrophin-positive fiber for every 5-8 thousand dystrophin-negative fibers. Nobody knows the precise explanation for it. There is genuine dystrophin in the membrane of those fibers demonstrable with either N- or C-terminal antibodies. I think it has been suggested that a second "reversing" mutation occurred in some muscle precursor cell(s) contributing to such fibers during early myogenesis. This putative "reversing" mutation somehow negated the deleterious effects of the original mutation, and the dystrophin gene was rendered competent. By clever experimental approach, these points could be further investigated!

Dr. Kunkel: Maybe I should address the question. Basically, I think it varies, depending on the mouse, the age of the mouse; you can see anywhere from less than 1% up to more. Dr. Hoffman had one mouse which had 80 positive fibers across the whole muscle.

Our assumption at the moment is that most of them represent revertence. That is Becker, deletion-like mutations then revert the phenotype back to that of a Becker. That is our working hypothesis. It also could be a cross reacting, embryonic dystrophin that is coming on in a rare fiber. Those are the only answers we have. But so far, as best we know, most of the antibodies do see that cross reacting.

Dr. Karpati: In our colonies the prevalence of these fibers is quite low. In some colonies it is much more. So some sort of environmental influence might play a role.

Dr. Kunkel: You have to be very careful with it because as I said 80 fibers per muscle is a lot. It would look like a successful take rate in a myoblast injection series. So you have to really look at the other leg very carefully and compare the two legs to each other.

Dr. Karpati: Your point is very well taken.

Dr. Partridge: Dr. Blau, did you have a comment?

Dr. Blau: Yes, I have a question for Dr. Partridge's group. You suggested that dystrophin has a direct effect on nuclear localization and localizing the nuclei to the periphery. On what do you base that?

Dr. Partridge: It is not a watertight suggestion still. It is based on the fact that mdx muscles that have been injected with normal muscle precursor cells, rapidly peripheralize their nuclei. The nuclei are to the outside of those fibers within 30 days. If you do the same thing by injecting mdx cells into mdx muscle, then the proportion of central nuclei increases progressively. This does not seem to be correlated to any great degree with regeneration. We know that in both instances the cells that we have put in have largely or, at least half, displaced the muscle cells of the host animal. So that most of those dystrophin-positive fibers have actually arisen by regeneration. They get their nuclei to the outside very quickly.

Dr. Blau: Which is quite unusual for regeneration.

Dr. Partridge: Yes. The only other instance in which we see it is in some of our very early experiments, where we grafted neonatal muscle. Then the neonatal muscle rapidly peripheralized its nuclei after regeneration.

Dr. Blau: So are you hypothesizing a function for dystrophin in development?

Dr. Partridge: Oh, no. I do not know that there is necessarily a direct effect but I do not think it is mediated by degeneration and regeneration.

Dr. Van Ommen: I have one comment for the suggestion of Dr. Ray on the minigene for the potential healing of the defect. I think one should realize that the gene is so enormously large that it would probably need a quite efficient promoter to get expressed at all and that if you try to truncate this gene to a cDNA, you have to be cautious. You may well end up in over-expression of the gene if you use the homologous promoter.

Dr. Ray: That is undoubtedly true but that seems to be the least of our concerns right at the moment.

Dr. Van Ommen: The question I had actually relates back to the previous question. This idea of back mutation is testable and we also had discussed that before the mdx transplantation into mdx should show a difference if you compare the irradiated and the non-irradiated specimen because then in the irradiated ones you would enhance mutations.

So you would see more dystrophin-positive fibers in those cases than if you did not irradiate? Has that experiment been done?

Dr. Morgan: X-irradiation should not cause a deletion and it must be a deletion in the mdx mouse that causes these back mutations so that it is not necessarily the case.

Dr. Whalen: I had a question for Dr. Law. The dystrophic dy mouse has a well characterized abnormality in the myelination of the peripheral nerve. Now, have you looked at the peripheral nerves to see whether in the animals which have regained such a clear functional performance have any improvements in the PNS? Otherwise, I would like to know what you think about how the animal can be so functionally, relatively normal, in the presence of these persisting PNS abnormalities.

Dr. Law: I was wanting to present some of the evidence earlier except that due to time constraints, I was not able to present the data. But way back in the 1970's, when the neural hypothesis was a very hot bandwagon, I did a parabiotic mouse experiment in which we demonstrated that dystrophic nervous system was normal in its ability to form synapses, to cause fiber type differentiation and to maintain the normal structure and functional integrity of the normal muscle.

This is a cross-parabiotic mouse experiment. And that, of course, provides very strong evidence which indicates that although the dystrophic nervous system is abnormal (after all the neurons contain the dystrophic gene), it is not bad enough to cause the normal muscle cells to undergo degeneration.

Dr. Whalen: I am not saying that the gene responsible for the mutation is necessarily uniquely and primarily expressed in neurons, nonetheless, there is pathology. I am still perplexed and I wonder if anyone else has a comment on this.

Dr. Partridge: I wonder if I could ask Dr. Karpati to comment on this.

Dr. Karpati: I see Dr. Peterson there. Are you going to talk about this, Alan? You worked on the amyelination of the lumbar roots in the dy mouse and I just want to remind the audience that although amyelination of lumbar roots must be somehow related to the fundamental gene defect (which is not known), you can correct amyelination simply by crushing the roots and then letting the roots regenerate! Is that correct, Dr. Peterson?

Dr. Peterson: That is generally true. That is correct. I specifically wanted to emphasize that we are well aware that the dy mutant in the mouse represents a very complex syndrome and I appreciated the fact that Dr. Whalen raised that point. I would find it extremely difficult to understand the dramatic improvement in the behavior that your film so elegantly showed without believing that there would have been some amelioration of the amyelination in the spinal roots of those mice. I think it would be an urgent experiment to complete because I think it underlies very much whether a particular mutation in the mouse is going to have general applicability as you suggested or perhaps be specifically related to its own defect.

I have one more question, if I may and that was, we all struggle very hard to find genotype markers and you mentioned in passing that your antibodies to GPI, if I understood you correctly, labelled a subpopulation of myonuclei. Is that correct?

Dr. Law: The antibodies against GPI were very universal for the normal cells. So we can have all of the normal cells labelled as the donor cells. Whereas for the host cells, they would not be labelled, as you can see, in our preparation. But in answering your question as raised about roots, obviously with all the dystrophic characteristics in the muscle, it is not animal model for peripheral neuropathy. Nonetheless, these show spontaneous and hereditary muscle degeneration which is what you are using it for.

Dr. Partridge: Dr. Peterson, could I ask you to be more specific on what you would expect from the amyelination defect and, for those of us who do not know the model well, if you could just indicate the implications of your question. I did not really follow you.

116

Dr. Peterson: The spinal roots of both dy and dy2J mice show a remarkable arrest of primary interaction between Schwann cells and axons, such that axons in that region traversing the longer spinal roots have never received appropriate Schwann cell ensheathment. There have been very elegant experiments, the electrophysiological consequences of this which appear to be both ephaptic and spontaneous generation of action potentials and much of the spontaneous dystonic movement of the legs in dy mice generated specifically from that defect.

Dr. Whalen: Your point being that as you watch some mice on a glass rod, they do not show any of those "dystonic" features, suggesting that the neural defect was corrected in some way or that there is a factor that we do not understand about the treatment corrected in some way, or that there is some factor that we do not understand about the treatment.

Dr. Peterson: I would think it would be extremely important to understand what the status of those nerve roots was.

Dr. Whalen: Dr. Law, have you looked at the nerve roots in the animals?

Dr. Law: Yes. In fact we were one of the earliest ones to actually look at some of those peripheral nerves, and also the nerve roots when I was with Dr. McComas who postulated a neurogenic hypothesis of dystrophy.

Dr. Whalen: I understand that you looked at them originally.

Dr. Law: Yes.

Dr. Whalen: But in response to the myoblast transfer treatment, have they changed?

Dr. Law: I have not looked into that yet. But my feeling is that has very little to do with it. In 1976, I published on the pseudomyotonia in the dystrophic mouse, in which we ascribe such pseudomyotonia not due to the demyelinated or amyelinated nerve roots, but rather it is due to the fact that here we have a muscle that is undergoing contracture. It is the firing of the Golgi tendon organ, as well as the firing of the muscle spindle, that is causing this sporadic flexion-extension of the thigh. To my mind, it is not originating from the amyelinated roots. What we have achieved here is a correction of the muscle, therefore, interrupting the firing of the Golgi tendon organs as well as the muscle spindle.

Dr. Whalen: Your answer to Dr. Peterson's question is that you are not interested in what the roots look like. Is that correct?

Dr. Law: What I am trying to say is that we are ascribing too much significance to amyelination in the animal model. We have done studies in which we demonstrated that such abnormal nervous systems could, in fact, support normal muscle fiber development.

Dr. Peterson: Perhaps all Dr. Whalen and I are perhaps saying is, that we are very interested in what the spinal roots of those particular mice would look like.

Dr. Watt: Could I ask Dr. Law one quick question? I was wondering if you had done control experiments when you actually implanted dystrophic muscle into the dystrophic hosts?

Dr. Law: No, that was proposed in our initial protocol, but eventually we did much more important experiments than putting dystrophic cells back into dystrophic muscle again.

Dr. Watt: It would seem to be a very important control.

Dr. Law: I think so, yes.

Dr. Worton: I just have a general question for the panel, or anyone who can answer it, and that is, is it known whether the mild phenotype in mdx is limited because of mouse as a species or whether it is the strain background that it is on? And has the mdx mutation been bred into other strains that may have different regenerative capacities?

Dr. Partridge: We have brought it on to our other strain background, which is I think a Swiss white background rat. We have also got it crossed on to the 129-REJ mouse, which was used to carry the dy dystrophy in our colony. In both cases the phenotype is the same.

Dr. Watt: I could perhaps just add to that. We have done some crush injury experiments in the mdx mice just to see whether the pattern of regeneration was that of an excellent regeneration or not. It suggests that it is not some kind of superior muscle regeneration associated with it.

Dr. Moxley: A couple of just quick questions to Dr. Blau. It is a follow-up on Dr. Partridge's question earlier. I am just curious about migration patterns of the injected myoblasts, whether or not you have evidence that they might migrate out of fascicles. In terms of just thinking ahead for practical concerns about what we could hope for in terms of our injectate. Then, as a corollary, do you have any sense of the absolute density required? We have heard of the number of cells per unit volume in the mice. Is it going to take a half million per cubic centimeter in a human?

Dr. Blau: To address the first question about how far the myoblasts are migrating. At this point, we know they are migrating to contribute to several fibers within their vicinity, and we are now looking to see what distances they can actually travel using different assays. So we do not have that answer yet. With respect to the myoblast number, that is really a question of injected myoblasts, and I would anticipate that the injected myoblasts, especially where degeneration and regeneration are going on, will migrate at least as well as the normal. That it bodes extremely well for cells being able to travel and contribute to multiple fibers when injected. We are definitely going to look at that question as well; labelling cells in vitro and introducing them into the mice, human cells and seeing how far they migrate and contribute to muscle fiber formation.

Dr. Partridge: And as a quick follow-up to Dr. Grounds, related to that, and based on the issue if there is something about the muscle environment that we could do in order to optimize, perhaps, in the future in our patients. In the not-too-distant future things like IGF-1 will become available that is human compatible. I am just curious if you had the opportunity in those Swiss compared to the BALBc mice, to look at differences in those muscles. So, for instance, in the density of the IGF-1 receptor, or something that might account for those differential responses that you saw in concert with your concern about the interface between the genetic predilection for improvement, along with other factors that may account for our ability to sustain or recover muscle fiber growth.

Dr. Grounds: Well, I think there is really an enormous number of possibilities you could look at. In our initial experiments, we are doing some breeding studies and looking at F2 mice to see how many genes might be involved. We are also doing cross-transplantation experiments to see as to what extent the host environment has a dominant effect. I really think it is a matter of just trying to get as much information as one can at this stage. I would be very happy if other people were prepared to run a battery of tests. There is very little information available as to what genetic factors really do control muscle regeneration in vivo.

Unidentified Participant: I just wanted to make a general comment about different methods for labelling myoblasts for transfer, and that some of those methods may minimize the amount of participation by the transferred myoblasts in that. It may well be that, particularly if the retroviral nuclear-marked myoblasts may not proliferate at the same rate than the unlabelled ones. The second thing is that when we have looked at the different ways of labelling the myoblasts, we could identify them. We find that very trivial amounts of treatment with thymidine incorporation markedly alter the ability of myoblasts to differentiate. Even though thymidine is a good marker, it does not only show proliferation but also interferes with differentiation. There is transfer of thymidine from one nucleus to another, although I do not have a real explanation for that.

I have a question for Dr. Law. Since there are many muscles in the hind limb, if you intentionally inject only one muscle versus injecting many muscles, what are the differences in outcome that you see?

Dr. Law: In fact, we have done both. In these, the initial experiment was done on one muscle of the calf; and later on we extended the injection into all of the major muscle groups.

I would say then, a small muscle would require less foci of injection as compared to the other major muscle groups. What I envision then when we apply this in the human, we probably will also be doing multi-focal injections of the major muscle groups that are badly involved. Although I do not believe that we need to inject all of the muscles of the patient.

Dr. Morgan: We find that when we inject precursor cells into the mdx mice, we frequently find donor cells and dystrophin-positive fibers in adjacent muscle, so the muscle precursor cells are moving from one muscle to another.

Dr. Partridge: Dr. Law, in your functionally, normal or semi-normal mice, if you inject into a single muscle, that animal then acquires functioning of what must involve many muscles in order to make a coordinated movement?

Dr. Law: In my belief our experiments were done more specifically. Our injections were made very specifically, intramuscularly. Also, you know, it is sometimes difficult to control. The fact is that some of the donor cells did leak out from the site of injection.

Unidentified Participant: The movement of those animals' legs obviously involved the coordinated movement of many, many muscles, not a single muscle and, therefore, I am curious about whether you saw those coordinated movements which had to involve other muscles when you injected a single muscle.

Dr. Law: In some animals a particular muscle group was injected, but the majority of them functioned very normally, because I think that the smooth functioning, smooth behavior is a function of not only the development of donor cells, but also the usage of the animal with the limb. So I think that the smooth movement, what you saw here, is one of those few successful experiments.

Unidentified Participant: Did the animal we saw in the movie have multiple muscles injected or one muscle?

Dr. Law: That was multiple muscle injection.

Unidentified Participant: How many muscles out of the panoply of mouse muscles did you inject?

Dr. Law: As I say, we grouped them into four major groups. The extensors would include the tibialis anterior, the flexors will include the triceps surae, the quadriceps, of course, and also the hamstring and the biceps.

Unidentified Participant: So there were four injections per mouse?

Dr. Law: No. There were more injections than four, but it is injection into these four major groups. We also injected into the external intercostal muscles.

Unidentified Participant: Into the external intercostals? Between each interspace?

Dr. Law: Not between the interspaces. The idea is to put in the needle and go sideways so that you get a lot of cell with minor damage of the external intercostal muscles, and in such a way a lot of cells were deposited as a film on top of the external intercostal muscles.

Dr. Partridge: On the matter of the relative effectiveness of single limited injections on mice, I think irrespective of the reason why it happens there is quite a lot of evidence that myogenic cells can move across the epimysium in mice. Most of the leg muscles have very thin epimysia and we have really got quite good evidence that cells can get across that. That does not seem to be the case in rats, where the epimysium is much thicker. So a widespread effect from single injections in the mouse may be something that is really quite special to the mouse, and may not work in other species.

Dr. Caskey: I just wanted to make a comment on the mouse model dy that Dr. Law presented. It is impressive that the effect is regional, and impressive that the effect is so dramatic. It makes me want to raise the possibility that in this particular mouse mutant the defect resides in the inability to make a functionally transacting factor.

Now if that speculation were correct, the data acquired from this mouse model would not be anticipated to be directly applicable to a disease such as Duchenne muscular dystrophy, or the mdx model, where the defect resides in a structural protein. So I just wanted to make those comments, and now ask the question. Have there been any studies carried out in vivo or in vitro to explore the possibility that the defect does indeed reside in a trophic factor?

Dr. Law: Yes. I was going to show this. Maybe I give it in the talk tomorrow. In 1972, we actually reported the first direct evidence to indicate that there was a functional abnormality in the dystrophic muscle cell membrane. We were able to demonstrate that the dystrophic muscle cell membrane was leaky to sodium and potassium ions. That implies a structural protein defect, and at that particular moment, since there were no mechanisms to correct such a spontaneous muscle degeneration, then obviously the cells to be degenerated have to be replenished in some way, and that started the transplanting experiments. Now the idea of the mouse experiment is to test out whether muscle degeneration in this dystrophic mouse model was secondary to a lack of neurotrophic substance. At that particular moment, we purposely put a normal fast nerve into a slow dystrophic muscle, and vice versa. To sum up that experiment, the slow muscles turned fast indicating that neurotrophic influence was exerted. However, the normal muscle stays normal, dystrophic muscle stays dystrophic, indicating that the nervous system was not able to control the spontaneous degeneration of the dystrophic muscle.

Dr. Kunkel: I really enjoyed today's session, and have a couple of questions. They are for the panelists. One of which is quite in the vein of Dr. Caskey's question. It is to Dr. Law, and I will ask a little bit more.

It seems like your take rate is fantastic, compared to even Dr. Morgan's rate, and I really wonder if you have considered first looking at the mdx mouse and looking for truly dystrophin-positive fibers in your transplantation experiments. And second, addressing maybe how well it has worked. I am not really aware, but is there knowledge about the dy mutation in various different strains? Specifically the Swiss-Webster strain, which you are using as your donor? Is it possible that the dy mutation is actually quite benign and that there is some trophic factor that Dr. Caskey is talking about that might have been capable of coming from myoblasts or compensating for this mutation?

Dr. Law: Incidentally, our transplantation were not only done with Swiss-Webster mice. It was also done with another strain of C57-Black-6 mice that Dr. Peterson was very kind to send us. As far as using the mdx mouse, we really have problems with the mdx mouse in a sense that I cannot see myself using an animal that does not show muscle weakness. I mean after all, muscular dystrophy is defined as a hereditary genetic defect causing muscle degeneration and weakening. I mean, Duchenne muscular dystrophy boys would not care if they do not have dystrophin, but they do care if they are going to become crippled and die because of muscle degeneration. Now as far as the dy mouse muscles are concerned, this animal model we have been working with it for about 20 years. We have every single thing documented in this animal model, except what its genetic defect is.

Dr. Kunkel: The specific question was the strain differences, and the Swiss-Webster.

Dr. Law: Right.

Dr. Kunkel: What is the variability of the phenotype?

Dr. Law: Variability of the genes?

Dr. Kunkel: Or is it constant? That is, you see, the exact same progressive loss of muscle wasting and loss of motility.

Dr. Law: Yes, I think it is an inbred strain of animal and the pathogenesis and pathology is extremely stereotyped and is highly predictable.

Dr. Kunkel: C57-Black-6, or onto a Swiss-Webster background, what is the phenotype of that mutation homozygote on those backgrounds, on those strain backgrounds?

Dr. Law: I do not think anybody has done that study. I would not be able to answer that question.

Dr. Kunkel: I think it would be a good idea to test that kind of an hypothesis.

Dr. Law: Yes, yes. Definitely.

Dr. Partridge: Some years ago I crossed the two dystrophic strains, the 129 and the C57-Black-6. The dystrophic dy mice from that cross were actually much longer lived than either of the individual parents.

Dr. Kunkel: Finally, I would like to close with this: I think the definition of Duchenne dystrophy entails the absence of dystrophin, and you should address the replacement of dystrophin if one is going to address the replacement or cure of this particular disease. That is the definition of this disease at the moment.

Dr. Law: Right. I will gladly do it, if one can actually demonstrate to me that the absence of dystrophin in a normal animal causes the pathology that we see in Duchenne muscular dystrophy. One should be able to take a normal mouse, get rid of its dystrophin, and show me then that the muscle pathology occurring is similar to Duchenne.

Dr. Kunkel: That is not necessarily true at all. There are other examples of genetic diseases, and cloned genes where the phenotype in mouse is extremely different than the phenotype in human beings. I would once again say that dystrophin deficiency is the defect that causes Duchenne muscular dystrophy. If one does not assay for the basic biochemical problem when one does transplantation and has the ability to do that, that is an absolute error, and that is all I will say about that.

Dr. Kaufman: I do not mean to pick on Dr. Law but I have to say this. In your film, I believe that you stated that the injection of fibroblasts led to detrimental consequences. Could you tell us, what was the source of those fibroblasts? How many were injected and what were the detrimental consequences?

Dr. Law: The fibroblasts were basically clones from our initial primary culture. We were injecting, if I remember correctly, about a million cells into the soleus muscle in which there were purely fibroblasts and no myoblasts there. It was only at a very high percentage of fibroblasts that we saw effects like that. Basically that is one of our control experiments.

Dr. Kaufman: Were the effects that you sought worse than what you started with? Or was there not any improvement upon the injection of fibroblasts?

Dr. Law: It was definitely much worse than the control dystrophic muscle. After injection of fibroblasts you will have a lot of connective tissue.

Dr. George-Weinstein: Question for Dr. Blau. Could the injection procedure itself have had a transient effect on the continuity of the basal lamina?

Dr. Blau: The sites at which we saw the migration of the satellite cells were often quite distant from the actual site of injection, which we knew, because we had marked it with the charcoal. So it is unlikely that in all the cases we saw, the basal lamina was disrupted, though it may have occurred in one or two. I think an important point that I wanted to make also from the migration story is that as we put myoblasts, transplant myoblasts into dystrophic muscles, and the hope is that we will make muscle fibers, the importance of finding that these cells can migrate is that if they could not migrate, one would form a fiber, and the satellite cell would be restricted in their movement, and simply be able to contribute to that fiber. What this suggests is that those muscle cells will be able to contribute even after the development of the new fibers to more new fibers.

Dr. Karpati: I should like to comment on the clever model of Dr. Morgan and co-workers, using irradiated mdx mouse muscle where the regeneration was practically arrested and myoblast transfer into such muscles produced a very high rate of dystrophin expression. One would assume that the X-irradiation, by negating the participation of the native (dystrophin-incompetent) satellite cells in regeneration, enhanced the fusion frequency of transferred donor myoblasts. Hence, most regenerated fibers must have become dystrophin-positive.

Dr. Morgan: In our irradiated mdx muscles there are no host satellite cells left, because we have knocked all those out by the X-irradiation. So that is the only real explanation I can think of, why and how these cells are becoming incorporated. They must be patching up the necrosing muscle fiber. I suppose we could look and see if that is happening.

Dr. Blau: Yes. I have a question for Dr. Ray. There is clinical evidence that steroids have a beneficial effect on patients with Duchenne muscular dystrophy. I was wondering whether your promoter construct for dystrophin responds to steroids and whether you have mapped a steroid responsive element?

Dr. Ray: No, we have not looked at that. There are a number of things like that we are intending to do, but we have not done them yet.

Unidentified Participant: I would like to make two comments. One is regarding the difference between mdx mouse and the other mouse presented by Dr. Law. I think that Dr. Kaufman mentioned an abstract that was in the recent Cell Biology Meeting, and that abstract showed that mixing of fibroblasts with myoblasts from dysgenic mice could cure phenotypically dysgenic mice. The fibroblasts fused with those myotubes and some of the morphological characteristics of dysgenic mice disappeared. I do not know more about his work, but the poster was very nicely presented, and maybe somebody else here can add something to it.

Now, regarding the movement of satellite cells from one area to another area: I think that Dr. Partridge should say something about his work from some years ago when he put two different muscles close to each other, and you could find the appearance of the isozyme in the other muscle. I think that you should mention it.

Dr. Partridge: We spent a long time showing just that I suppose. But the methods we were using were really comparatively insensitive. I never felt terribly happy about them. I would feel much happier if I could see little blue cells zipping along muscle fibers.

Dr. Strohman: I had just a question on the irradiation model. I think it is a picky question but I do not know. I was sort of surprised that the brief irradiation that you gave was so successful in eliminating practically all of the satellite cells. Would you not have expected a more random stochastic effect from that? That there would be some survivors? Or am I not getting something?

Dr. Morgan: I think it was quite a heavy dose of X-irradiation.

Dr. Strohman: It was?

Dr. Morgan: It was 1800 rads just to one leg of the mouse.

Dr. Strohman: To one leg? For how long?

Dr. Morgan: So it would knock out all the proliferating cells.

Dr. Strohman: All the proliferating cells. But then that would suggest they are all synchronous. That does not seem to be.

Dr. Morgan: Not all the proliferating cells, but any cells capable of proliferating would have been stopped from being able to ever proliferate.

Dr. Fischman: I think that the radiation model is fascinating. I would just like to urge a little caution. We have shown that if you irradiate myoblasts in monolayer, they are still capable of fusion. They did not have to go through a round of replication. If you irradiated satellite cells, they are still capable of cell fusion. With the transplanted myogenic cells you could still get a hybrid myotube without replication of the host satellite cells. So do not assume that irradiation prevents fusion. It will prevent replication but they still might be competent to fuse. I just urge a little caution on it.

Dr. Morgan: Well, we have no evidence about being competent to fuse. In that instance, if you just leave those muscles to their own devices, you get no repair, no signs of regeneration in them. So they are not fusing with one another and they are not fusing with the surviving bits of the muscle fibers.

Dr. Fischman: I am not trying to quibble. I mean, I would just be a little cautious still. That's all. I think the data are fascinating, and it would suggest that the myogenic cells are fusing with myofibers, which is interesting. I would just like to be sure that the host satellite cells are really gone.

Dr. Stedman: I have two related questions to put to Drs. Morgan and Grounds. Both involve the model of phosphorylase deficiency in a mouse, which is relevant to something that we are doing in another related metabolic myopathy. One subtlety on one of the slides that you presented appeared to be that if you look at those three bands, the topmost and the bottommost bands in the group to the left on the one slide were much stronger than the heterodimeric band. Whereas, to the right, in the phosphorylase deficient following the myoblast transfer, it looks as if the hybrid band was actually stronger than the above. I wonder if you have ever found that gives you any kind of indication as to comparative amounts of homofusion, if you will, versus heterofusion amongst the cells in that situation. As I recall from the slide the cells to the left were all mdx and the ones to the right were all for phosphorylase. Correct me if I am wrong.

Dr. Morgan: No. If you are just thinking of the phosphorylase kinase side.

Dr. Stedman: Right.

Dr. Morgan: The ones to the left were where I injected normal cells into autographed phosphorylase-kinase deficient muscle, and the ones on the other side were where I injected normal cells into growing phosphorylase-kinase deficient muscle.

Dr. Stedman: Okay.

Dr. Morgan: So it really looks as if when I injected into regenerating phosphorylase-kinase muscle, that there were probably quite a few fibers purely of donor origin forming more so than when I injected into the growing muscles.

Dr. Stedman: Okay. The second question for you and Dr. Grounds. Have you thought of coupling an X-irradiation to marcaine injection in that same animal to see whether you can get a situation a bit more like the mdx in terms of your total uptake?

Dr. Morgan: Well, I mean in the irradiated mdx mouse, putting in marcaine probably would not help much. The reason why I am really proposing marcaine in the human situation is that it is such a huge scaling up from that seen in the mouse.

Dr. Stedman: Right.

Dr. Morgan: I mean the efficiency in a mouse is very impressive. I think it will be interesting to try it. But I think the problem arises when you start to look at a human situation; you are talking about a very big distance to get the cells across. But I certainly think in the human situation it is worth entertaining the idea to see whether it would sufficiently make a significant increase in the efficiency of the myoblast transfer.

Dr. Karpati: In the natural history of Duchenne dystrophy, at any given moment, there is a certain number of muscle fiber segments that undergo necrosis. That percentage varies; at an early stage it is a higher percentage than in the later stages of the disease. I guess, at an early stage when perhaps most of these transfer experiments are contemplated, or should be contemplated, one could make a very crude estimate that approximately 5 to 8% of the fibers at any given moment in the vulnerable muscles are actually necrotic. It is fair to say that only those fibers would be therefore available for the immediate mosaicization. But if you consider the possibility that the large number of the injected myoblasts will probably also assume satellite cell position, and will revert to Go, you do not really need to increase artificially the necrotic rate in the Duchenne muscle above the natural prevalence by a technique which could be considered objectionable. This is because as time goes on, many different fiber segments will become necrotic from the natural disease and will pull in those donor myoblasts that are waiting as satellite cells to come in and participate in the regeneration. Thus, I am not sure about the wisdom of creating additional artificial necrosis to increase the mosaicization rate in Duchenne dystrophy.

Dr. Morgan: Sorry. You have made the comment about the implanted cells going into satellite position, and I wonder what that is based on.

Dr. Karpati: Well, it is based on the fact that I saw some of them in such a position. Presently, I cannot give you quantitative data on that.

Dr. Morgan: Have you looked under the EM? Because I think that was only the way you could identify a satellite cell.

Dr. Karpati: You could also identify them by N-CAM cytochemistry and subsequent performance of radioautography; thus you can verify that it is both a donor cell and a satellite cell.

Dr. Morgan: One of the reasons I am raising this point is that, if these cells are persisting outside the muscle fiber, they can pose immunological problems as mononuclear cells. Whereas, if you can get them inside the muscle fiber with a great efficiency, this is immunologically more attractive because they are lying under the plasmalemma and therefore silent. Therefore, if you can actually incorporate large numbers of them at one time, that would seem to be advantageous.

Dr. Karpati: Yes, if you did not use immunosuppression that might be a plus, but most people would use some immunosuppression to cover allogeneic myoblast transfer.

SECTION 4

IN SITU FUSION: NUCLEAR DOMAINS AND mRNA/PROTEIN MIGRATION

FIBROBLASTS FUSE WITH MYOTUBES DEVELOPING IN CULTURE

Nirupa Chaudhari and Kurt G. Beam

Department of Physiology
Colorado State University
Fort Collins, CO 80523

INTRODUCTION

Skeletal muscle cells are formed during development in vivo as well as in vitro by the fusion of mononucleate, undifferentiated precursors, termed myoblasts. Over the last year, we have observed that, although rarely, fibroblasts also are able to fuse with developing myotubes in tissue culture (Chaudhari et al., 1989). The experiments which we present here were carried out primarily on muscle cells from the muscular dysgenesis (mdg) strain of mice (Pai, 1965). This strain carries a mutation in the gene for a skeletal muscle specific calcium channel (Tanabe et al. 1988), rendering skeletal muscle totally paralyzed. When dysgenic myoblasts fuse in culture, these myotubes are unable to twitch, either spontaneously or in response to electrical stimulation. However, dysgenic myotubes can be induced to regain twitch responses by the introduction of either a normal myoblast nucleus (Peterson and Pena, 1984) or a normal gene to replace the altered calcium channel (Tanabe et. al. 1988). Thus, dysgenic myotubes allowed us to visualize (through their altered phenotype), rare events: the incorporation of single, normal, fibroblast nuclei.

METHODS

Animals

A colony of mice, carrying the muscular dysgenesis (mdg) locus, is maintained at Colorado State University. Homozygous normal and heterozygous mice from this strain are phenotypically normal by all examined anatomical and physiological criteria (Pai, 1965, Beam et al., 1986). Non-paralyzed neonatal mice from this colony are designated +/mdg? and were used as a source of normal myoblasts in this study.

Tissue Culture

Primary cultures of normal or dysgenic muscle were established by dissociating neonatal skeletal muscle with 0.15%

trypsin at 37°C, and plating liberated myoblasts in DMEM (Gibco Laboratories, Grand Island, NY) containing 20% fetal calf serum (Hyclone Laboratories, Logan, UT), 100 units/ml penicillin and 100 µg/ml streptomycin. Approximately 10^5 cells were plated per 35 mm Falcon Primaria dish. One day later, cultures were re-fed with plating medium. When the cultures were 2 to 4 days old, they were transferred to maintenance medium (DMEM with 10% horse medium with penicillin and streptomycin as above). In some experiments, cytosine arabinoside (araC, 10 µM) was added to the culture medium between days 3 and 5 and was removed 24-48 hours later. The schedule for addition of fibroblasts to myoblast cultures is described in the text.

Staining

Fixation and staining for ß-galactosidase was according to Sanes et al. (1986). Hoechst 33342 was obtained from Sigma Chemical Company (St. Louis, MO), was stored as a 10 mg/ml stock and was diluted to 50 µg/ml in culture medium for use. Cells were stained for 15 minutes at 37°C. A mouse monoclonal antibody, Mab 1406, against rat Thy1.1 was obtained from Chemicon International (El Segundo, CA), was diluted 1:100 in maintenance medium, and was used to stain living cells for 15 minutes at 37°C. Cultures were washed with several changes of saline and then were incubated with a TRITC-secondary antibody, (rabbit anti-mouse from Sigma), diluted 1:50, for 15 minutes at 37°C. After extensive washing, cells were fixed in 2% paraformaldehyde in phosphate-buffered saline.

RESULTS

Myotubes derived from mice affected with muscular dysgenesis develop in culture but remain paralyzed (i.e. never contract spontaneously or in response to stimulation), whereas primary myotubes derived from normal mice twitch spontaneously. About a year ago, we learned that Drs. J. Koenig and J. Powell, working at the University of Bordeaux, had added normal fibroblasts to primary cultures of dysgenic myotubes and had observed that a fraction of such myotubes recovered twitch responses (Courbin et al., 1989). In order to determine the mechanism of this functional rescue, we considered two categories of mechanism. One was that normal fibroblasts provide an essential component for the development of dysgenic myotubes that is not provided by dysgenic fibroblasts. As one test of this, we treated dysgenic myotubes in culture with media conditioned by either normal primary dermal fibroblasts or 3T3 fibroblasts. All dysgenic myotubes remained paralyzed following such treatment. We also considered that a factor(s) effecting rescue of dysgenic myotubes might be associated with the plasma membrane of normal fibroblasts. In order to test this possibility, we cultured monolayers of 3T3 fibroblasts, lysed the cells in situ with distilled water or with ultraviolet irradiation, and then established primary cultures of dysgenic myoblasts on this substratum. The dysgenic myotubes which developed remained unable to twitch.

An alternative mechanism for the functional rescue of dysgenic myotubes was that fibroblasts were able to fuse with the myotubes. The fibroblast nuclei presumably would then be subjected to factors within the myotube cytoplasm, which would activate muscle-specific genes, thereby producing functional calcium channels within the dysgenic myotubes. Such gene activation is well documented in the case of artificially induced fusions between muscle and non-muscle cells (Blau et al., 1985). An initial test of this hypothesis was carried out using, as a marker, the enzyme ß-galactosidase. The bacterial gene for ß-galactosidase has been incorporated into a clonal line of 3T3 fibroblasts designated C1W (Sanes et al., 1986). The enzyme is produced constitutively within these cells and can be detected cytochemically. C1W fibroblasts were added to primary cultures of dysgenic or normal mouse myoblasts. The cultures were maintained for 5 to 15 days and then were fixed and stained for ß-galactosidase activity. As shown in Fig. 1, myotubes were observed in such cultures that stained brightly for enzyme activity; the majority of myotubes did not stain for enzyme. C1W cells cultured alone never formed myotube-like structures. Further, homogenates of C1W cells, when added to myotube cultures, did not induce ß-galactosidase activity in myotubes. These results implied that C1W fibroblasts could fuse with myotubes.

We wished to examine whether fibroblasts could fuse with myotubes of all ages. As detailed in Figure 2, myoblasts and myotubes at different stages of maturity were exposed to C1W fibroblasts for limited periods. In three experiments of this type, we consistently saw that fibroblasts were able to fuse most efficiently with myoblasts as the latter were just beginning to fuse with each other.

Having shown that fibroblasts can fuse with phenotypically normal, developing myotubes, we next questioned whether the fusion of fibroblasts of normal genotype is a necessary prerequisite for the functional rescue of dysgenic myotubes in culture. Thus, we added C1W fibroblasts to one day old primary

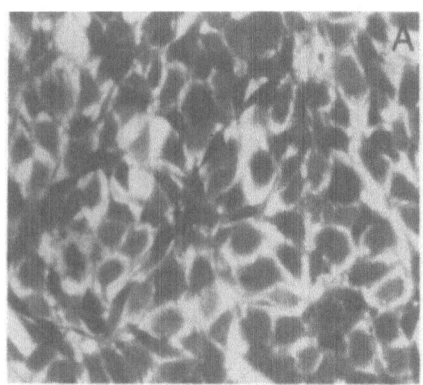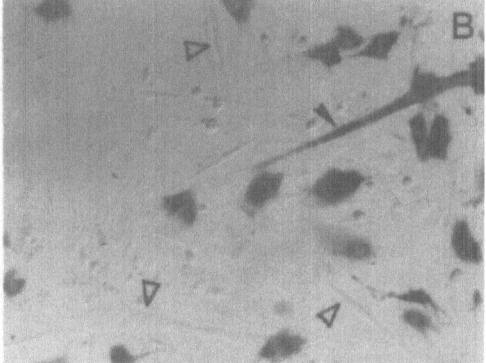

Figure 1. C1W fibroblasts can fuse with growing myotubes in culture. (A) C1W fibroblasts possess a typical 3T3-like morphology and also stain for ß-galactosidase activity. (B) Occasional myotubes are seen which contain ß-galactosidase activity (filled arrowheads) although most do not (open arrowheads). In this culture, phenotypically normal mouse myoblasts (+/mdg?) were plated and C1W cells were added 1 day later. The culture was fixed and stained at 8 days.

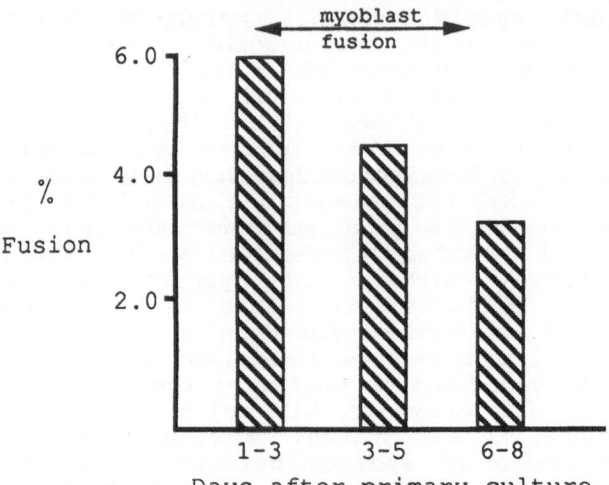

Figure 2. A primary culture of phenotypically normal (+/<u>mdg</u>?) mouse myoblasts was established on Day 0. 10^4 C1W fibroblasts were added to pairs of dishes on Day 1, 3 or 6. Fibroblasts and myoblasts/myotubes were co-cultured for 2 days (i.e. day 1-3, 3-5 or 6-8) after which, the majority of fibroblasts were removed by treatment with araC for two days. In the case of addition of C1W cells on day 6, araC had to be applied before addition of C1W fibroblasts in order to remove excess primary fibroblasts which would have prevented growth of C1W fibroblasts. 9 days after establishment of the culture, all dishes were fixed and stained for ß-galactosidase activity and the fraction of myotubes which stained brightly were recorded.

cultures of dysgenic myoblasts. Dysgenic myotubes were allowed to develop in culture for a total of 6 days in the absence, or 5 days in the presence, of 10 µM tetrodotoxin before testing for restoration of contractile function. We monitored the twitching of individual myotubes in response to extracellularly applied current pulses (Tanabe et al., 1988) and then fixed and stained the cultures for ß-galactosidase activity. In separate experiments, 85 to 88% of myotubes which had demonstrated twitch responses, also stained brightly for enzyme activity, implying that the fusion of normal C1W fibroblasts was necessary for restoration of function.

Since the observed fusion of fibroblasts with myotubes was previously undocumented in the literature, we felt it was necessary to confirm the phenomenon using additional, independent markers. Toward this end, we used the difference in nuclear morphology between mouse and rat, and also a plasma membrane marker (Thy1.1), as described below. When stained with Hoechst 33342, mouse nuclei display a punctate fluorescence whereas rat nuclei stain homogeneously (Fig. 3A). Thy1.1 is an antigen expressed on the surface of fibroblasts (Stern, 1973) and also on myotubes (Schweitzer et al., 1987). A monoclonal antibody, specific for the rat antigen, was used to stain mouse and rat fibroblast cell lines. As expected, all the rat fibroblasts demonstrated a bright fluorescence at their plasma membranes whereas mouse fibroblasts did not (Fig. 3B). Mouse myotubes also did not immuno-stain for rat Thy1.1 (not

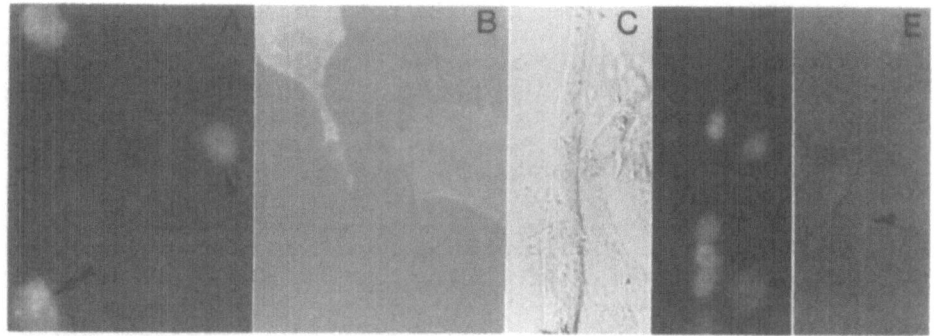

Figure 3. Fibroblast nuclei are contained within dysgenic myotubes. (A) Mouse fibroblasts, stained with Hoechst 33342 show punctate fluorescence in their nuclei (double arrowhead); rat fibroblasts, similarly stained show homogeneously stained nuclei (arrowhead). (B) The same field as in (A), stained with a monoclonal antibody against rat Thy1.1 is viewed for rhodamine. Rat fibroblast nuclei are surrounded by stained plasma membrane while mouse nuclei are not. (C) A dysgenic mouse myotube, rescued by Rat2 fibroblasts (D) The same field as in (C), viewed for Hoechst shows a single rat nucleus (arrow). (E) The same field as in (C), stained and viewed for rhodamine, shows that the rat nucleus is not surrounded by rat fibroblast plasma membrane.

shown). Rat2 cells (a clonal line of fibroblasts) were added to one day-old primary cultures of dysgenic myoblasts and then were removed after 2-3 days of co-culture by treatment with araC. Following another 5-6 days of culture in maintenance medium, spontaneous twitching could be observed in some myotubes whereas the majority of myotubes remained paralyzed (ie. true to the dysgenic phenotype). Myotubes which had displayed twitches contained at least one rat nucleus within them (Fig. 3D). Since no Thy1.1 immunoreactivity was localized surrounding these rat nuclei (Fig. 3E), we concluded that they

Table 1

Fibroblast Type	Source	Percent Functional Rescue of Dysgenic Myotubes
Cloned dermal	Rat skin	1 - 5 %
Dermal	Rabbit skin	4 %
* Rat2, clonal	Rat embryo	2 - 25 %
* 3T3, clonal	Mouse embryo	20 %
* NRK-52E, clonal	Rat kidney	2 - 20 %
# C1W	Mouse embryo, contains gene for ß-galactosidase	5 - 20 %

* Obtained from American Type Culture Collection, Rockville, MD.
# Obtained from Dr. J. Sanes, Washington University Medical School.
Remaining cell lines produced in our laboratory.

were derived from Rat2 fibroblasts but had become incorporated into mouse myotubes. Six independent experiments each demonstrated that 100% of the examined myotubes which displayed twitches (6 to 20 rescued myotubes per experiment) contained at least one normal (rat) fibroblast nucleus. By contrast only 15 to 35% (in separate experiments) of randomly examined myotubes contained rat nuclei.

In order to determine if there was a specificity to the type of fibroblast which could fuse with developing myotubes, we added a variety of fibroblast types to one day-old primary cultures of dysgenic myotubes. The fraction of dysgenic myotubes which developed twitch responses varied between experiments and fibroblast types.

DISCUSSION

We have shown (using primarily dysgenic mouse myoblasts), that a variety of fibroblast types are able to fuse spontaneously with very early myotubes. Qualitatively similar results were obtained using dysgenic (mdg/mdg) mouse myoblasts, phenotypically normal (+/mdg?) mouse myoblasts, and also normal rat myoblasts, when the myoblasts were grown with ß-galactosidase-expressing (C1W) fibroblasts. The frequencies of fibroblast-fusion for the normal myoblast cultures are lower (1-10% of myotubes are stained) than seen for dysgenic myoblasts. We do not presently know whether normal fibroblasts are able to fuse with developing human myotubes, nor whether such fusions ever occur during development in vivo.

REFERENCES

Beam, K.G., Knudson, C.M. and Powell, J.A. Nature 320, 168-170 (1986).
Blau, H M., Pavlath, G.K., Hardeman, E.C., Chiu, C.-P., Silberstein, L., Webster, S.G., Miller, S.C. and Webster, C. Science 230, 758-766 (1985).
Chaudhari, N., Delay, R. and Beam, K.G. Nature, in press (1989).
Courbin, P., Koenig, J., Ressouches, A., Beam, K.G. and Powell, J.A. Neuron, 2, 1341-1350 (1989).
Pai, A.C. Devel. Biol. 11, 82-92 (1965).
Peterson, A. and Pena, S. Muscle and Nerve 7, 194-203 (1984)
Sanes, J.R., Rubenstein, J.L.R. and Nicolas, J.-F. EMBO J. 5, 3133-3142 (1986).
Schweitzer, J.S., Dichter, M.A. and Kaufman, S.J. Exp. Cell Res. 172, 1-20 (1987)
Stern, P. L. Nature New Biol. 246, 76-78 (1973).
Tanabe, T., Beam, K.G., Powell, J.A. and Numa, S. Nature, 336, 134-139 (1988).

Discussion of Dr. Chaudhari's paper

Dr. Fischman: Dr. Chaudhari, there is a large literature, particularly in the chick-quail chimera studies, indicating that in the developing limb, fibroblasts have a different embryological origin from myoblasts. The former derive from the lateral plate mesoderm, the latter from the paraxial mesoderm (somites). In vivo, these two cell populations never fuse. What is different about your in vitro system that permits myoblast-fibroblast fusion? And why has fibroblast fusion with myoblasts not been observed by others working with cell cultures?

Dr. Chaudhari: Let me answer the last question, since I am more familiar with in vitro systems. In my experience, there is a fairly narrow time window during which fibroblasts can fuse with myoblasts. Such fusion only occurs during the earliest stages of myoblast fusion. If fibroblasts are added to myotubes more than 4-5 days old, no fusion occurs. I do not know if there any unique features of our culture system nor do I believe that our procedures differ from those used routinely elsewhere. I do not how to comment about the in vivo studies; perhaps there is some inhibition related to myotube age but that would be sheer conjecture. It is too premature to compare these results with other in vivo studies.

Dr. Fischman: May I ask one further question related to the efficiency of your myoblast/fibroblast fusion events? Is this a common event? Is there a difference in frequency of fusion between fibroblasts and myoblasts or myoblasts and myoblasts? The implication from your presentation is that you can obtain 100% rescue, is that correct?

Dr. Chaudhari: I am sorry if I implied that rescue is complete. In different experiments it can vary widely; usually it ranges between 2 and 25% rescue, with a typical experiment giving 10% of myotube rescue. There is usually only a small percentage of fibroblast fusion with myotubes; perhaps one fibroblast nucleus for forty myoblast nuclei. Regarding the efficiency of fibroblast fusion, I would estimate an efficiency of 10^{-4}, i.e. one fibroblast fusion event for every 10,000 fibroblasts added to the cultures.

PHENOTYPIC AND FUNCTIONAL REVERSION OF *MUSCULAR DYSGENESIS* BY HETEROTYPIC

FIBROBLAST-MYOTUBE FUSION *IN VITRO*

Luis Garcia, Patrick Dreyfus, Martine Pinçon-Raymond, Albert
Villageois, Olivier Chassande, Georges Romey*, Michel Lazdunski*
and François Rieger

Groupe de Développement, Pathologie et Régénération du Système
Neuromusculaire, INSERM U.153, 17 rue du Fer-à-Moulin 75005
Paris
* Institut de Pharmacologie Moléculaire et Cellulaire UPR 411
CNRS, 660, route des Lucioles, Sophia Antipolis 06560 Valbonne

INTRODUCTION

Muscular dysgenesis in the mouse (Glueksohn-Waelsch 1963; Pai
1965[a,b]) is a genetic disease with autosomal recessive inheritance
characterized by an immature internal organization of foetal muscle
(Glueksohn-Waelsch 1963; Bowden-Essien 1972; Pinçon-Raymond *et al.*,
1985). It is now well established that dysgenic skeletal muscles display
important decreases in both the level of the 1,4 dihydropyridine (DHP)
receptor (Pinçon-Raymond *et al.*, 1985) and the level of the L-type Ca^{2+}
channel activity (Romey *et al.*, 1986; Beam *et al.*, 1986). These molecular
defects have been correlated with the total lack of excitation-
contraction coupling in the mutant skeletal muscles (Rieger *et al.*, 1987;
Tanabe *et al.*, 1988).

Coculture of diseased mouse muscle myotubes with spinal cord cells
from normal mice was shown to restore contractile activity (Koenig *et al.*,
1982) and several features of a normal ultrastructure including a normal
triadic organization with regularly spaced densities (Rieger *et al.*,
1987). In such conditions, the restoration of the missing excitation-
contraction coupling was found to be always accompanied by the apparition
of the L-type Ca^{2+} channel. In an effort to understand the cellular and
molecular mechanisms underlying such a spectacular phenotypic reversal
to normal of mutant myotubes, experiments have been conducted first on
the possible action of trophic factors and secondly on the possible
interaction of cell types other than neuronal cells which could be
present in the spinal cord suspension. During the course of these studies
we observed, as it has been recently reported by Courbin *et al.*, (1989),
that the presence of normal fibroblasts can induce by itself the
restoration of normal function in the dysgenic myotubes. We show here
that normal primary fibroblasts, among other cells, are able to transfer
cytoplasmic markers into mutant myotubes - suggesting fusion - and impose
upon the recipient myotubes new features suggestive of a quasi normal

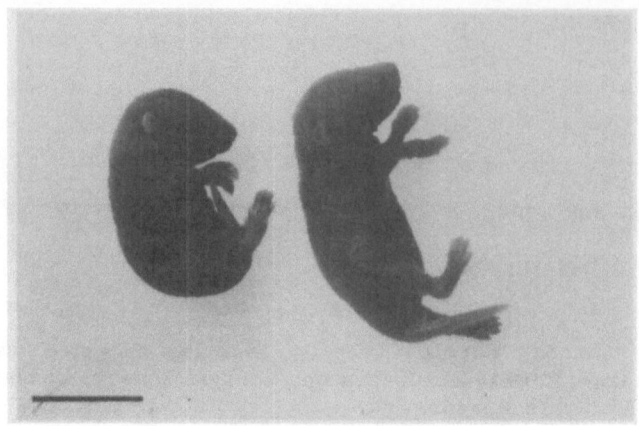

Fig. 1. *Muscular dysgenesis* (*mdg*/*mdg*) *(left)*
and normal new-born mice. (bar = 1cm).

differentiation with normal ultrastructure, slow Ca^{2+} currents and
accumulation of the α1 subunit of the DHP receptor.

MATERIAL AND METHODS

A stock of mice carrying the *mdg* mutation has been maintained as a
closed colony in our laboratory in Paris since 1978. *Mdg/mdg* mice never
move during intra-uterine life and die at birth. A normal and a *mdg/mdg*
neonate are shown on Fig. 1. The mutant new-born is characterized by no
muscle activity, but normal heart beat, a cleft palate, a fixture of
joints and a characteristic body curvature.

Primary muscle cell cultures were derived from limb muscles of newborn mice as described in Rieger *et al.*, (1987). Cocultures with nonmuscle
cell types were performed using plating densities of 10^4 myoblasts per
35 mm plate and adding at the time of myoblast fusion (7th day of culture)
10^4 non muscle cells per plate, namely primary fibroblasts prepared from
neonatal mouse sciatic nerve. Cocultures were studied on a 2 week period
and observed daily to score contractile activity.

In most experiments prior to coculture, nonmuscle cells were exposed
to 0.5 µm fluorescent (FITC) latex beads (PolySciences Inc.) which are
readily phagocytized by the cells and can be used as a cytoplasmic
marker. Immunofluorescent studies in coculture conditions were performed
using either single staining provided by the FITC labeled cells or a
double staining obtained by a combination of this same fluorescein
staining and a rhodamine staining of a goat polyclonal antibody (raised
against the rabbit DHP receptor - a gift from Dr M. Hosey) revealing the
α1 subunit of the DHP receptor.

RESULTS AND DISCUSSION

Internal disorganization at the ultrastructural level is the allmark
of *mdg/mdg* skeletal muscle myotubes at embryonic day 18. There are no
clear muscle masses in the limbs, abundance of myoblasts, nuclei in a
peripheral location and no clear sarcomeric organization (Fig. 2).
Mdg/mdg myotubes *in vitro* also display the same basic abnormal features
(data not shown).

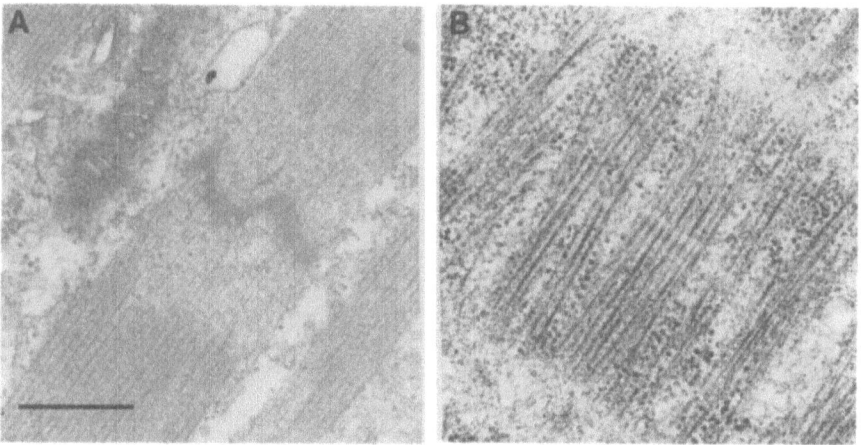

Fig. 2. *Ultrastructural disorganization in mdg/mdg diaphragm muscle at embryonic day 18 (x38,000 - bar = 0.5μm). (A) Normal muscle has already well organized Z lines and sarcomeres. (B) Dysgenic muscle is characterized by the absence of Z striation.*

The two main abnormalities directly related to the uncoupling of excitation-contraction (E-C) in *mdg/mdg* skeletal muscle are:

1) abnormal ultrastructural features of the triadic junction. Fig. 3 shows normal triads with regularly spaced junctional densities and their mutant counterparts with no spaced densities and dilated sarcoplasmic reticulum.

2) An important decrease of DHP binding to skeletal muscle membranes (Pinçon-Raymond et al 1985), although it is noteworthy that binding is not totally absent.

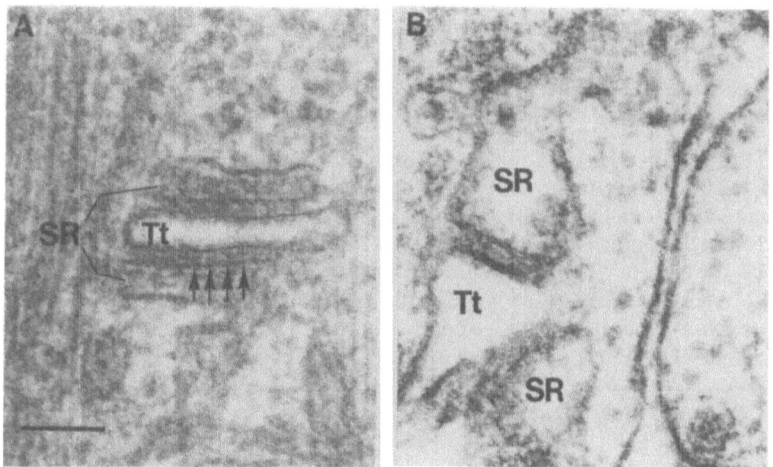

Fig. 3. *(A) Triads in normal muscle are constituted by the apposition of two terminal cisternae from the sarcoplasmic reticulum (SR) against a portion of T-tubule (Tt). The cleft between SR and Tt is occupied by regularly spaced densities bridging the membranes (arrows). (B) Abnormal triadic junction in mdg/mdg muscle with dilated cisternae and absence of spaced densities (x137,500 - bar = 0.1μm).*

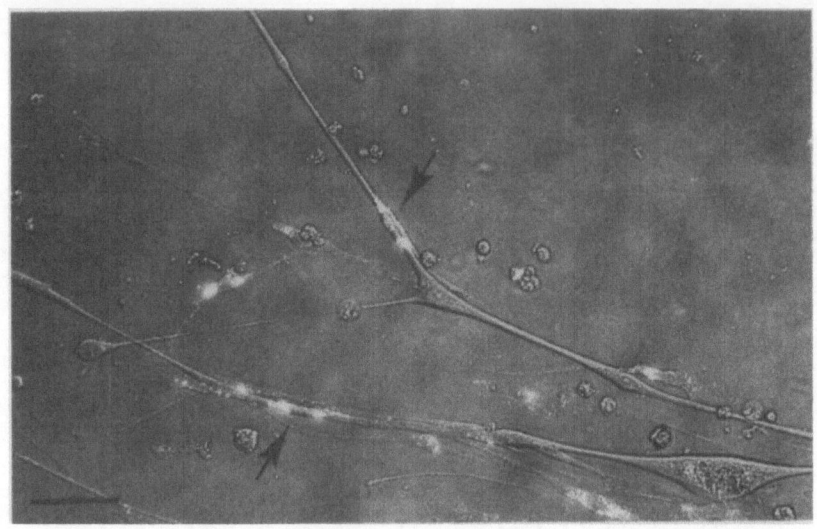

Such an abnormal phenotype can be corrected *in vitro* by coculturing mutant myotybes with normal spinal cord cells (Koenig *et al.*, 1982; Rieger *et al.*, 1987). We demonstrated in these conditions the recovery of both normal triadic ultrastructure and slow Ca^{2+} currents in such myotubes (Rieger *et al.*, 1987).

In view of the possible involvement of nonmuscle, nonneural cells in such an effect, we cocultured mutant myotubes with normal cells of fibroblastic or epithelial origin. In order to distinguish between trophic action and genetic complementation by fusion, we loaded normal fibroblasts with fluorescent latex beads. Coculture of these labeled fibroblasts and *mdg/mdg* muscle cells at the time of fusion (7th day in culture) resulted after 4-5 days into what seemed to be a rather frequent internal myotube labeling around nuclei (Fig. 4 - using a combined Nomarski and fluorescence technique). In almost all of the fluorescent areas (75%), we observed intense contractile activity and voltage-clamp technique showed recovery of slow Ca^{2+} currents (data not shown). Frequent fusion events between fibroblasts from primary cultures were also observed at the ultrastructural level (Fig. 5).

Using a goat polyclonal antibody directed against the α1 subunit of the rabbit DHP receptor in a double staining technique (Fig. 6), we found that in myotube regions rich in latex beads, generally accumulated around specific nuclei, a very significant α1 staining could be detected. The frequency of what appears to be fibroblast fusion in cocultures of primary fibroblasts and *mdg/mdg* early myotubes was found to be around 1 per 100 muscle nuclei in our conditions. In contrast, in normal muscle cultures exposed to the fibroblasts in the same conditions, fibroblast fusion was extremely rarely observed in agreement with all other recently published data (Blau *et al.*, 1985). One fusion event could be scored every 35 mm culture dishes and thus its frequency was 10^2-10^3 times less than in *mdg/mdg* cultures. The reason for this difference between normal and mutant myotubes in their permissivity for heterologous fusion is not known. However, there is one important defect which could be related to the ability of fibroblasts to fuse with *mdg/mdg* myotubes. *Mdg/mdg* myotubes (Fig. 7) have little or no basal lamina in contrast to normal myotubes, which may favor adhesion and subsequent fusion.

Fig. 4. *Fusion of normal fibroblasts with* mdg/mdg *myotubes. Fluorescent beads labeled fibroblasts were added into young* mdg/mdg *myotubes in culture. After two days, fluorescent beads surrounding nuclei are clearly visible into the myotubes (combined fluorescence and Nomarski observation - bar = 50μm).*

The DHP receptor is mainly located in transverse tubules (Fosset *et al.*, 1983) and is thought to have a dual function as a voltage sensor for E-C coupling and as a slow Ca^{2+} channel (Schwartz *et al.*, 1985; Rios and Brum 1987). The DPH receptor defect may be the primary defect of the mutation, a suggestion strengthened by recent data by Tanabe *et al.*, (1989) showing recovery of E-C coupling *in vitro* after microinjection of a cDNA coding for the α1 subunit of the DHP receptor into mutant myotubes. In the studies reported here we most probably induced a near

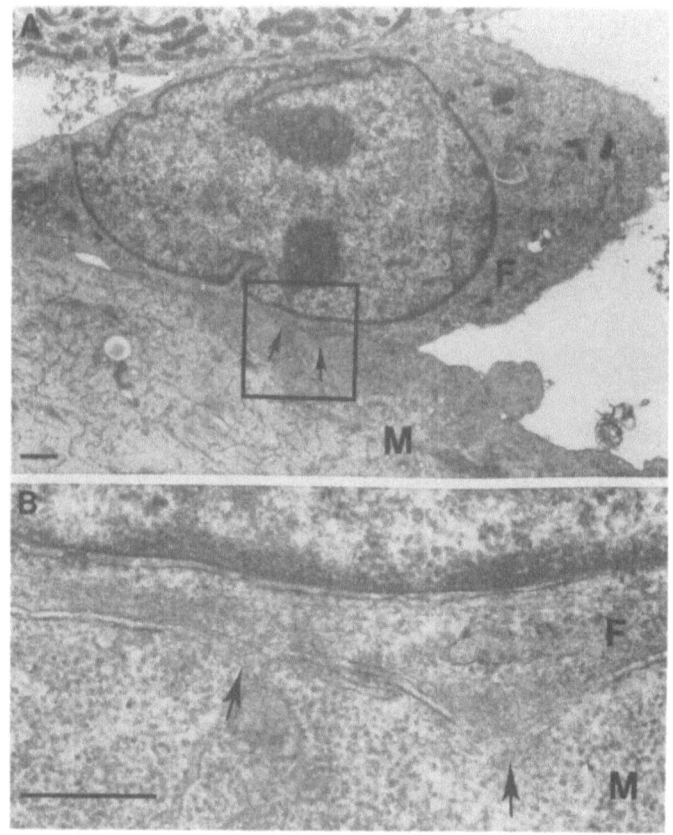

Fig. 5. *Ultrastructural aspects of normal fibroblast (F) and mutant myotube (M) fusion. (A) Low magnification (x6,000 - bar = 1μm). (B) High magnification of the region designated by a box demonstrating cell fusion (arrows) (x45,000 - bar = 0.5μm).*

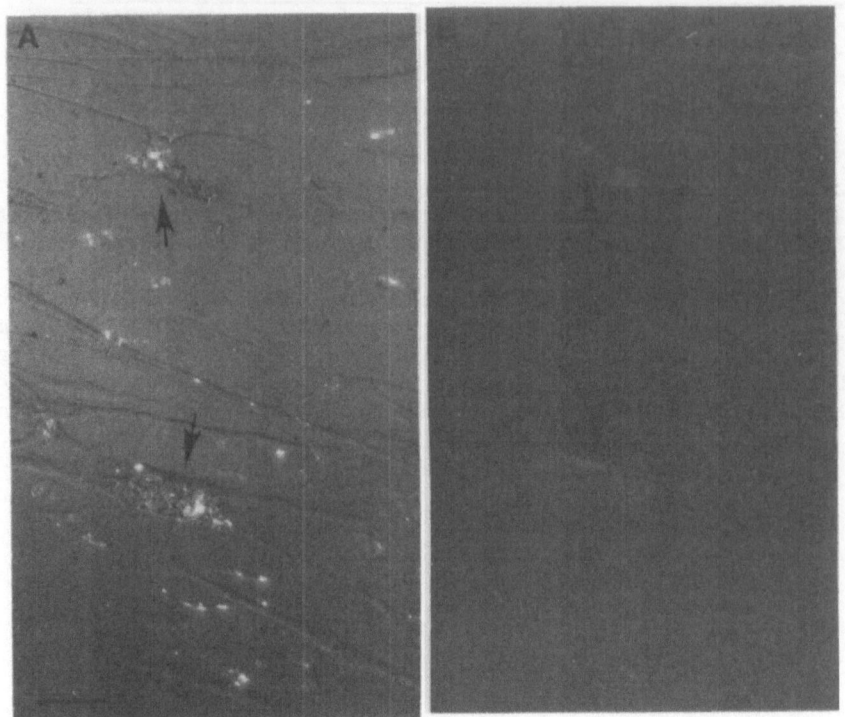

Fig. 6. *Phenotypic and functional reversion of* <u>*muscular dysgenesis*</u>.
(A) Coincidence of latex beads accumulations and foci of contractile activity were observed in live cultures.
(B) After fixation of the same cultures, detection of the α1 subunit of the L-type Ca²⁺ channel was performed by immunocytochemistry. The α1 subunit labeled regions were systematically colocalized with the accumulations of fluorescent beads and the contracting areas (bar = 50μm).

total recovery of features essential for E-C coupling in the areas of fibroblast fusion, including normal triads with regularly spaced densities - which are thought to correspond to the ryanodine receptor or sarcoplasmic reticulum Ca^{2+} release channel -and normal DHP receptor levels, at least in the regions near the sites of fibroblast fusion. It has been demonstrated through artificially induced fusion (Blau *et al.*, 1985) of muscle and non muscle cells that muscle cytoplasm induces the nucleus of the heterologous cell to express muscle-specific genes. Such a mechanism may be active here after fusion of the fibroblast with the mutant myotube, leading locally to the synthesis of near-normal levels of the α1 subunit of the DHP receptor. Our observations, together with those of others (see Chaudhari and others in this volume), strongly suggest that one way (although not the only way) of rescuing *mdg/mdg* myotubes is to incorporate normal genetic information by fusion of non muscle cells into the diseased myotubes. Such observations may be relevant to the search for techniques leading to the incorporation of foreign normal proteins or gene elements into diseased muscle cells in order to correct genetic defects and particularly perform gene transfer therapy.

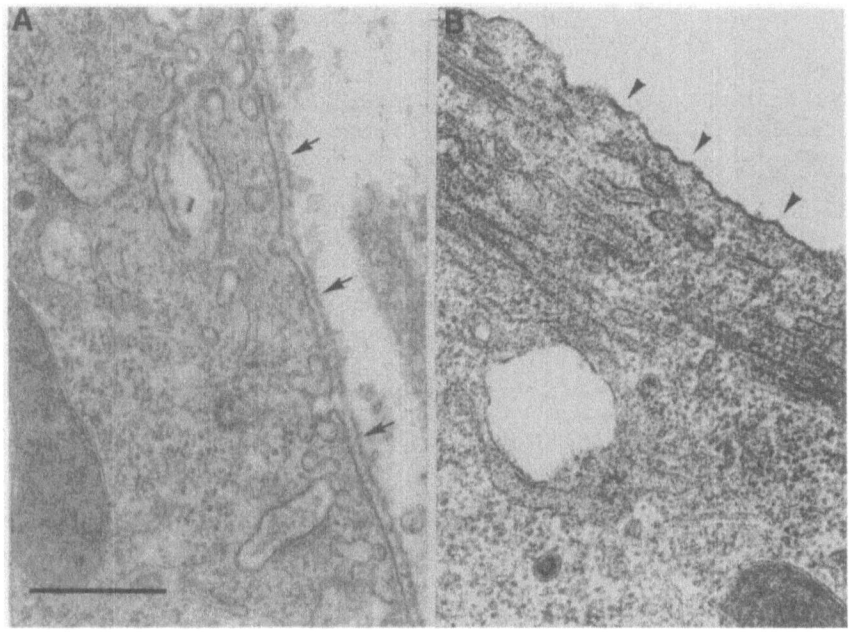

Fig. 7. *Lack of a basal lamina in* mdg/mdg *myotubes.*
(A) 2 days after fusion, normal myotubes display contractile activity
and a continuous basal lamina sheath (arrows).
(B) In mdg/mdg *myotubes, no basal lamina is visible at the cell*
surface. (x45,000 - bar = 0.5µm).

Acknowledgements

This work has been partially supported by the Association Française
contre les Myopathies, by CNRS and INSERM.

References

Beam, K.G., Knudson, G.M. & Powell, J.A. (1986) *Nature* 320, 168-170
Blau, H.M., Paulat, G.K, Hardeman, E.C., Chiu, C.P., Silberstein,
    L., Webster, S.G., Miller, S.C. & Webster, C. (1985) *Science* 230,
    758-766
Bowden-Essein, F. (1972) *Devl. Biol.* 27, 351-364
Courbin, P., Koenig, J., Ressouches, A., Beam, K.G. & Powell, J.A.
    (1989) *Neuron* 2, 1341-1350
Inui, M., Saito, A. & Fleischer, S. (1987) *J. Biol. Chem.* 262,
1740-1747
Gluecksohn-Waelsch, S. (1963) *Science* 142, 1269-1276
Koenig J., Bournaud R., Powell J.A. & Rieger F. (1982) *Devl. Biol.*
    92, 188-196
Pai, A.C. (1965[a]) *Devl. Biol.* 11, 82-92
Pai, A.C. (1965[b]) *Devl. Biol.* 11, 93-109
Pinçon-Raymond, M., Rieger, F., Fosset, M. & Lazdunski, M. (1985)
    Devl. Biol. 112, 458-466
Rieger, F., Bournaud, R., Shimahara, T., Garcia, L., Pinçon-Raymond,
    M., Romey, G. & Lazdunski, M. (1987) *Nature* 330, 563-566
Rios, E. & Brum, G. (1987) *Nature* 325, 717-720

Romey, G., Garcia, L., Dimitriadou, V., Pinçon-Raymond, M., Rieger, F. & Lazdunski, M. (1989) *Proc. Natl. Acad. Sci. USA* **86**, 2933-2937

Romey, G., Rieger, F., Renaud, J.F., Pinçon-Raymond, M. & Lazdunski, M. (1986) *Biochem. Biophys. Res. Commun.* **136**, 935-940

Schwartz, L.M., McCleskey, E.W. & Almers, W. (1985) *Nature* **314**, 747-751

Tanabe, T., Beam,K.G., Powell, J.A. & Numa, S. (1988) *Nature* **336**, 134-139

Discussion of Dr. Rieger's paper

Dr. Fischman: Dr. Rieger, is there a possibility that either in your work or that of Dr. Chaudhari's that not all of the cells termed fibroblasts are truly fibroblasts? Could some be myogenic precursors?

Dr. Rieger: I cannot rule out that possibility in our own work since most of the transplanted cells are from primary cultures.

Dr. Chaudhari: We were quite concerned about that point and for that reason we selected long term fibroblast cell lines and also isolated fibroblasts from the kidney, a source unlikely to have any myogenic precursors.

CONTROL OF SATELLITE CELL PROLIFERATION

Richard Bischoff

Department of Anatomy and Neurobiology
Washington University School of Medicine
St. Louis, Missouri

Satellite cells are reserve stem cells of adult skeletal muscle and thus represent a potential source of myogenic cells to use in myoblast replacement therapy of neuromuscular disease. Optimal application of this strategy requires knowledge of factors that control the growth and differentiation of satellite cells. Control of satellite cell proliferation is a complex phenomenon determined by the contribution of both positive and negative factors (Fig. 1).

Among possible effectors are general wound mediators derived from the vascular system during early inflammation following tissue injury. Platelet derived growth factor released from activated platelets may stimulate proliferation (Jodczyk, et al. 1986) or migration (Venkatasubramanian and Solursh 1984) of myogenic cells, while transforming growth factor beta (TGFβ) from the same source may inhibit myoblast growth (Allen and Boxhorn 1989; Florini and Ewton 1988). Several observations imply a relationship between innervation and cell cycle status of satellite cells. Denervation of muscle results in increased proliferation of satellite cells and connective tissue (McGeachie and Allbrook 1978) suggesting that nerve-derived factors promote withdrawal from the cell cycle. Also, substances released from nerve terminals may stimulate mitosis of myoblasts during development (Ross, et al. 1987), and nerve-mediated muscle activity may modulate the response of satellite cells to growth factors (this report). Heparin-binding growth factors stored in the muscle extracellular matrix may stimulate satellite cells to proliferate following injury (DiMario, et al. 1989). Finally, contact with the surface of a mature myofiber can suppress satellite cell proliferation. Conversely, death of the myofiber leads to release of contact inhibition and also liberate soluble growth factors for satellite cells (Bischoff 1986a; Bischoff 1989b; this report). These observations point out many possible factors controlling satellite cell growth, but the details of the regulatory system have yet to be determined.

To study these factors, we have utilized a system in which satellite cells are maintained in culture under conditions designed to mimic their environment in the animal, i.e., beneath a basal lamina and in contact with a normal myofiber (Bischoff 1986a). Single intact myofibers with attached satellite cells are isolated from the toe muscle of adult rats using purified enzymes under conditions that free the

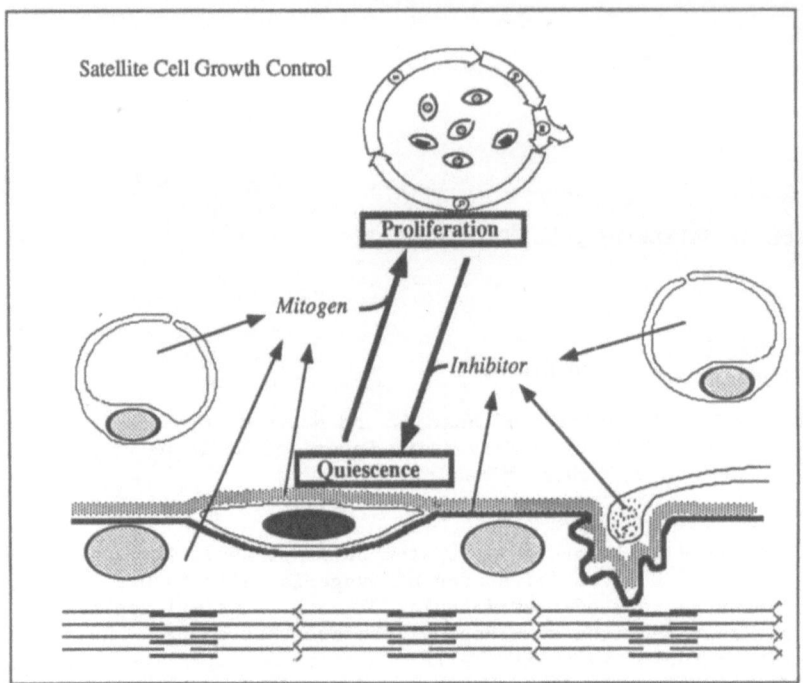

Fig. 1. Possible sources of effectors for control of satellite
cell growth. Both mitogens and inhibitors may be involved
in shifting satellite cells between proliferation and cell
cycle quiescence. See text for discussion.

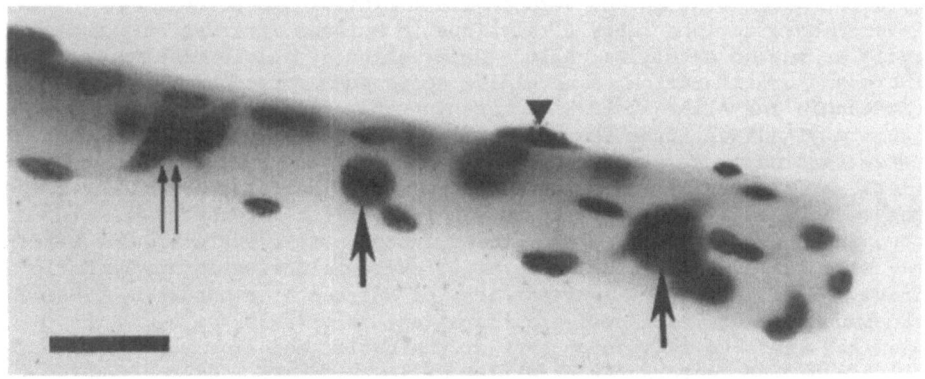

Fig. 2. Radioautograph of single myofiber cultured for 2 d in
basal medium (Eagle's MEM with 10% horse serum and
antibiotics) containing 0.5 mg/ml muscle extract.
For the final 4 h tritiated thymidine ([3H] TdR, 0.5
μCi/ml) was added to the medium. The satellite cells
can be identified by the basophilic cytoplasm
surrounding them while the myonuclei lack this.
Shown are satellite cells in mitosis (arrows).
labeled satellite cells in interphase (arrowhead),
and unlabeled satellite cells (double arrow).
Hematoxylin stain. Bar = 25 μm.

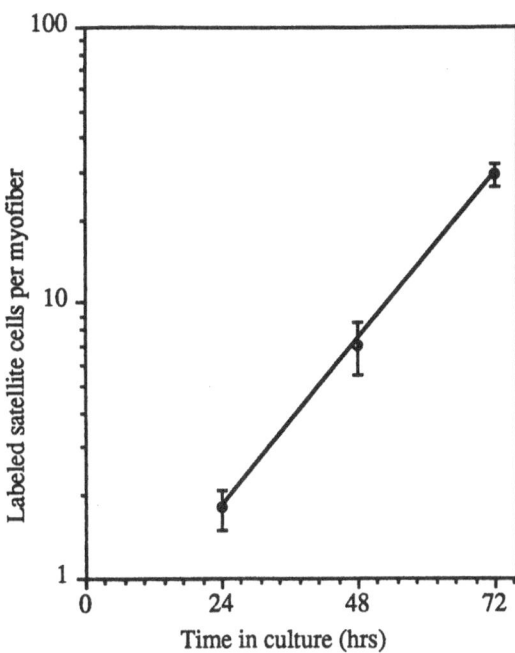

Fig. 3.   Exponential growth of mitogen-stimulated
          satellite cells on single myofibers.
          Myofibers with attached satellite cells were
          cultured in basal medium plus 0.5 mg/ml
          muscle extract and labeled with [3H] TdR at
          the times indicated. The slope of the
          growth curve is 12 h and the intercept is
          about 16 h. During the first 3 d all satel-
          lite cells incorporate [3H] TdR in
          continuous labeling experiments (not shown).
          Points are the mean 1 standard error of
          counts from at least 50 myofibers.

myofibers from connective tissue attachments but do not remove the basal
lamina. Each myofiber is about 1 mm long and contains about 125
myonuclei and 2-3 satellite cells beneath its basal lamina. Most of the
satellite cells remain mitotically quiescent when the myofibers are
cultured in medium containing 10% serum, but can be stimulated to
proliferate when exposed to a saline extract of adult muscle (Bischoff
1986b) (Fig. 2). The myofibers remain viable in culture but do not give
rise to mononucleated cells.

    Satellite cells exposed to muscle extract begin to proliferate only
after a lag period during which time the nucleus changes from
heterochromatic to euchromatic and the cell enlarges. The first mitotic
figures are seen about 24 h after exposure to muscle extract. Labeling
with tritiated thymidine [3H] TdR reveals that the cells enter the DNA
synthetic phase (S) of the cell cycle at about 16 h after exposure to
muscle extract and proliferate with a doubling time of 12 h (Bischoff
1986a) (Fig. 3).

    Since all the satellite cells are proliferating, the growth
fraction is 1 and the doubling time is equal to the cell cycle time.
Because the lag period before S is longer than the entire cell cycle

time, it can be concluded that the satellite cells were in the Go phase of the cell cycle prior to exposure to muscle extract. Exponential growth continues for 3 days in mitogen-containing medium but declines after that as cells withdraw from the cell cycle and fuse with each other forming multinucleated myotubes. The satellite cells do not fuse with their associated myofiber at this time (Bischoff 1989b).

The satellite cell mitogen from crushed muscle extract exhibits tissue-source and target cell specificity (Bischoff 1986b) and is active with both satellite cells in culture and in newborn and adult rats (Bischoff 1989a). Recent experiments suggest that the mitogen acts during the lag period prior to DNA synthesis to commit satellite cells to enter the cell cycle and is not needed continuously for progression through the cell cycle (Bischoff, unpublished).

Efforts to purify the mitogen suggest that multiple factors may be involved. Since fibroblast growth factor (FGF), a heparin-binding polypeptide first isolated from nervous tissue (Westall, et al. 1974) stimulates proliferation of myoblasts and satellite cells (Allen, et al. 1984; Clegg, et al. 1987; Gospodarowicz, et al. 1975; Kardami, et al. 1985), a purification method designed for FGF was applied to the muscle extract. The extract was mixed with heparin-agarose beads and eluted sequentially with 1M and 2M NaCl (Fig. 4).

Mitogenic activity for satellite cells was found only in the 1M salt peak. Full activity comparable to that obtained with unfraction-ated extract was achieved only after adding low molecular weight mater-ial previously removed by ultrafiltration (Bischoff 1989b). Combining the 1M and 2M NaCl peaks did not enhance the activity. About 10% of the total extract protein was present in the 1M NaCl peak and the major molecular species in SDS gels was a 37 kD polypeptide (Fig. 4). Further

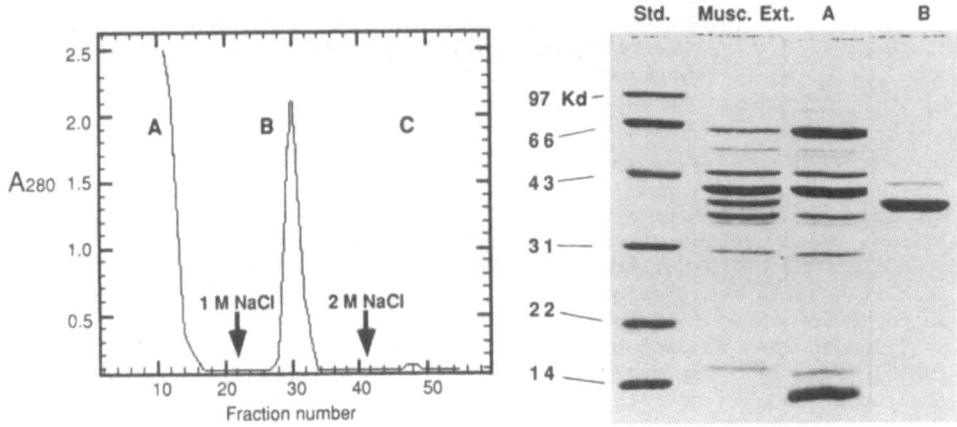

Fig. 4. Heparin-agarose affinity chromatography of muscle extract. About 150 mg of muscle extract protein in starting buffer (10 mM Tris-HCl, pH 7.0) containing 0.2 M NaCl was mixed with 7 ml of packed heparin-agarose (Sigma) for 3 h at 4b C. The gel was poured into a column and eluted sequentially with starting buffer containing 0.2M NaCl (peak A), 1M NaCl (peak B), and 2M NaCl (peak C). Fractions including the peaks were pooled, dialyzed, concentrated, and electrophoresed on 12% polyacrylamide gels containing SDS. The gels were loaded with 5 μg/well (whole extract, peak A) or 2 5g/well (peak B) and stained with Coomassie Blue. No stained bands were detected in the material from peak C (not shown).

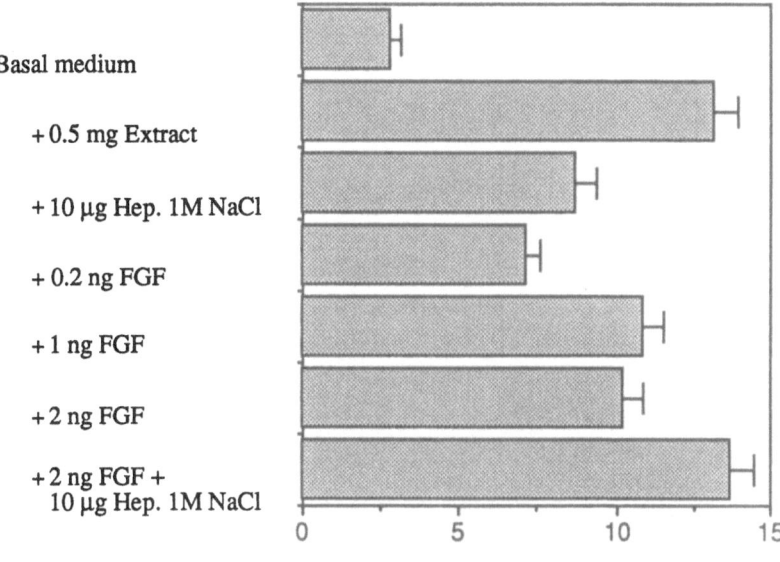

Labeled satellite cells per myofiber

Fig. 5. Additive effects of FGF and purified muscle extract
on proliferation of satellite cells. Single
myofibers were cultured for 48 h and labeled with
[3H] TdR for the final 4 h. Additions to basal
medium were crude muscle extract, muscle extract
purified by elution from heparin-agarose beads with
1 M NaCl, or fibroblast growth factor
(Collaborative Research). Counts of labeled
satellite cells represent the mean 1 standard error
of at least 30 myofibers.

fractionation of the 1M NaCl peak on calibrated Sephadex G150 columns
showed that the satellite cell mitogen elutes at a molecular size of
about 100 kD. This peak contained only the 37 kD polypeptide in SDS
gels (not shown) and still required low molecular weight material for
full activity with quiescent satellite cells.

The relation of this mitogen to FGF remains unclear. Most forms of
FGF have a molecular size of about 16 kD and elute from heparin agarose at
1.5-2 M NaCl (Gospodarowicz, et al. 1984). The 2M NaCl heparin column
fraction of muscle extract contains no mitogenic activity for satellite
cells on single fibers, but this fraction is a potent mitogen for endothelial cells in monolayer culture (Bischoff 1989a). Although purified
FGF is mitogenic for satellite cells on single fibers (Bischoff 1986a),
its activity, even at high doses, does not equal that obtained with
unfractionated muscle extract. Furthermore, the effects of an optimal
dose of FGF and the 1M NaCl heparin column fraction are additive (Fig. 5).

Taken together, these results suggest that although FGF may be
involved in providing a positive mitogenic signal for Go satellite cells,
other factors yet to be identified are also involved.

Thus far I have discussed positive-acting signals that trigger
quiescent satellite cells to begin cycling, but muscle growth is a complex
process that is controlled by growth stimulatory and inhibitory signals.
Since the satellite cells are in close contact with both the plasmalemma
and the basal lamina of the myofiber, the effect of these surfaces on

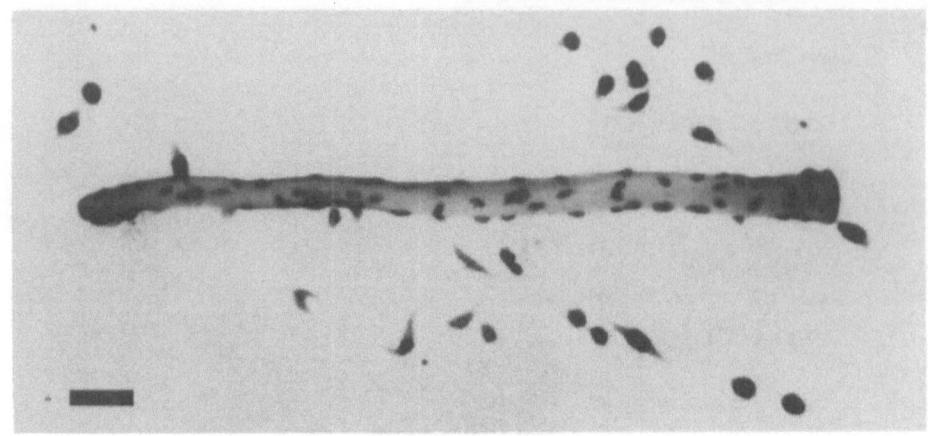

Fig. 6. Radioautographs of single myofiber centrifuged in the
culture dish at 1500 xg for 30 min and grown in basal
medium with 0.5 mg/ml muscle extract for 2 d. All
satellite cells have been detached from the myofibers
by centrifugal force and are growing nearby on the
culture dish. Many of the satellite cells are labeled.
Hematoxylin stain. Bar = 50μm.

satellite cell proliferation was tested. For these experiments the
satellite cells were cultured under three conditions (Bischoff 1989b):

1. On intact myofibers surrounded by basal lamina
2. Removed from both the myofiber and basal lamina
3. Attached to the basal lamina alone

In the first experimental condition, satellite cells were in contact
with both the plasmalemma and basal lamina of the myofiber and these
served as controls for the other two groups. Satellite cells in the
second condition were detached from the myofibers by centrifugal force and
deposited on the culture dish (Fig. 6). In the third condition, the

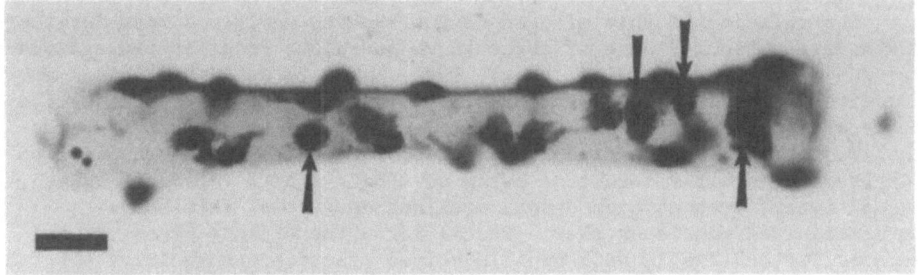

Fig. 7. Radioautograph of myofiber killed by exposure to
0.05% Marcaine (Winthrop-Breon) for 10 min at
0-time. The cultures were washed and grown in basal
medium containing 0.5 mg/ml muscle extract for 2 d
and labeled with [3H] TdR as in Fig. 2. The
myonuclei have lysed and the satellite cells are
growing within the basal lamina tube of the dead
myofiber. Some of the labeled satellite cells are
indicated (arrows). Hematoxylin stain. Bar = 25μm.

TABLE 1
Effect of Contact on Satellite Cell Proliferation

Satellite Cell Contact With Myofiber	Labeled Satellite Cells Per Myofiber	p*
Expt. #1		
Plasmalemma+ basal lamina	17.2 1 1.1**	
None	23.9 1 1.7	<.001
Expt. #2		
Plasmalemma+ basal lamina	14.8 1 0.8	
Basal lamina	21.0 1 0.7	<.001

In the first experiment isolated myofibers were centrifuged
at 1500 xg for 30 min in 35 mm dishes containing coverslips
coated with type I collagen then treated with polylysine to
promote cell adhesion. All satellite cells were detached
and adhered to the substratum adjacent to the myofiber
where they remained during subsequent proliferation. The
cultures were grown for 48 h in medium containing 0.5 mg/ml
muscle extract and labeled for the final 4 h with [3H] TdR.
Radioautographs were prepared and labeled satellite cells
in contact with both the plasmalemma and the basal lamina
from control cultures centrifuged at low speed were
compared with satellite cells removed from contact with the
myofiber by high speed centrifugation.

In the second experiment isolated myofibers were killed by
exposure to 0.05% Marcaine for 10 min. The myofibers
rapidly undergo hypercontraction leaving the satellite
cells attached to the inner surface of the basal lamina.
Cultures were treated with muscle extract and [3H] TdR as
for experiment #1. Labeled satellite cells on intact
viable myofibers from control cultures were compared with
satellite cells on the basal lamina alone from the
Marcaine-treated cultures.

*The p value for the difference between the means was
obtained from Student's t-test.
**Mean ± Standard Error

myofibers were killed by brief exposure to the myotoxic drug Marcaine,
leaving the satellite cells attached to the inner surface of the basal
lamina (Fig. 7).

In all cultures the satellite cells were stimulated to proliferate by
a limiting concentration of muscle extract so as to avoid overriding
possible surface contact effects. The results of these experiments show
that contact with the myofiber plasmalemma, but not the basal lamina,
inhibits the response of the satellite cells to the mitogens in muscle
extract (Table 1).

Taken together, these experiments show that contact with the plasmalemma depresses satellite cell proliferation by about 30% compared to
contact with the basal lamina alone or with a neutral surface, the
culture dish substratum. Although the degree of suppression is relatively modest, additional dose response studies shows that inhibition of
satellite cell proliferation by the intact myofiber extends over a wide

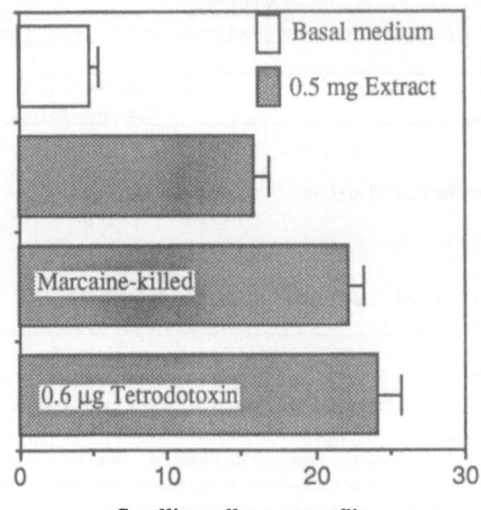

Satellite cells per myofiber

Fig. 8. Effect of myofiber activity on mitogen-
stimulated proliferation of satellite
cells. Single myofibers were cultured in
basal medium (A), or basal medium plus 0.5
mg/ml muscle extract (B,C,D) for 48 h.
Some cultures were treated at 0-time for
10 min with 0.05% Marcaine (Winthrop-
Breon) to kill myofibers (C), or incubated
with 0.6 μg.ml tetrodotoxin (Sigma) to
block contractile activity (D). To
facilitate counting of total satellite
cells per myofiber, all myofibers were
killed with Marcaine four hours before
sacrifice. The myonuclei lyse within this
time leaving only satellite cells within
the basal lamina tubes. This concentration of tetrodotoxin abolishes spontaneous
contractions of the myofibers and prevents
induced contractions driven by 30 volt
pulses delivered from a stimulator (Grass,
Mod. SD9D) to the culture via platinum
electrodes in the medium.

range of mitogen concentration (Bischoff 1989b). At low concentration
of mitogen satellite cells proliferate on dead myofibers but remain
quiescent on living myofibers. This type of push-pull control may be
important in the fine tuning of regeneration. For example, satellite
cells may be stimulated to proliferate on single myofibers, or small
groups, following necrosis induced by overwork, while satellite cells on
neighboring undamaged myofibers remain quiescent (Darr and Schultz 1987;
Giddings, et al. 1985; Irintchev and Wernig 1987).

The mechanism of contact inhibition in controlling satellite cell
proliferation is not yet understood, but various properties of the myofiber plasmalemma may be involved. Among these are physical masking of
receptor sites, biochemical modulators on the surface, and electrical
activity of the myofiber. Preliminary experiments suggest that activity
may be involved. Isolated myofibers twitch spontaneously in culture and
this activity can be blocked by adding tetrodotoxin to the medium. There

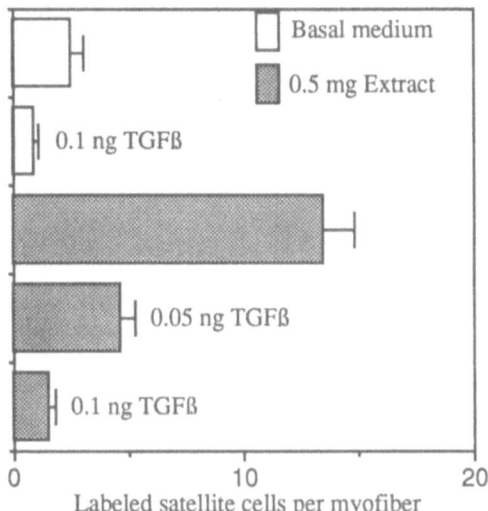

Fig. 9. Effect of transforming growth factor beta on
mitogen-stimulated proliferation of satellite
cells. Isolated myofibers were grown for 48 h and
labeled with [3H] TdR for the last 4 h. Supplements to basal medium consisted of 0.5 mg/ml crude
muscle extract and 0.05-0.1 ng/ml TGFβ
(Collaborative Research). At least 50 myofibers
from each treatment were scored in radioautographs.

was significantly greater proliferation of satellite cells on myofibers
in tetrodotoxin-treated cultures than in control cultures (Fig. 8).

Cell growth on paralyzed but viable myofibers was comparable to that
found in Marcaine-killed myofibers. This suggests that electrical
activity at the myofiber surface may, in part, be responsible for modulat-
ing the response of attached satellite cells to soluble growth factors.

In addition to growth suppression mediated by cell contact, soluble
inhibitors may also be involved. Since TGFβ has been reported to inhibit
proliferation of myogenic cells in monolayer culture (Allen and Boxhorn
1989; Florini and Ewton 1988), the effect of this factor was tested in the
satellite cell-myofiber system. TGFβ proved to be a remarkably potent
inhibitor of extract-stimulated satellite cell proliferation (Fig. 9).

As little as 0.05 ng/ml severely depressed satellite cell growth and
0.1 ng/ml blocked it completely. After 2 d exposure to 0.1 ng/ml, the satellite cells resumed proliferation when the myofibers were washed with
saline and cultured in normal medium. Even the low level of "spontaneous"
satellite cell proliferation in basal medium was reduced by TGFβ (Fig. 9).
This background level of proliferation in basal medium varies between experiments and may arise from release of mitogen from a few damaged myofibers in the culture or from endogenous growth factors present in the serum.

The combination of stimulation and inhibition in satellite cell
growth control may explain how the extent of the regenerative response is
adjusted to fit the degree of injury. For example, death of only a few
myofibers elicits a focal regeneration of the necrotic myofibers, while a
massive injury may cause regenerative changes in adjacent, non-necrotic
muscle tissue (Grounds and McGeachie 1987; McGeachie and Grounds 1987).

Release of low levels of mitogen combined with escape from contact
inhibition following death of a myofiber may be sufficient to trigger its
attached satellite cells into the cell cycle while neighboring satellite
cells are kept quiescent by contact with a viable myofiber.

These results may also be useful in helping to obtain optimal
growth of implanted myoblasts. Since culture-derived myoblasts will
already be cycling, there should be no further need for competence
factors such as those thought to be present in muscle extract.
Progression factors present in serum (or plasma) will allow the cells to
continue proliferation after implantation. Individual progression
factors have not yet been identified for myoblasts. If the dividing
myoblasts contact the plasmalemma of a mature myofiber however, further
proliferation may be suppressed. For maximum growth of implanted cells
it may be advantageous therefore, to produce some local injury so as to
kill myofibers and remove the source of inhibition.

REFERENCES

Allen, R. E. and Boxhorn, L. K., 1989, Regulation of skeletal-muscle
    satellite cell-proliferation and differentiation by transforming
    growth factor-beta, insulin-like growth factor-I, and fibroblast
    growth-factor, J. Cell. Physiol., 138:311.
Allen, R. E., Dodson, M. V., and Luiten, L. S., 1984, Regulation of
    skeletal muscle satellite cell proliferation by bovine pituitary
    fibroblast growth factor, Exp. Cell Res., 152:154.
Bischoff, R., 1986a, Proliferation of muscle satellite cells on intact
    myofibers in culture, Dev. Biol., 115:129.
Bischoff, R., 1986b, A satellite cell mitogen from crushed adult muscle,
    Dev. Biol, 115:140.
Bischoff, R., 1989a, Analysis of muscle regeneration using single
    myofibers in culture, Med. Sci. Sports Exer. (In Press).
Bischoff, R., 1989b, Interaction between satellite cells and skeletal
    muscle fibers, Development (In Press).
Clegg, C. H., Linkhart, T. A., Olwin, B. B., and Hauschka, S. D., 1987,
    Growth factor control of skeletal muscle differentiation:
    commitment to terminal differentiation occurs in G1 phase and is
    repressed by fibroblast growth factor, J. Cell Biol., 105:949.
Darr, K. C. and Schultz, E., 1987, Exercise-induced satellite cell acti-
    vation in growing and mature skeletal muscle, J. Appl. Physiol.,
    63:1816.
DiMario, J., Buffinger, N., Yamada, S., and Strohman, R. C., 1989,
    Fibroblast growth factor in the extracellular matrix of dystrophic
    (MDX) mouse muscle, Science, 244:688.
Florini, J. R. and Ewton, D. Z., 1988, Actions of transforming growth
    factor-beta on muscle cells, J. Cell. Physiol., 135:301.
Giddings, C. J., Neaves, W. B., and Gonyea, W. J., 1985, Muscle fiber
    necrosis and regeneration induced by prolonged weight-lifting
    exercise in the cat, Anat. Rec., 211:133.
Gospodarowicz, D., Cheng, J., Lui, G.-M., Baird, A., and Bohlent, P.,
    1984, Isolation of brain fibroblast growth factor by
    heparin-Sepharose affinity chromatography: Identity with pituitary
    fibroblast growth factor, Proc. Nat. Acad. Sci. USA, 81:6963.
Gospodarowicz, D., Weseman, J., and Moran, J., 1975, Presence in brain
    of a mitogenic agent promoting proliferation of myoblasts in low
    density culture, Nature, 256:216.
Grounds, M. D. and McGeachie, J. K., 1987, A model of myogenesis in
    vivo, derived from detailed autoradiographic studies of
    regenerating skeletal muscle, challenges the concept of quantal
    mitosis, Cell Tissue Res., 250:563.

Irintchev, A. and Wernig, A., 1987, Muscle damage and repair in voluntarily running mice: strain and muscle differences, Cell Tissue Res., 249:509.

Jodczyk, K. J., Bankowski, E., and Borys, A., 1986, Stimulatory effect of platelet-breakdown products on muscle regeneration, Zentralbl. Allg. Pathol., 131:357.

Kardami, E., Spector, D., and Strohman, R. C., 1985, Myogenic growth factor present in skeletal muscle is purified by heparin-affinity chromatography, Proc. Nat. Acad. Sci. USA, 82:8044.

McGeachie, J. and Allbrook, D., 1978, Cell proliferation in skeletal muscle following denervation or tenotomy. A series of autoradiographic studies, Cell Tissue Res., 193:259.

McGeachie, J. K. and Grounds, M. D., 1987, Initiation and duration of muscle precursor replication after mild and severe injury to skeletal muscle of mice. An autoradiographic study. , Cell Tissue Res., 248:125.

Ross, J. J., Duxson, M. J., and Harris, A. J., 1987, Formation of primary and secondary myotubes in rat lumbrical muscles, Development, 100:383.

Venkatasubramanian, K. and Solursh, M., 1984, Chemotactic behavior of myoblasts, Dev. Biol., 104:428.

Westall, F. C., Lennon, V. A., and Gospodarowicz, D., 1974, Brain-derived fibroblast growth factor: identity with a fragment of the basic protein of myelin, Proc. Natnl. Acad. Sci USA, 75:4675.

Discussion of Dr. Bischoff's paper

Dr. Fischman: Dr. Bischoff, is the active factor present in the muscle fiber or must it be produced during the 1 hr incubation after crushing?

Dr. Bischoff: The factor seems to be present in muscle since we can obtain activity by homogenizing muscle and using this directly. However, the specific activity appears to be less than in the crush and extraction protocol.

Dr. Fischman: But does the activity level increase with time after muscle crush? In other words, is being released from muscle by some proteolytic process or does it require biosynthetic events after muscle injury?

Dr. Bischoff: We have attempted these experiments in the presence of protease inhibitors and the results are unchanged. It appears that the factor is not created by proteolysis after muscle injury.

Dr. Fischman: Is the factor active across species lines? Will a rat extract stimulate mouse or human satellite cell proliferation?

Dr. Bischoff: A rat extract is active in cultures of chick embryo myoblasts but we have not tested this extensively. We have not tested the effects on human or mouse satellite cells.

Dr. Fischman: I was struck by the fact that the mitogen only has to be present for 10-12 hours. Is geometric growth maintained after removal factor.

Dr. Bischoff: For at least 48 hr the satellite cells do grow logarithmically, even though they have seen the factor for only 10-12 hr. We have not looked beyond 48 hr since the cells begin to fuse and form new myotubes and it is difficult to count satellite cell numbers.

REGENERATION OF SKELETAL MUSCLE INDUCED

BY SATELLITE CELL GRAFTS

Hala S. Alameddine and Michel Fardeau

INSERM U. 153 / CNRS URA 614
17 rue du Fer-à-Moulin
75005 Paris
France

INTRODUCTION

Trauma to a skeletal muscle may be experimentally induced by various types of injury. This results in the necrosis of the injured muscle fibers followed by the degradation of cell debris and subsequent regeneration. Revascularization and the presence at the site of injury of myogenic stem cells, the "satellite" cells, are essential for the regeneration to take place (Allbrook 1981; Carlson and Faulkner, 1983). This sequence of degeneration - regeneration is observed in many genetically inherited myopathies among which the most dramatic is Duchenne Muscular Dystrophy. Conspicuous foci of necrosis and regeneration are common in young patients and although active regeneration occurs particularly in the earlier stages of the disease, it is ineffective in preventing fiber loss and progression of the disease (Cullen and Mastaglia 1980). This leads to the replacement of the muscle fibers by a fibro-fatty tissue at more advanced stages of the disease and is accompanied by progressive weakness due to necrosis and loss of muscle fibers.

Long before the DMD gene and its protein product "dystrophin" have been described (Kunkel et al., 1985; Monaco et al., 1985; Hoffman et al., 1987), trials were made on experimental animals to investigate the fusion of myoblasts into regenerating muscles with the aim of correcting genetic defects (Sloper and Partridge, 1980). With the same aim in mind, we have undertaken, since 1983, a study to test the ability of in vitro multiplied satellite cells to regenerate new muscle fibers in an irreversibly injured muscle (Alameddine at al., 1987).

EXPERIMENTAL REQUIREMENTS

A number of conditions should be met in order to clearly demonstrate the ability of satellite cells multiplied in vitro to form new muscle fibers:

1) induce an irreversible injury to a given muscle without affecting its revascularization. This should insure proper degradation of cell debris by invading cells and provide nutritional requirements for the development of the grafted cells.

2) label the cells in order to follow their fate after grafting. The label(s) should not affect viability of the cells, and allow long term labeling even if they are diluted during cell division of the grafted cells. They should be internalized by the mononucleated cells and should not be physiologically liberated outside the cytoplasm.

3) prevent immunological rejection of the grafted cells. We choose to circumvent rejection by use of autologous grafts as a first step of our demonstration.

RESULTS

The experimental model has been performed on *extensor digitorum longus* (EDL) of adult male rats. It is derived from the free autotransplantation model described by Carlson and Gutmann 1975. The experimental procedure consists in sectioning the proximal tendon of the muscle and reflecting it distally until it is completely removed from its bed; the muscle remains attached by the distal tendon. It is then replaced in its original site and the cut tendon ends sutured with no attempt to perform neurovascular anastomosis.

Free Autotransplantation

This model is based on ischemic injury of the most central fibers composing a muscle and is characterized by the survival of a thin peripheral rim 5-6 muscle fibers thick. Degradation and phagocytosis of the necrotic muscle fibers progress centripetally following revascularization. Regeneration progresses in the same centripetal manner; so after a few days we observed three concentric zones within the muscle: a peripheral rim of intact muscle fibers, a very dense intermediate zone composed of newly formed myotubes and necrotic muscle fibers invaded by macrophages, and a central zone made up of necrotic muscle fibers not yet degraded. By the end of the second week all muscle fibers have regenerated

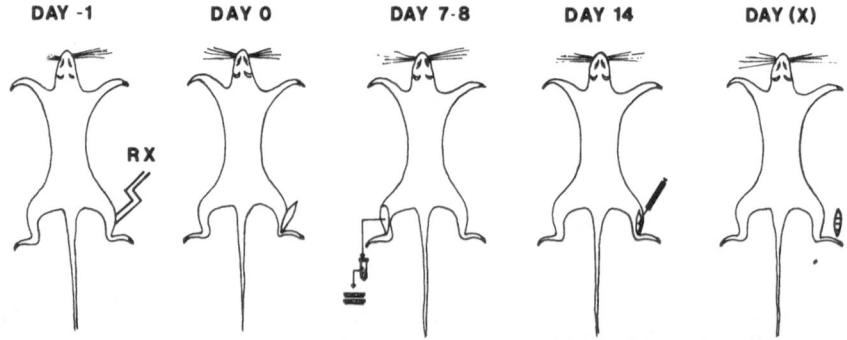

Fig. 1. Schematic representation of the experimental model.

and at 30th day the muscle has recovered a subnormal appearance with differentiated muscle fibers, except the persistence of centrally located nuclei. Stabilization of the graft occurs at the end of the third month after the operation.

Association of X-Irradiation to Free Autotransplantation

It is possible to prevent this spontaneous regeneration by an X-irradiation. Since we wanted to determine the optimal dose of X-ray that would irreversibly inhibit spontaneous regeneration without affecting revascularization, we tested three different doses of X-irradiation before and after the ischemia. Our observations indicate that degradation of cell debris is not affected by irradiation when 1500 and 2500 rad X-ray doses were administered; it was only slightly retarded when the muscles received 3500 rad. X-irradiation is more efficient to prevent regeneration when applied before than after the surgical procedure, and spontaneous regeneration was completely inhibited after a single X-ray exposure of 2500 rad performed before the surgery. EDL muscle are transformed, after macrophage elimination of the necrotic fibers, to a cystic structure formed by the peripheral rim of intact muscle fibers surrounding a central space composed by a loose connective network containing a few cellular elements (Fig. 3a). This transformation was stable up to 8 months after the surgical procedure. This site appeared to us as an ideal zone for the grafting of autologous satellite cells that have been multiplied *in vitro* to investigate their ability to regenerate new muscle fibers.

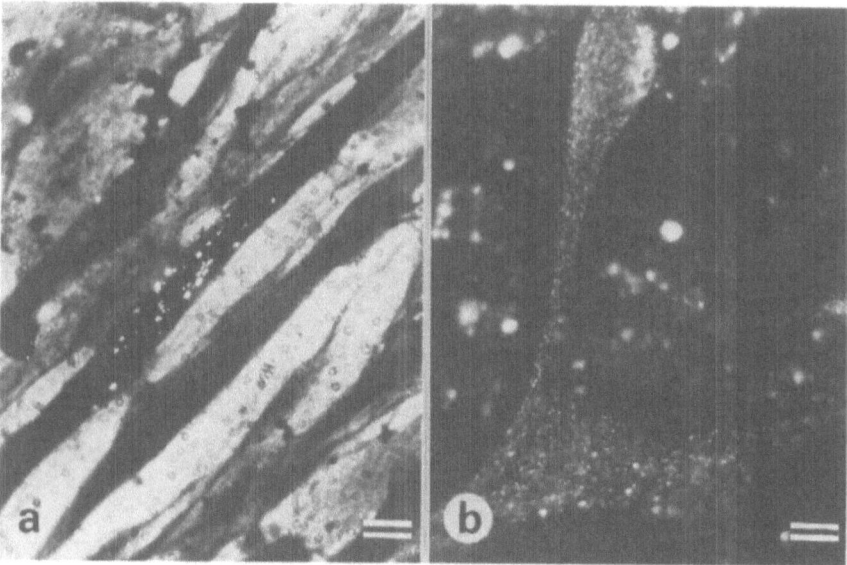

Fig. 2. *In vitro labeling of myogenic cells with a) FITC-latex beads and b) Carbocyanine dye (DiI). Fluorescence microscopy, bar = 25 μm. Mononucleated cells internalize the labels and transport them into myotubes during differentiation. These labels do not affect viability of the cells and remain for a long time in the myotubes (10 days of culture in a and 21 days in b).*

Grafting Autologous Satellite Cells

We thus designed the following experimental procedure: right hind-limb received an X-Ray dose of 2500 rad 24 hours before autotransplantation of EDL muscles was performed. Seven days after surgery, muscles from the contralateral leg were removed and satellite cells enzymatically dissociated and grown *in vitro*. These cells were labeled *in vitro* and collected after 7 days of growth. The cells were injected in the central cystic zone of the injured muscles and histological controls were performed at 3, 7, 14, 30, 60, and 90 days after cell grafting (Fig. 1).

Two kinds of fluorescent probes were used to label the satellite cells *in vitro* : FITC-latex beads and carbocyanine dyes (Dil). The latter emit a yellow fluorescence when examined with fluorescein filter and an orange-red fluorescence when examined with rhodamine filter. These labels were administered to the mononucleated cells in the culture medium. All mononucleated cells incorporate the labels into their cytoplasm and myogenic cells transport it into the myotubes during the fusion process (Fig. 2).

The incorporation of one or both labels within the cytoplasm of the mononucleated cells allowed to follow the fate of the grafted cells.

Immediately after cell grafting, dense aggregates of fluorescent cells occupied the center of the cystic structure obtained 14 days after autotransplantation of a previously irradiated EDL. The introduced cells disaggregate and disperse within the loose fibrous network. Myotubes were first identified 3 days after the implantation of the cells (Fig. 3b). They colocalized with the cells emitting the fluorescent signal. The number as well as the diameter of the regenerated myotubes increase with survival time. At 14 days after cell grafting, the number of regenerated myotubes have reached its maximum. A "no fiber land" clearly separates the newly formed fibers from the peripheral fibers that survived the ischemia. Thirty days after the grafting, the central zone was completely reorganized according to the longitudinal axis of the muscle and muscle fibers were perfectly well organized among each other. Most of the regenerated muscle fibers were reinnervated as evidenced by histochemical differentiation. For longer survival periods (Fig. 3c), structural observations remained mainly similar to the ones reported here above with the exception of the appearance of a few hypotrophic muscle fibers in the central zone probably reflecting a lack of reinnervation of these muscle fibers.

Fig. 3. Transverse sections of EDL muscles stained with hematein-eosin, bar= 200μm. a- Cystic structure obtained by X-irradiation and autotransplantation 14 days after surgery. b- Cellular masses composed of mononucleated cells and myotubes fill part of the central space 3 days after cell grafting. c- Numerous myofibers have formed in the central space. Most of them have normal diameter while a few others are of small size 2 months after cell grafting.

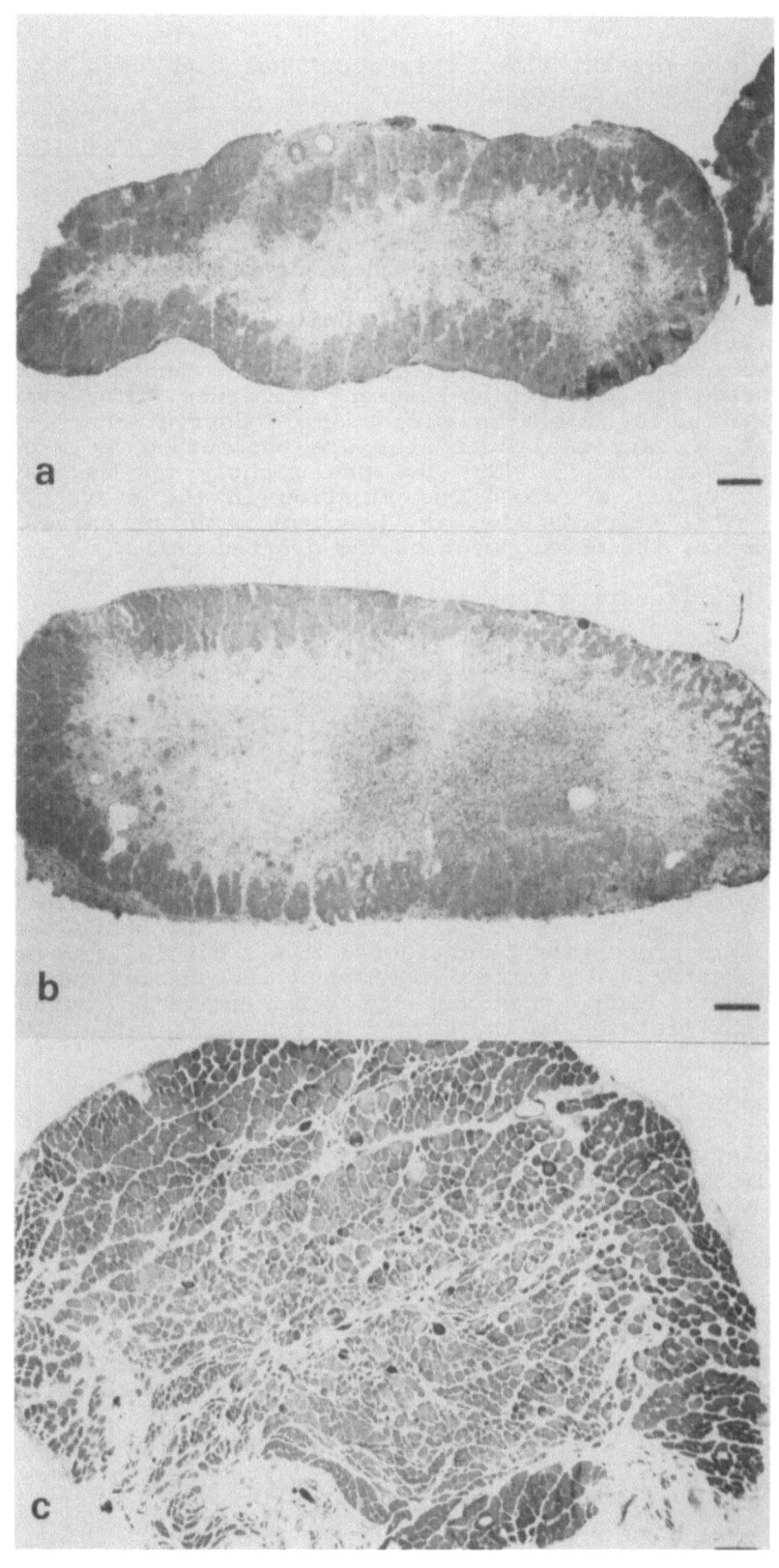

COMPLEMENTARY STUDIES

At the present time, this model has been used to check three important points:

1) the Role of the Extracellular Matrix in the Architectural Reorganization

The ischemic necrosis associated to X-irradiation preserves the basement membrane of the old muscle fibers. Immunoreactivity to major components of the basement membrane i.e. fibronectin, laminin, type IV collagen, and to heparan sulfate proteoglycan is present in the central space of the cystic structure. The regenerated muscle fibers evidenced by the presence of fluorescent latex beads in their sarcoplasm form inside the preexisting basement membranes. This evidence is supported by electron microscopic observations. Young myotubes of very small diameter are ensheathed by a double basement membrane. The new one tightly apposed to the myotubes within a folded one floating in the extracellular space. This suggests that the basement membrane serves as a template for the development of the grafted cells.

2) the Proliferative Capacity of the Grafted Cells

The ability of grafted satellite cells to proliferate has been investigated by the injection of tritiated thymidine at different intervals after the grafting. Our results indicate that myogenic cells retain their proliferative capacity at least for 7 days after grafting. Labeled nuclei are found within myotubes while others are found in satellite position.

3) the Functionality of the Regenerated Muscle

Functional recovery induced by satellite cell grafts has been explored by electrical stimulation of the sciatic nerve after sectioning of other interfering elements. The contractile properties investigated show a partial recovery of all parameters. The force developed by the grafted muscles is significantly higher when compared to the ungrafted control and is directly related to the weight of the regenerated muscle.

CONCLUSION

This work shows that grafting myogenic cells may serve not only to complement the genome of abnormal muscle fibers, but also to reconstruct morphologically and functionally normal muscle fibers.

ACKNOWLEDGMENTS

This work was supported by Association Française contre les Myopathies. The authors acknowledge the contribution of Michèle Dehaupas, Daniel Hantaï, Yolaine Joubert, Jean-Pierre Louboutin and Alain Sébille in some of the studies.

REFERENCES

1. Alameddine H., Dehaupas M., Fardeau M., Régénération musculaire par autogreffe de cellules satellites multipliées in vitro. C. R. Acad. Sc. 304:493, 1987.

2. Allbrook D., Skeletal muscle regeneration. Muscle and Nerve 4:234, 1981.

3. Carlson B.M., Faulkner J.A., The regeneration of skeletal muscle fibers following injury: a review. Med. Sci. Sport Exer. 15:187, 1983.

4. Carlson B.M., Gutmann E., Regeneration in grafts of normal and denervated muscle in the rat: morphology and histochemistry. Anat. Rec. 183:47, 1975.

5. Cullen M.J., Mastaglia F.L., Morphological changes in dystrophic muscles. Brit. Med. Bull. 36:145, 1980.

6. Hoffman E.P., Brown R.H., Kunkel L.M., Dystrophin: the protein product of the Duchenne Muscular dystrophy locus. Cell 51:919, 1987.

7. Kunkel L.M., Monaco A.P., Mieddlesworth W., Ochs H.D., Latt S.A., Specific cloning of DNA fragments absent from the DNA of a male patient with an X chromosome deletion. Proc. Natl. Acad. Sci. USA 82:4778, 1985.

8. Monaco A.P., Bertelson C.J., Middlesworth W., Colleti C.A., Aldridge J., Fischbeck K.M., Bartlett R., Pericak-Vance M.A., Roses A.D., Kunkel L.M., Detection of deletions spaning in the Duchenne muscular dystrophy gene. Nature 323:646, 1985.

9. Sloper J.C., Partridge T.A., Skeletal muscle: regeneration and transplantation. Brit. Med. Bull. 36:153, 1980.

Discussion of Dr. Fardeau's paper

Dr. Fischman: Dr. Fardeau, have you attempted heterologous transplantation in your system?

Dr. Fardeau: No, not yet.

Dr. Fischman: What tests have been performed to prove heterokaryon formation in your new myotubes? Do the new myotubes contain nuclei from both host and donor sources?

Dr. Fardeau: We believe that hybrid myotubes have formed but have not performed rigorous tests to prove that.

Dr. Fischman: Do you have any idea how far the transplanted satellite cells can migrate in vivo?

Dr. Fardeau: At present it is not possible to estimate the distance traveled. The satellite cells appear to travel along the longitudinal axis of the irradiated fibers, perhaps along the basal lamina but we do not know exactly how far they move from the site of insertion.

Dr. Fischman: Do you inject the satellite cells at one or multiple sites?

Dr. Fardeau: We insert the cells at one site with a single injection. The number of cells injected is about 5×10^6 and this can regenerate about 60 mg of tissue. For a human, we estimate that about 10^{12} cells will be required to regenerate all muscles of the body. This is an enormous number.

LOCALIZATION OF MUSCLE GENE PRODUCTS IN NUCLEAR DOMAINS:

Does this Constitute a Problem for Myoblast Therapy?

Helen M. Blau, Grace K. Pavlath, Kevin Rich and Steven G. Webster

Department of Pharmacology
Stanford University School of Medicine
Stanford, CA 94305

A question of major interest in considering myoblast therapy is whether the gene product dystrophin, provided by the introduced myoblasts, can contribute to the function of the myofiber into which the cells fuse. As shown in several papers in this volume, there is now ample evidence that injected myoblasts can fuse with myofibers and produce dystrophin, but the question of whether that dystrophin can restore muscle function still remains. In this regard, a major consideration is whether dystrophin can gain access to distant sites within the myofiber or remains localized in the vicinity of the nucleus that encoded it.

We have been addressing this question by examining to what extent products within myofibers are mobile. These studies have shown that some products are distributed throughout the myofiber whereas others remain localized in nuclear domains. As we have shown in another paper in this volume, when β-galactosidase is introduced into a myofiber using a retroviral construct, it is distributed over a great distance, often encompassing as many as 10 to 20 nuclei in vitro and 1,000 nuclei in vivo. This is the case not only for the cytoplasmic β-galactosidase but also for nuclear β-galactosidase. These results indicate that a foreign protein such as β-galactosidase is readily distributed over large distances within myofibers, a finding in good agreement with that of Ralston and Hall.[1]

Is this also true for a protein that is native to muscle? In order to address this question, we used heterokaryons, a system which has proved useful in addressing a range of questions regarding muscle development.[2,3] Heterokaryons are produced by fusing mouse muscle cells with human non-muscle cells, a procedure which leads to the activation of human muscle genes (Figure 1). Heterokaryons are routinely produced with polyethylene glycol and are cultured so that no mitosis occurs. There is no nuclear division, no chromosome loss, and no chromosome rearrangement, in contrast to traditional hybrids that are obtained following extensive growth during genetic selection. As a result, the effect of the components of one cell type on the components of another cell type can be followed over a period of time: immediately after fusion, and for up to two weeks thereafter. With heterokaryons, we have shown that in response to trans-acting factors muscle genes in a wide range of non-muscle cell types, including keratinocytes from ectoderm and hepatocytes from endoderm, in addition to mesodermal cells, can be induced to express muscle genes that they normally never would express.[3]

An example of a heterokaryon is shown in Figure 2. By fluorescence microscopy, the nuclear composition of the heterokaryon shown in phase contrast (top panel) is revealed. The mouse muscle nucleus is punctate and the human fibroblast nucleus is uniformly stained (middle panel). Upon fusion of the two parent cells with polyethylene glycol the expression of human 5.1H11, a cell surface antigen that is muscle-specific, was detected using a single cell assay (bottom panel). This antibody recognizes N-CAM (neural cell adhesion molecule).[4] Thus, in response to muscle trans-acting factors, human muscle N-CAM is being produced by the human fibroblast nucleus that normally would not produce it. Species-specific reagents have been used to establish that an additional 10 muscle genes encoding products as diverse as enzymes, contractile proteins, and membrane components are activated.[3]

Myoblast Transfer Therapy
Edited by R. Griggs and G. Karpati
Plenum Press, New York, 1990

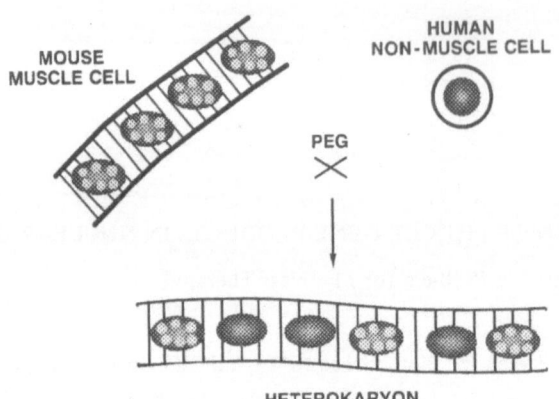

Figure 1. Production of a multinucleated heterokaryon containing mouse and human nuclei. Mouse muscle cells which are naturally multinucleated are fused with human non-muscle cells using polyethylene glycol. Trans-acting factors from the mouse muscle cell can gain access to the human nonmuscle nuclei and result in the activation of expression of human muscle genes in such heterokaryons. When stained with the fluorophore Hoechst 33258, mouse nuclei appear punctate, whereas human nuclei are uniformly stained.

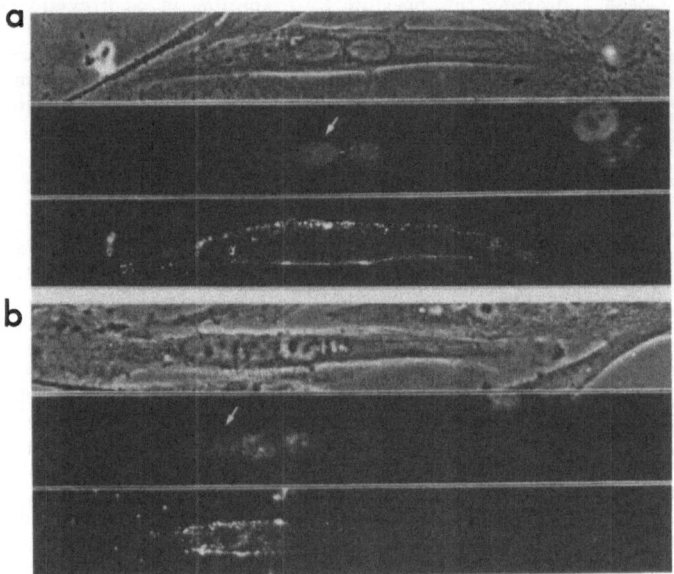

Figure 2. Heterokaryon with a non-localized distribution of N-CAM (a) and heterokaryon with N-CAM localized in the vicinity of the human nucleus (arrows) encoding it (b). Triplicate panels for the two different heterokaryons show boundaries of individual multinucleated cells in phase contrast (top), nuclei stained with Hoechst 33258 (middle) to distinguish uniformly stained human nuclei (arrows) from punctate mouse nuclei, and distribution of N-CAM by immunofluorescence microscopy (bottom). N-CAM was visualized by reaction with monoclonal antibody 5.1H11 in a hybridoma supernatant at 37°C followed by biotin-conjugated anti-mouse IgG and finally avidin-conjugated Texas Red. Heterokaryons were produced by PEG-fusion of mouse C2C12 myotubes and human MRC-5 fetal lung fibroblasts and analyzed 4 days after fusion. The 5.1H11 antibody was generously provided by Dr. Frank Walsh. *Reprinted by permission from Reference 5.*

Using species-specific markers, we have examined the products encoded by specific nuclei in heterokaryons and determined to what extent they are localized in the vicinity of the nucleus that encoded them or gain access to distant sites within the myotube.[5] We observed that in some heterokaryons, muscle cell surface N-CAM was uniformly distributed over the entire surface (Figure 2A), whereas in other cases it appeared to be restricted to the domain

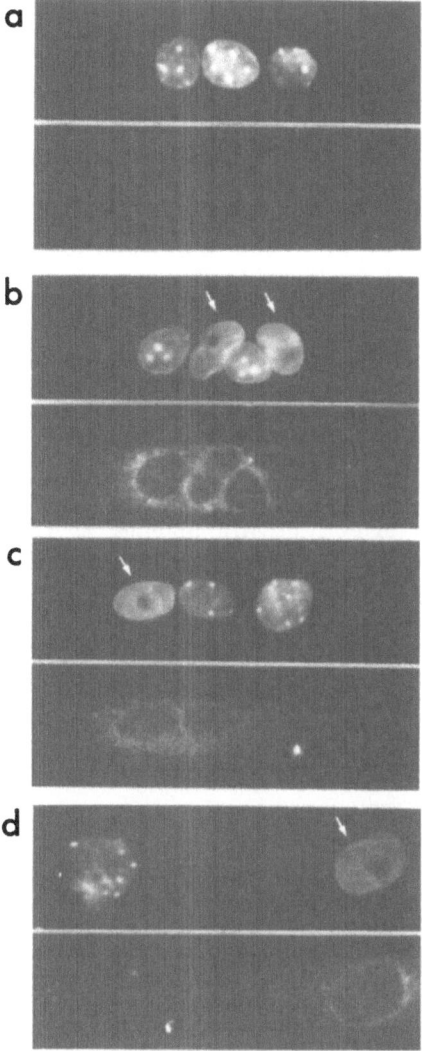

Figure 3. Localized distribution of Golgi apparatus in heterokaryons. (A) An affinity-purified polyclonal antibody (O_{12}) to the human Golgi enzyme galactosyl transferase is a human-specific reagent at a concentration of 1:10 and does not react with the mouse Golgi enzyme in a myotube containing only mouse muscle nuclei. In (B) and (C,left), when human (arrows) and mouse nuclei are closely juxtaposed, the human Golgi is shared. In (C,right) and (D), when human and mouse nuclei are separated, the human Golgi is not shared and remains distinct. In contrast to the Golgi of most cell types, the Golgi shown here is circumnuclear, a distribution typical of striated muscle cells that is induced in non-muscle cells after fusion with muscle cells to form heterokaryons.[14] The O_{12} antibody was generously provided by Dr. Eric Berger. *Reprinted by permission from Reference 5.*

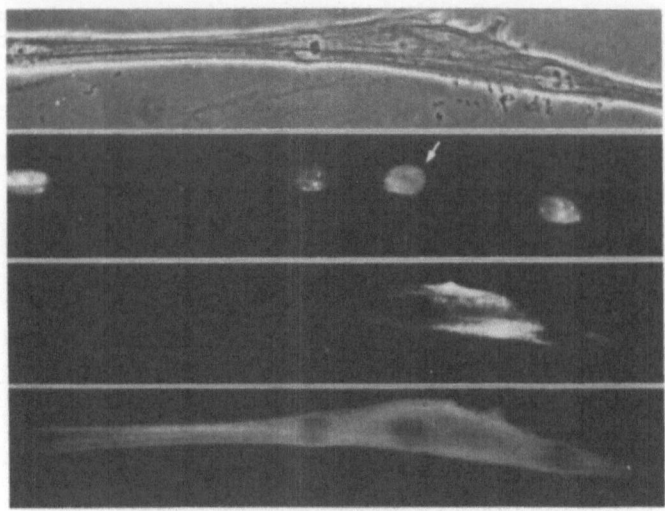

Figure 4. Localized distribution of myosin heavy chain in heterokaryons. A heterokaryon is shown with myosin heavy chain localized in the vicinity of the human nucleus (arrow) reponsible for encoding it. The same field is shown in four views, from top to bottom: phase contrast to define myotube boundaries, Hoechst 33258 to distinguish mouse and human nuclei, monoclonal antibody 4A.951 to reveal human slow SMHC isoforms, and monoclonal antibody 4A.1025 to reveal all SMHC isoforms.[15,16] The heterokaryon was formed by PEG fusion and analyzed 4 days after fusion. Cell types are described in Figure 2. A partial view of the heterokaryon is shown. The total number of mouse and human nuclei (mouse, human) in this heterokaryon is 5,2. Cells were fixed and then reacted in four sequential steps with 4A.951 hybridoma supernatant for 90 min, with fluorescein isothiocyanate -conjugated goat anti-mouse IgG (1:100) for 60 min, then with biotinylated monoclonal antibody 4A.1025 (1:200) for 45 min, Texas red-avidin (1:800) for 15 min, then stained with Hoechst 33258 and visualized. *Reprinted by permission from Reference 5.*

of the nucleus that encoded it (Figure 2B). This result is in marked contrast with that observed by others.[6-8] In most cases, when different cell types were fused together, the cell surface antigens readily diffused throughout the cell within minutes. In addition to N-CAM, the Golgi apparatus, which mediates its transport to the cell surface, was localized. The Golgi was monitored in heterokaryons using an antibody to galactosyltransferase. As shown in Figure 3A, under the conditions used, this antibody is human-specific and does not recognize the mouse Golgi in a myotube containing only mouse nuclei. In a heterokaryon, the Golgi apparatus is not shared unless the mouse and human nuclei are closely juxtaposed (Figure 3B,D). Thus, in a multinucleated muscle fiber, organelles can be maintained in a discrete distribution in the vicinity of the nucleus that encoded them. Again, this finding contrasts with results obtained in most cell fusion experiments in which the organelles contributed by the two cell types rapidly intermixed.[9] Sarcomeric myosin heavy chain (SMHC) also remained localized in nuclear domains in heterokaryons (Figure 4). The antibody used in the bottom panel detects all SMHCs (human and mouse), demonstrating that myosin was distributed throughout the heterokaryon myotube. However, as shown in the third panel, human SMHC remains in the vicinity of the human nucleus that encoded it.

Experiments with spontaneous heterokaryons indicated that nuclear domains are an intrinsic property of the muscle cell. Such heterokaryons were formed by the fusion of mouse myoblasts with human myoblasts in culture. The localization of gene products in spontaneous heterokaryons ruled out the possibility that nuclear domains were a result of fusing fibroblasts into myotubes that had already formed a contractile apparatus. Nuclear domains were also not due to the use of PEG or the artificial combination of disparate cell types.

The frequency of localization in PEG fused heterokaryons increased with nuclear ratio: the more muscle nuclei relative to fibroblast nuclei, the more localized was the product, suggesting that localization is a muscle-specific property. To determine whether N-CAM and SMHC were initially localized within a heterokaryon, but then dispersed, we monitored the distribution of these products over time. One distribution did not give rise to another distribution, but remained stable.

In contrast to the products discussed above, some native muscle products are not localized in domains. For instance, the subunits of creatine kinase, an enzyme involved in generating the energy for contraction, readily intermix.[2] Thus, more heterodimers are detected than homodimers in heterokaryons. Similar results were obtained for the enzyme glucosephosphateisomerase by Partridge[10] and by Mintz.[11] In addition, the trans-acting factors that mediate the activation of muscle genes gain access to distant sites within the myotube. Taken together, these experiments suggest that there are mechanisms within the myotube for recognizing products and either restricting their movement and maintaining them in localized domains or allowing them to distribute freely.

What is the significance of this finding for Duchenne muscular dystrophy? First it helps explain the finding that two-thirds of carriers exhibit some muscle defects such as elevated creatine kinase in their serum.[12] Muscle fibers of carriers are natural heterokaryons. The tissues of females are mosaic for the expression of the X-chromosome. Thus, in accordance with the Lyon hypothesis, some of their cells express one X-chromosome due to random inactivation, and the remainder express the other X-chromosome. If the normal nuclei could complement the defective nuclei within heterokaryons, muscle defects would be expected only rarely. However, our results suggest that dystrophin, which appears to be a membrane associated cytoskeletal protein,[13] may be localized in nuclear domains, resulting in focal lesions along a fiber that could lead to elevated serum creatine kinase.

Is this going to be a problem for myoblast transplantation? That localization occurs in myotubes suggests the possibility that dystrophin cannot gain access to distant sites and is maintained in nuclear domains. To determine whether this constitutes a problem, experiments need to be designed to examine the rescue of muscle function, not merely the rescue of product expression following myoblast transplantation in vivo. It will be important to determine the necessary ratio of nuclei of introduced myoblasts to existing myoblasts. One consideration that suggests optimism is that although carriers exhibit elevated creatine kinase in their serum, most carriers have relatively little impairment of muscle function.

References

1. Ralston, E. and Z.W. Hall, Transfer of a protein encoded by a single nucleus to nearby nuclei in multinucleated myotubes, Science, 244:1066 (1989).
2. Blau, H.M., C.-P. Chiu and C. Webster, Cytoplasmic activation of human nuclear genes in stable heterocaryons, Cell 32:117 (1983).
3. Blau, H.M., G.K. Pavlath, E.C. Hardeman, C.-P. Chiu, L. Silberstein, S.G. Webster, S.C. Miller and C. Webster, Plasticity of the differentiated state, Science 230:758 (1985).
4. Walsh, F.S., G. Dickson, S.E. Moore, C.H. Barton, Unmasking N-CAM, Nature 339:516 (1989).
5. Pavlath, G.K., K. Rich, S.G. Webster and H.M. Blau, Localization of muscle gene products in nuclear domains, Nature 337:570 (1989).
6. Watkins, J.F. and D.M. Grace, Studies on the surface antigens of interspecific mammalian cell heterokaryons, J. Cell Sci., 2:193 (1967).
7. Harris, H., E. Sidebottom, D.M. Grace and M.E. Bramwell, The expression of genetic information: A study with hybrid animal cells, J. Cell Sci., 4:499 (1969).
8. Frye, L.D. and M. Edidin, The rapid intermixing of cell surface antigens after formation of mouse-human heterokaryons, J. Cell Sci., 7:319 (1970).
9. Tassin, A.-M., B. Mara and M. Bornens, Fate of microtubule - organizing centers during myogenesis in vitro, J. Cell Biol., 100:35 (1985).
10. Partridge, T.A., M. Grounds and J.C. Sloper, Evidence of fusion between host and donor myoblasts in skeletal muscle grafts, Nature, 273:306 (1978).
11. Gearhart, J.D. and B. Mintz, Clonal origins of somites and their muscle derivatives; evidence from allophenic mice, Develop. Biol. 29:27 (1972).
12. Hausmanowa- Petrusewicz, I., I. Niebroj- Dobosz, J. Borkowska, E. Lukasik and D. Liszewska-Pfejfer, Carrier detection in Duchenne dystrophy in: "Pathogenesis of Human Muscular Dystrophies," L.P. Rowland, ed., Excerpta Medica, Oxford (1977).

13. Koenig, M., A.P. Monaco and L.M. Kunkel, The complete sequence of dystrophin predicts a rod-shaped cytoskeletal protein, <u>Cell</u>, 53:219 (1988).
14. Miller, S.C., G.K. Pavlath, B.T. Blakely and H.M. Blau, Muscle cell components dictate hepatocyte gene expression and the distribution of the Golgi apparatus in heterokaryons, <u>Genes Devel.</u>, 2:330 (1988).
15. Silberstein, L., S.G. Webster, M. Travis and H.M. Blau, Developmental progression of myosin gene expression in cultured muscle cells, <u>Cell</u>, 46:1075 (1986).
16. Webster, C., L. Silberstein, A.P. Hays and H.M. Blau, Fast muscle fibers are preferentially affected in Duchenne muscular dystrophy, <u>Cell</u>, 52:502 (1988).

Discussion of Dr. Blau's paper

Dr. Fischman: Could the restricted distribution of N-CAM be a unique property of this molecule because of its interaction with extracellular matrix components or with other transmembrane components? Has the diffusion of other molecules besides N-CAM been examined in your heterokaryon system?

Dr. Blau: I think that the localization of specific molecules will correlate with functional properties of each. For example the localization of myosin heavy chain could relate with mechanisms of targeting and sarcomere assembly and the co-translational assembly of sarcomeric proteins. For N-CAM I think its specific function may relate to it is localized and the cytoskeleton may mediate that localization.

MOUSE CHIMERAS AND GENETIC RESCUE OF MOSAIC MUSCLE

Alan Peterson[1] and David Cross[2]

[1]Ludwig Institute   [2]Department of Pediatrics
687 Pine Ave. W.   The National Jewish Center
Montreal, Quebec   1400 Jackson St.
Canada H3A 1A1   Denver, CO 80206, USA

INTRODUCTION

The precise developmental mechanisms leading to myoblast fusion and the subsequent functional relationships that exist amongst the myonuclei contained within each syncytium are for the most part ill-defined. In contrast, extraordinary advances have been achieved in defining the precise molecular deficits causing several inherited diseases of muscle. In such diseases, functional rescue of muscle might be achievable by the introduction of genetically normal myonuclei but this strategy ultimately depends upon the interrelationships that exist between the myonuclei and the syncytial cytoplasm. If the mRNA encoded by each myonucleus was translated into protein molecules that were subsequently restricted to a "nuclear territory", rescue of such mosaic fibers might depend upon not only the number of transplanted myonuclei but also their intrafiber distribution. Alternatively, if proteins encoded by each myonucleus were distributed uniformly throughout the syncytium, the spatial organization of the transplanted myonuclei could be irrelevant.

Here we review evidence on the nature of the nuclear-cytoplasmic relationships that exist *in vivo* in the skeletal muscle of both normal and mutant mouse chimeras. Dystrophia muscularis (dy and dy^{2J}) and muscular dysgenesis (mdg) chimeras have been analysed in detail, while only preliminary evidence is available for chimeras containing cells homozygous for the mdx allele of the dystrophin locus.

METABOLIC ENZYMES AND THE ABSENCE OF NUCLEAR TERRITORIES

Whether myoblast fusion occurred *in vivo* or only as an artifact of tissue culture was definitively resolved by one of the earliest observations made on chimera muscle.

Myoblast Transfer Therapy
Edited by R. Griggs and G. Karpati
Plenum Press, New York, 1990

Chimeras derived by aggregating embryos homozygous for different alleles of the dimeric enzyme isocitrate dehydrogenase (Id-1), revealed heteropolymer isozyme in skeletal muscle samples. Such heteropolymers could only have formed if the cytoplasm of individual fibers contained a mixture of the two genetically different monomer types and that in turn could only be achieved by the *in vivo* fusion of the genetically different myoblasts into a common syncytium[1].

This observation also revealed that the Id-1 monomers encoded by the genetically different myonuclei must be mixed in the cytoplasm of muscle fibers prior to dimer formation. Thus, the products encoded by each myonucleus could not be restricted to a domain of the fiber subserved by only the local myonucleus. Similar findings were subsequently reported for additional multimeric enzyme systems including glucosephosphate isomerase (GP1-1)[2] and malic enzyme (Mod-1)[3], and for a nuclear-encoded mitochondrial enzyme, mitochondrial malic enzyme (Mod-2)[4].

While the results of these early investigations clearly demonstrated that the cytoplasm of mosaic muscle contained a mixture of proteins encoded by multiple myonuclei, they did not address how far the proteins encoded by a single myonucleus could distribute within the fiber. With the advent of techniques sensitive enough to detect and measure the ratios of GP1-1 isozymes contained within single cells[5], it became possible to directly address that question. In chimera muscle, the GP1-1 isozyme distribution was identical along the length of individual mosaic muscle fibers. Not only was the proportion of each isozyme type the same throughout each fiber, but also, the ratios of dimer types were precisely those predicted if all monomers synthesized within each fiber had aggregated randomly[6]. At least for this one cytoplasmic enzyme, the entire muscle fiber contains a homogeneous mix of the products encoded by multiple myonuclei. Thus, GP1-1 isozyme ratios detected in any sample of a chimera muscle fiber should provide an accurate measurement of the GP1-1 isozyme proportion existing throughout the remainder of the fiber.

This finding has been exploited as the basis of a genotype marking system in a wide array of muscle culture, transplant, and chimera investigations. However, this approach has specific limitations and requires certain assumptions. First, if all myonuclei within a fiber transcribe GP1-1 mRNA at an equivalent rate, GP1-1 isozyme proportions would accurately reflect the proportion of myonuclear genotypes within each mosaic fiber. This hypothesis has not been tested and at least one protein, the α subunit of the acetylcholine receptor, appears to be synthesized by only a subset of myonuclei[7]; should this also occur for GP1-1 the presence of some myonuclei could go undetected by this method. Second, if the muscle samples analysed for GP1-1 contained multiple fibers, any between-fiber differences in GP1-1 content would yield a result in which the actual myonuclear proportions existing within individual fibers might not be accurately represented.

MUTANT ↔ NORMAL CHIMERAS AND THE EXPRESSION OF MUSCLE
DISEASE

Dystrophia muscularis (dy and dy^{2J})

dy was the first mutant recovered in the mouse that was
thought to represent a primary myopathy[8]. As a potential
model of human muscle disease dy has received extraordinary
attention, but the molecular defect underlying the disease
phenotype remains enigmatic. Moreover, it is now known that
both dy and dy^{2J} (a second mutation at the dy locus) result
in a remarkable syndrome of abnormalities characterized by
developmental arrest of Schwann cell maturation[9,10]
associated with a later occurring, but progressive,
degeneration of skeletal muscle.

All attempts to convincingly define the precise cell
type or types in which the dy locus is expressed have so far
failed. Evidence against a primary abnormality in either
neural or muscle structures has been obtained using various
chimera[11], transplant, culture[12], and regeneration[13,14]
preparations. Muscle in dy^{2J}/dy^{2J}↔ +/+ chimeras, in which
no normal myonuclei could be detected using the GP1-1 assay
system, failed to reveal any pathology[16]. On this basis, the
muscle disease must occur either as a secondary consequence
of a primary defect expressed in some different cell type, or
else a primary defect within dy^{2J}/dy^{2J} muscle can be
alleviated by some diffusible factor. Schwann cell
abnormalities, which include a failure to appropriately
associate with spinal root axons and to elaborate intact
basal laminas[10], are similarly rescued in chimeras. In such
preparations, the majority of the dy^{2J}/dy^{2J} Schwann cells are
capable of ensheathing axons projecting in spinal roots[11] and
throughout the peripheral nervous system they elaborate
morphologically normal basal laminas[15]. Such a phenotypic
rescue could only be mediated by an extracellular component
provided in the chimera environment by the coexisting
genotypically normal cells and, at this time, a candidate for
the cell type that expresses dy is the fibroblast[12]. These
results suggest that the pathogenesis of the dy mutant has
little or no relevance as an experimental model of a primary
myopathy consequent upon dystrophin deficiency.

Muscular dysgenesis (mdg)

From its initial description, the mdg mouse has
presented a fascinating array of phenotypic abnormalities[16].
With the recent identification of a molecular defect
affecting the α1 subunit of the dihydropyridine calcium
channel[17,18], many seemingly paradoxical observations will
likely find a rational explanation.

Homozygous mdg/mdg mice die at birth from asphyxiation
caused by the complete failure of all skeletal muscle to
contract. At birth, all skeletal muscle has an immature

myotube-like appearance. Remarkably, when mdg/mdg myoblasts are placed into suitable culture conditions they succeed in producing multinucleated myotubes at a rate and to a size that rival both their own *in vivo* capabilities and those of similarly cultured myoblasts obtained from normal mice. However, such myotubes do not contract despite their ability to propagate normal action potentials and their elaboration of a caffeine-sensitive myofibrilary apparatus[19].

In an attempt to establish whether the introduction of normal myonuclei could correct the *in vitro* failure of excitation-contraction coupling and to determine what proportion of normal myonuclei might be required to achieve such rescue, normal and mdg/mdg myoblasts were cocultured. It was our expectation that mosaic myotubes containing myonuclei of both genotypes would form in such cocultures and that, by assessing both their *in vitro* contractile phenotype and their genotypic composition, we could address both of the above issues.

In these cocultures, spontaneously contracting myotubes were encountered frequently and many of the larger quiescent myotubes contracted vigorously in response to acetylcholine stimulation. Subsequent analysis of myotube genotypes, using the GP1-1 assay on individually recovered myotubes, revealed that normal myonuclei were in fact present in all contracting fibers. Surprisingly, the proportion of normal myonuclei required to achieve this *in vitro* functional rescue was very small. Based upon a visual estimate of the number of myonuclei present in the largest myotubes and the GP1-1 isozyme proportions observed, it appeared likely that a single normal myonucleus could restore apparently normal physiological function to the entire myotube[20]. These results were consistent with an intrinsic deficiency within the mdg/mdg myotube and defined a situation in which phenotypic rescue *via* myoblast transfer was highly successful.

On the basis of these *in vitro* investigations, we predicted that chimeras derived by aggregating mdg/mdg ↔ ±/± embryos would develop and function as normal mice; ie. that the majority of muscle fibers in chimeras would in fact be mosaic and such fibers would be expected to have a normal phenotype. Moreover, as the primary defect in mdg is now known to be restricted to muscle, mdg cells in the remainder of the chimera host would be expected to contribute normally to the environment supporting myogenesis.

By correlating the genotype and phenotype of the muscle in such chimeras, it is possible to demonstrate directly whether such rescue of mosaic fibers occurs *in vivo* and to determine what proportion of genetically normal myonuclei are required.

Chimeras were generated by aggregating normal 8 cell embryos from the C57BL/6J strain with 8 cell embryos obtained from crosses between proven mdg/± mice. C57BL/6 mice are gp1-1[bb] and the mdg stock is homozygous for the gpi-1[a] allele. From the mdg/± x mdg/± cross, 25% of the embryos should be mdg/mdg, 50% mdg/±, and 25% ±/±. Thus, only 25% of the resulting chimeras were expected to contain homozygous mdg/mdg cells. As it was our expectation, based upon the in vitro observations, that such chimeras would have a clinically normal phenotype, we devised a mating scheme to distinguish those chimeras containing mdg/mdg cells from the rest. Viable chimeras were raised to maturity and mated to normal 129/J mice. Pigmentation differences in the offspring allowed us to distinguish between those derived from the normal C57BL/6 gametes produced in the chimera and those derived from gametes originating from the albino (c/c)-bearing mdg cells. If the mdg derived cells were homozygous for the mdg allele then all albino progeny of the chimera x 129 cross would be obligate mdg/± heterozygotes and when crossed to known mdg/± mice, every one would generate affected pups in the expected 25% frequency. Should any of the albino offspring from a chimera fail to generate such affected progeny, the mdg derived component of that particular chimera would be identified as either mdg/± or ±/±.

A total of 23 viable chimeras were successfully analysed by the above strategy (>12 albino progeny each producing > 12 offspring) and all proved negative for cells bearing the mdg/mdg genotype. In contrast, 5 chimeras either died at birth (Cesarian section on day 19 of pregnancy) or shortly thereafter. The majority of these non-viable chimeras demonstrated many if not all of the abnormalities typical of homozygous mdg/mdg newborns including microagnathia, cleft palate and an edematous appearance. For every such potential chimera, pigmentation mosaicism of the eyes and subsequent GP1-1 analysis confirmed that both C57BL/6J-and mdg-derived cells were present in each. Based upon the striking mdg/mdg like appearance of these chimeras we concluded that they were the 25% (5:23) expected to contain homozygous mdg/mdg cells. Our original predictions of this preparation could not have been more erroneous.

Throughout this investigation, chimeras were delivered by Cesarian section, making it possible to appropriately process samples from non-viable chimeras for subsequent analysis of phenotype and genotype. Non-viable chimeras were dissected either just prior to imminent death or immediately thereafter. Representative non-muscle tissues were recovered and frozen at -40°C for subsequent GP1-1 analysis while limbs were mounted on cryostat chucks and frozen in liquid nitrogen-cooled isopentane for cryostat sectioning. Tissue samples were homogenized in an equal volume of water and the GP1-1 isozyme proportions were determined as previously described[5]. For the limb samples, serial cryostat sections were prepared and processed as follows: the first section was mounted and stained with heamatoxylin and eosin; the second section was stained by the Gomori-trichrome method; and the third section was placed upon a polyvinylpyrollidone-covered microscope slide at -20°C. A 15mm diameter perspex ring was placed around this section and immediately filled with silicone oil to permit subsequent dissection and recovery of cross sections of muscle fiber fascicles and fibers within morphologically recognizable muscles for GP1-1 analysis[6]. Polaroid micrographs of the histological preparations were taken and the samples recovered were identified and outlined on the photograph. The muscle phenotype and its genotype, assessed by GP1-1 isozyme proportions, were then determined.

As all samples recovered for GP1-1 analysis contained cross sections of multiple muscle fibers (myotubes) it was possible to assign only a rough estimate of morphological maturity to each sample. A 5 point scale was devised with the following criteria: mdg-like; partial mdg; inter-mediate; almost normal; normal. Two observers in-dependently assigned one of the above morphological maturity scores to each sample by examining the histological preparations and marked photographs. A summary of these results is presented in figure 1.

An apparently linear relationship existed between the proportion of normal C57BL/6 myonuclei and the morphological maturation score assigned to each muscle sample. Most striking was the fact that unambiguously normal muscle morphology was assigned only to that sample group in which the mean proportion of C57BL/6J myonuclei was subsequently determined to be on the order of 50%.

As seen in figure 1, the genotype of these chimera muscle samples was dominated by the mdg/mdg component. Previous investigations of C57BL/6J ↔129/J chimeras had revealed that 129 cells tend to dominate in such chimera combinations. The mdg stock we used was originally derived from multiple backcrosses to the 129/J strain, and therefore the present result likely reflects only this between-strain difference rather than any specific preferential advantage of mdg/mdg myoblasts *per se*. In support of this interpretation, the majority of the non-muscle tissues within these chimeras also expressed a preponderance of mdg/mdg cells (data not shown). In every muscle sample analysed, GP1-1 heterodimers were prevalent, demonstrating that the majority of the muscle was mosaic, containing normal myonuclei. Thus, a striking disparity between the requirements for successful rescue of *in vitro* function and the restoration of a normal *in vivo* phenotype appears to exist in mdg/mdg ↔ +/+ mosaic muscle fibers.

Prior to the characterization of the molecular deficit in mdg, several rational hypotheses could account for this discrepancy. However, the only tenable explanation of these findings must be sought considering only the mosaic muscle within these chimeras. An attractive hypothesis would predict that the α1 subunit of the Ca^{++} channel becomes localized in close proximity to each normal myonucleus within the mosaic syncytium. Any fiber that had a domain in which only mdg/mdg myonuclei existed would lack normal excitation-contraction coupling within that region and, as a consequence, demonstrate a localized failure to appropriately mature. In fact, individual teased muscle fibers were found that expressed a relatively mature muscle fiber morphology

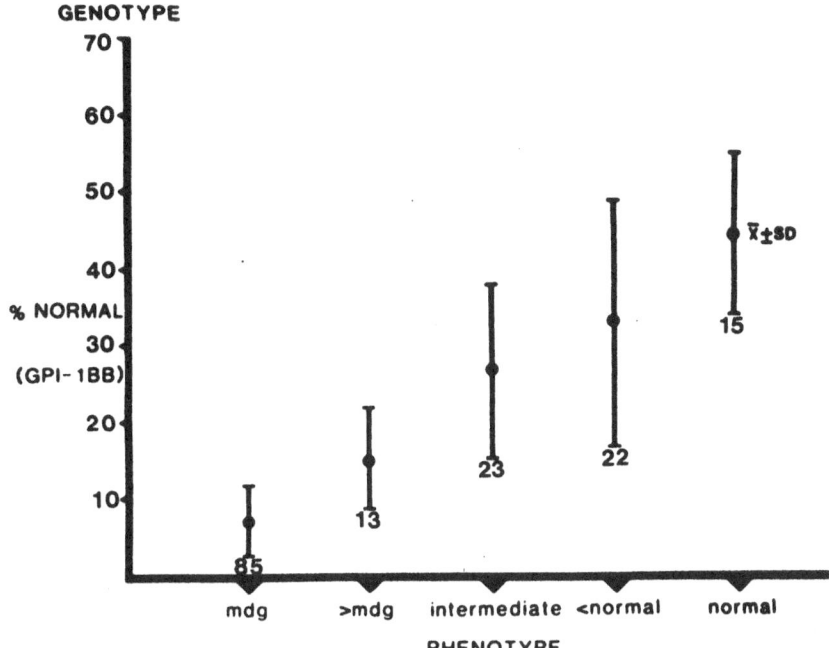

Figure 1. Relationship between proportion of genotypically normal myonuclei and maturation of skeletal muscle in mdg/mdg ↔ +/+ chimeras. A significant correlation (0.84) exists between increasing proportions of normal myonuclei and normalization of phenotype. Number of samples is given on bottom of error bars.

along part of their length while other parts of the fiber had a characteristic mdg/mdg-like myotube appearance[21].

Such regional disparity within myotubes *in vitro* could escape detection by the visual analysis of gross contraction. Alternatively, the apparent sharing of Golgï by multiple nuclei in immature myotubes could result in a more homogeneous distribution of the mdg gene product in the cultured myotubes[22]. A further possibility is that requirements for rescue of diseased myofibers *in vivo* are more quantitatively stringent than they are *in vitro*. The phenotypes that we assessed *in vitro* and *in vivo* are quite different and a proportion of normal myonuclei sufficient to achieve *in vitro* contractility may not be adequate to allow full morphological maturation. It is also important to note that the degree of *in vitro* maturation falls far short of that seen *in vivo* and this may produce a lesser need for the normal mdg gene product.

The "nuclear territory" hypothesis would also explain additional observations made on the patterns of diaphragm innervation in these chimeras. During development, the diaphragm has a diffuse end plate region and many fibers have multiple end plates. In normal mice, multiple innervation of fibers is lost prior to birth and the end plate region of the diaphragm becomes condensed. In mdg/mdg, the immature pattern persists. In the chimera diaphragms, regional differences in the relative maturity of the innervation pattern were observed but these differences did not correlate with the overall genotype of the diaphragm muscle[21]. It is therefore probable that the intrafiber spatial arrangements of the normal myonuclei contributed to this result.

This hypothesis should become directly testable using immunocytochemistry to detect the intrafiber distribution of the $\alpha 1$ Ca^{++} channel subunit in further mdg/mdg ↔ ±/± preparations. Nonetheless, the present results demonstrate clearly that for one characterized deficiency intrinsic to skeletal muscle, the *in vitro* behavior of mosaic myotubes did not accurately predict the behavior of presumably similar mosaic fibers differentiating *in vivo*.

Muscular dystrophy (mdx)

The characterization of dystrophin deficiency in Duchenne and Becker muscular dystrophy[23] and the subsequent recognition that the mdx mouse[24] has the same molecular defect provides a basis for testing the design of therapeutic strategies, the most obvious being myoblast transfer[25,26,27,28,29].

Both the female carrier of Duchenne dystrophy and the heterozygous mdx/± mouse are essentially spared a disease phenotype[30,31]. Because of random X-chromosome in-activation, female carriers are thought to contain muscle that is an equivalent mix of functionally normal and functionally mutated myonuclei. Despite the fact that cells with a normal active X chromosome are present throughout fetal development, both mouse and human carriers typically produce elevated serum levels of muscle-specific enzymes[32]. On this basis alone, it is reasonable to predict that experimentally-imposed muscle fiber mosaicism might restore a fully normal phenotype to a fiber only when the proportion of normal myonuclei approaches or exceeds a value of 50%.

There is however additional information, at least for the mouse, that would suggest that naturally occurring X-inactivation mosaicism represents a special circumstance that could be difficult if not impossible to achieve experimentally. Specifically, X-linked markers in heterozygous females reveal a finely variegated mosaicism[33]. Therefore a relatively uniform spatial distribution of dystrophin-expressing myonuclei should exist throughout the majority of the muscle fibers within such carriers. As the existing evidence strongly suggests that dystrophin is a protein localized in close proximity to the myonucleus that synthesizes its message[28], further understanding of the myonuclear distribution in carrier females will be critical when designing any future therapeutic strategies. The mouse chimera preparation may also yield additional insight.

As seen for the mdg/mdg ↔ ±/± chimeras, such mice can have a genetic constitution that is significantly skewed from an equal representation of both cell lines. This would be expected to lead to intrafiber domains, nucleated solely by mutant myonuclei, that are larger than those occurring in the carrier female[34]. Such chimera fibers should therefore provide an opportunity to determine both the _minimum_ proportion and spatial distribution pattern of normal myonuclei that is required to prevent the initial degeneration of the fiber. To this end _mdx_/_mdx_ ↔ _+_/_+_ chimeras have been produced and preliminary histological evidence reveals that such muscles have a mosaic phenotype containing apparently normal, regenerated and regenerating fibers, Figure 2.

This result suggests that within individual chimera muscles some fibers have a complement of normal myonuclei adequate to confer levels of dystrophin sufficient for normal maturation and function. Determination of the genotype proportions within such rescued fibers is in progress.

Within these chimera muscles the presence of fibers with centrally located nuclei indicates that some muscle fibers did undergo degeneration[35]. Genotype analysis of such fibers will, however, be less informative as their final nuclear proportions should reflect only the genotype of the satellite population secondarily recruited to regenerate the fiber, and not the original genetic composition of the fiber that degenerated.

The presence of apparently actively regenerating fibers in these muscle samples was unexpected. In the particular example shown, the chimera was over 6 months old, well past the age when active degeneration and regeneration of large caliber fibers normally occurs in the _mdx_/_mdx_ mouse[35,36]. This observation may indicate that fibers with a normal genetic compliment close to threshold requirements may survive for an extended period but eventually succumb. Should this be a common finding, experiments addressing the effectiveness of myoblast transfer therapy may yield overly optimistic results if only short time periods are examined.

SUMMARY

The nuclear-cytoplasmic relationships existing within mosaic muscle will likely determine whether myoblast transfer can effectively rescue diseased muscle. The mouse chimera preparation is one source of such mosaic muscle in which that _in vivo_ relationship can be investigated in the complete absence of complicating immunological or surgical trauma.

For several metabolic enzymes, the mature muscle fiber appears to contain a homogeneous mix of the proteins encoded by multiple myonuclei. This relationship is clearly not representative of all muscle proteins, as several examples of proteins highly localized to "nuclear territories" have now been described. Nonetheless, the intrafiber distribution of certain enzymes, particularly GP1-1, is appropriate for the basis of a genotype marking system applicable to mosaic fibers.

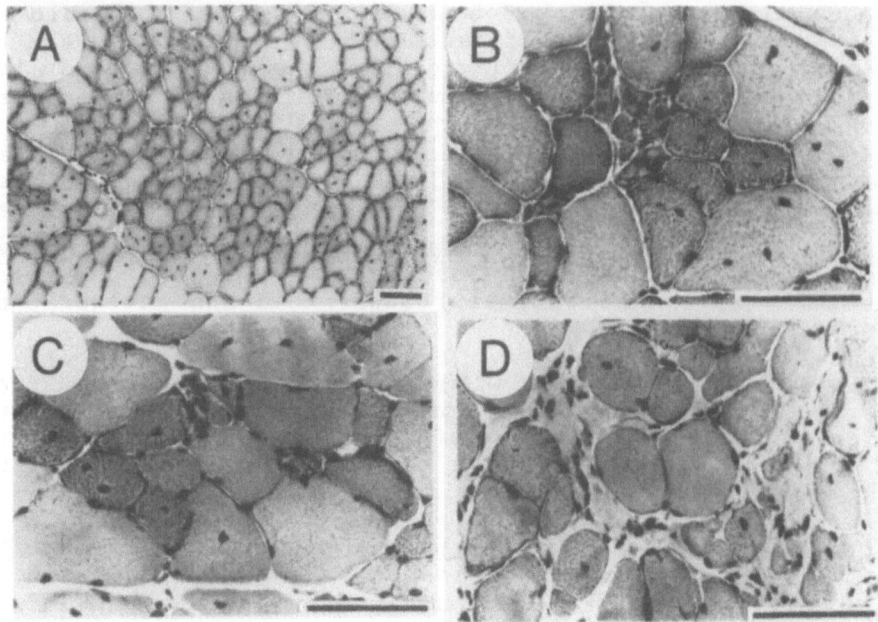

Figure 2. Muscles from a mdx/mdx ↔ ±/± chimera; hematoxylin and eosin stain of 10μm thick cryostat sections. A and B: right anterior tibialis. A: Region containing a mixture of centrally-nucleated and normal-appearing fibers. B: Focus of small apparently regenerating fibers surrounded by centrally- nucleated fibers (right) and normal fibers (left). C: Right gastrocnemius; centrally-nucleated and normal fibers surrounding small apparently regenerating fibers. D. Left anterior tibialis; fibrotic region containing profiles of both centrally-nucleated and normal-appearing fibers. Calibration bars = 100μm.

In vitro rescue of mdg myotubes is readily achievable by incorporation of few normal myonuclei and possibly by only one. *In vivo* requirements are apparently far more stringent and an hypothesis in which the mdg gene product, a Ca^{++} channel subunit, is restricted to nuclear territories would be consistent with the disparate results obtained *in vitro* and *in vivo*.

Finally, chimeras containing mdx/mdx cells can show a partial amelioration of muscle pathology and may provide a means of determining the minimum genetically normal myonuclear compliment required to prevent degeneration of dystrophin-deficient fibers.

ACKNOWLEDGEMENTS

The authors thank I. Tretjakoff and P. Valera for excellent technical assistance. Grant supporting parts of this work were provided by the MRC and MDAC.

REFERENCES

1. B. Mintz, and W. Baker, Normal Mammalian Muscle Differentiation and Gene Control of Isocitrate Dehydrogenase Synthesis. PNAS 58:592-598 (1967).
2. A. Peterson, P. Frair, H. Rayburn, and D. Cross, Development and Disease in the Neuromuscular System of Muscular Dystophic ↔ Normal Mouse Chimaeras. In Society for Neuroscience Symposia. Vol.4: Aspects of Developmental Neurobiology, pp. 258-273 (1979), Society for Neuroscience, Bethesda, MD.
3. A. Peterson, Chimeras Mouse Study Shows Absence of Disease in Genetically Dystrophic Muscle. Nature 248: 561-564 (1974).
4. P. Frair, P. Strasberg, K. Freeman, and A. Peterson, Mitochondrial Malic Enzyme in Mosaic Skeletal Muscle of Mouse Chimeras. Biochem.Genet. 17:693-702 (1979).
5. A. Peterson, P. Frair, and G. Wong, A Technique for Detection and Relative Quantitative Analysis of Glucosephosphate Isomerase Isozymes from Nanogram Tissue Samples. Biochem. Genet. 16:681-690 (1978).
6. P. Frair, and A. Peterson, The Nuclear-Cytoplasmic Relationship in 'Mosaic' Skeletal Muscle Fibers from Mouse Chimaeras. Exp. Cell Res. 145:167-178 (1983).
7. D. Harris, D. Falls, and G. Fischbach, Differential Activation of Myotube Nuclei Following Exposure to an Acetylcholine Receptor-Inducing Factor. Nature 337: 173-176 (1989).
8. A. Michelson, E. Russell, and P. Harman, Dystrophia muscularis: A Hereditary Primary Myopathy in the House Mouse. Proc. Nathl. Acad. Sci. U.S.A. 41:1079-1084 (1955).
9. W. Bradley, and M. Jenkison, Abnormalities of Peripheral Nerves in Murine Muscular Dystrophy. J. Neurol. Sci. 18: 227-247 (1973).
10. R. Madrid, E. Jaros, M. Cullen, and W. Bradley, Genetically Determined Defect of Schwann Cell Basement Membrane in Dystrophic Mouse. Nature 257:319-321 (1975).
11. A. Peterson, Peripheral Nerves in Shiverer ↔ Dystrophic Mouse Chimeras: Evidence that a Non-Schwann Cell Component is Required for Axon Ensheathment in vivo. J. Neurosci. 5:1740-1754 (1985).
12. C. Cornbrooks, F. Mithen, J. Cochran, and R. Bunge, Factors Affecting Schwann Cell Basal Lamina Formation in Cultures of Dorsal Root Ganglia from Mice with Muscular Dystrophy. Dev. Brain Res. 6:57-67 (1983).
13. C. Stirling, Experimentally Induced Myelination of Amyelinated Axons in Dystrophic Mice. Brain Res. 87:130-135 (1975).
14. G. Bray, S. David, T. Carlstedt, and A. Aguayo, Effects of Crush Injury on the Abnormalities in the Spinal Roots and Peripheral Nerves of Dystrophic Mice. Muscle Nerve 6:497-503 (1983).
15. A. Peterson, and G. Bray, Normal Basal Laminas are Realized on Dystrophic Schwann Cells in Dystrophic ↔ Shiverer Chimera Nerves. J. Cell Biol. 99:1831-1837 (1984).

16. S. Gluecksohn-Waelsch, Lethal Genes and Analysis of Differentiation. _Science_ 142:1269-1274 (1963).
17. T. Tanabe, K. Beam, J. Powell, and S. Numa, Restoration of Excitation-Contraction Coupling and Slow Calcium Current in Dysgenic Muscle by Dihydropyridine Receptor Complementary DNA. _Nature_ 336:134-139 (1988).
18. C. Knudson, N. Chaudhari, A. Sharp, J. Powell, K. Beam, and K. Campbell, Specific Absence of the α1 Subunit of the Dihydropyridine Receptor in Mice with Muscular Dysgenesis. _J. Biol. Chem_. 264:1345-1348 (1989).
19. J. Powell, and D. Fambrough, Electrical Properties of Normal and Dysgenic Mouse Skeletal Muscle in Culture. _J. Cell Physiol_. 82:21-38 (1973).
20. A. Peterson, and S. Pena, Relationship of Genotype and _in vitro_ Contreactivity in _mdg_/_mdg_ ↔ _+_/_+_ "Mosaic" Myotubes. _Muscle and Nerve_ 7:194-203 (1984).
21. F. Rieger, D. Cross, A. Peterson, M. Pinçon-Raymond, and I. Tretjakoff, Disease Expression in +/+ ↔ _mdg_/_mdg_ Mouse Chimeras: Evidence for an Extramuscular Component in the Pathogenesis of Both Dysgenic Abnormal Diaphragm Innervation and Skeletal Muscle 16 S Acetylcholinesterase Deficiency. _Dev. Biol_. 106:296-306 (1984).
22. G. Pavlath, K. Rich, S. Webster, and H. Blau, Localization of Muscle Gene Products in Nuclear Domains. _Nature_ 337:570-573 (1989).
23. E. Hoffman, R. Brown, and L. Kunkel, Dystrophin: The Protein Product of the Duchenne Muscular Dystrophy Locus. _Cell_ 51:919 (1987).
24. G. Bulfield, W. Siller, P. Wight, and K. Moore, X-Chromosome-Linked Muscular Dystrophy (MDX) in the Mouse. _Proc Natl Acad Sci_ 81-1189 (1984).
25. G. Karpati, Y. Pouliot, S. Carpenter, and P. Holland, Implantation of Nondystrophic Allogenic Myoblasts into Dystrophic Muscles of MDX Mice Produces "Mosaic" Fibers of Normal Microscopic Phenotype. In _Cellular and Molecular Biology of Muscle Development_ Alan R. Liss, Inc. 973-985 (1989).
26. D. Watt, J. Morgan, and T. Partridge, Use of Mononuclear Precursor Cells to Insert Allogeneic Genes into Growing Mouse Muscles. _Muscle & Nerve_ 7:741 (1984).
27. D. Watt, K. Lambert, T. Partridge, and J. Sloper, Incorporation of Donor Muscle Precursor Cells into an Area of Muscle Regeneration in the Host Mouse. _J. Neurol Sci_ 57:319 (1982).
28. G. Karpati, Y. Pouliot, E. Zubrzycka-Gaan, S. Carpenter, P. Ray, R. Worton, and P. Holland, Dystrophin is Expressed in mdx Skeletal Muscle Fibers After Normal Myoblast Implantation. _Am J. Pathol_. 135:27-32 (1989).
29. T. Partridge, J. Morgan, G. Coulton, E. Hoffman, and L. Kunkel, Conversion of mdx Myofibres from Dystrophin-Negative-to-Positive by Injection of Normal Myoblasts. _Nature_ 337:176-179 (1989).
30. Y. Tanabe, K. Esaki, and T. Nomura, Skeletal Muscle Pathology in X-Chromosome-Linked Muscular Dystrophy (MDX) Mouse. _Acta Neuropathol (Berl)_ 69:91 (1986).

31. F. Torres, and L. Duchen, The mutant MDX: Inherited Myopathy in the Mouse. Brain 110:269 (1987).

32. V. Chapman, D. Miller, D. Armstrong, and T. Caskey, Recovery of Induced Mutations for X Chromosome-Linked Muscular Dystrophy in Mice. Proc. Natl. Acad. Sci. 86: 1292-1296 (1989).

33. S. Garther, and R. Andina, Mammalian X-Chromosome Inactivation. Adv. Hum. Genet. 7:99-140 (1976).

34. M. Nesbitt, Chimeras vs X-Inactivation Mosaicism in the Mouse. Dev. Biol. 26:252-263 (1974).

35. J. Dangain, and G. Vrobova, Muscle Development in MDX Mutant Mice. Muscle & Nerve 7:700 (1984).

36. G. Karpati, S. Carpenter, and S. Prescott, Small-Caliber Skeletal Muscle Fibers do not Suffer Necrosis in MDX Mouse Dystrophy. Muscle & Nerve 11:795-803 (1988).

Discussion of Dr. Peterson's paper

Dr. Fischman: Your work on random mixing of the glucose phosphate isomerase would suggest wide diffusion of protein subunits in the sarcoplasm. I would like to ask if this widespread diffusion is a property only of soluble enzymes of intermediary metabolism or is also true of proteins which are part of the cytoskeleton or of the plasma membrane, where one might anticipate more restricted free diffusion.

Dr. Peterson: The only example that we have examined which might fit that description is a nuclear encoded, mitochondrial enzyme. The rationale being that here is an organelle that might incorporate its protein quickly from the cytoplasmic site of synthesis and this would reveal possible nuclear domains. The data that we have obtained indicate that heterodimers of mitochondrial malic dehydrogenase are fairly rare in mosaic muscle fibers. The theme that seems to be emerging is that some cytoplasmic metabolic enzymes are widely distributed while other proteins of more specialized function may in fact have more restricted domains.

GENERAL DISCUSSION

IN SITU FUSION: NUCLEAR DOMAINS AND mRNA/PROTEIN MIGRATION

     Unidentified Participant: Dr. Rubin, do you know of any
data on local variations of calcium concentrations which might
bear on the domain question?

     Dr. Rubin: We have been trying to analyze that problem
for some time in collaboration with Dr. Shelanski.   It turns
out to be far more difficult than we anticipated because of the
time resolutions required.   The Ca^{++} transient following muscle
contraction is very brief and the measurements are not
completely straightforward.   Also, cultured muscles may not be
the best system to examine these transients.   In fact, we have
been using the flexor digitorum brevis for some of these
experiments and now using fura 2.   We are now able to load the
cells, measure the transient but we have not looked at
different regions of the muscle cytoplasm.   There are also
other possibilities that must be considered.   For example,
there could be regional variations in a hypothetical calcium-
binding protein.

     Same Participant: The other comment that I wanted to make
concerns the possibility that fibroblasts from primary cultures
might be contaminated with myogenic cells or the possibility
that an established non-muscle cell line might convert to one
capable of myogenesis.   Based on the strategies used by
Woodring Wright one might attempt the following type of
experiment to answer this possibility.   If only one cell per
hundred were myogenic it probably would not encounter another
myogenic cell and would not fuse and express muscle-specific
gene products.   This is basically what Woody Wright did when
he was looking for fusion-defective myoblasts and measuring
myogenic potential.   With the large number of monoclonal
antibodies available today it should be fairly straightforward
to score for potentially myogenic cells in the total population
when co-culturing with mdg myotubes.

     Dr. Blau: I think that one of the cell lines used by Dr.
Chaudhari was 3T3 and there might be some ambiguity about their
fibroblast phenotype.   I want to raise this point since I am
intrigued by her results.   We have been studying heterokaryons
for a number of years and have made control cultures and cocultures of myoblasts with fibroblasts from a number of sources
and we have never observed spontaneous fusion of the type
reported by Dr. Chaudhari.   We have used C2 muscle cells from
mice, at least four different fibroblastic cell types and
normal human neonatal myoblasts or myoblasts from Duchenne
patients.   In these combinations, we have not seen myoblast-

187

fibroblast fusions. Thus, I am a bit perplexed by the findings reported here. One possibility that strikes me is that perhaps this phenomenon is specific for the mdg myogenic cells.

Dr. Chaudhari: I concur with Dr. Blau about the perplexing nature of these observations. We do not know what is different about these experiments from those reported by others. It is puzzling that this has not been reported before because we felt that other than the fact that we had a functional assay, we did not specifically look for such heterotypic fusion. Since spontaneous contractions occurred in these co-cultures, we could not help but see it. We do not know whether there is something unique about the dysgenic myotubes which permits such fusion. We recognize that we have not performed as many co-culture experiments as that of Dr. Blau; to date, we have done three or four experiments with normal cell cultures where we have looked for fibroblast-muscle fusion. The normal cell cultures were typically normal homozygotes or normal heterozygotes from our mdg colony. These normal mice have been considered normal based on physiological and morphological criteria widely used in this field. I assume that they are normal but there is the potential that there is something unique about this mouse line that was not recognized before.

Dr. Blau: I would suggest that it is really important to look at a number of mouse lines and one should recognize that such fusion may not pertain to human muscle.

Unidentified Participant: Is it possible that the media may contain a fusigen in it that might vary from lot to lot?

Dr. Chaudhari: That may be a possibility since we do observe variability in the frequency of myotubes which contain fibroblast nuclei.

Dr. Cooper: Dr. Rubin, are you aware of work being done by Rick Steinhardt at Berkeley showing that there are calcium transient abnormalities in mdx mice? Would you comment about how such might relate to your system? Also, I was wondering if the ACh clustering that you see is symmetrical enough to suspect if this represents the first step in organizing the Golgi apparatus for the export of membrane cytoskeletal proteins such as dystrophin.

Dr. Rubin: With respect to your first comment, I know of Rick Steinhardt's work but I cannot say anything more than you have already. With respect to your second comment, I say that you are correct. That is precisely what is happening. Basically, we initiate receptor clustering in the absence of nerves but the nerve terminals probably secrete something which has the effect of clustering receptors beneath the ending. This membrane event must alter the associated cytoskeleton in a profound manner. The alteration at the membrane must extend from the cell surface to the nucleus. It is a very interesting possibility that dystrophin is involved in this process. I once asked Lou Kunkel about this possibility and he said that others were using dystrophin antibodies to examine this question. We have not done so ourselves.

Dr. Kaufman: Getting back to the discussion of fibroblast fusion, I would like to reiterate what Helen Blau said. In our experience, we have never been able to see the insertion of fibroblast cell lines into myoblast lines. In fact, we find that various cell lines were quite inhibitory on myogenesis. One suggestion I would like to make is that the fusigenic capacity of cells can be mediated by endogenous and exogenous viruses and since your finding is, or seems to be, restricted or at least more prevalent in this particular line of cells, there might be the possibility that the cells or cultures may have exogenous viral contamination.

Dr. Chaudhari: Yes, that is a good point and we will repeat these experiments with a completely independent line and strain of mice.

Dr. Tapscott: Dr. Chaudhari, do you ever see fusion of fibroblasts in cultures not expressing lacZ?

Dr. Chaudhari: Yes.

Dr. Tapscott: Dr. Blau, you made the point that in female carriers of Duchenne, there are higher CPK levels and by the Lyon hypothesis, some X-chromosomes are inactivated and some are not. To determine what fraction of nuclei are expressing dystrophin and what fraction are not, it should be possible to look at the mixture of muscle fibers that survive in carriers and, by using markers for one or the other X-chromosome, to determine what percentage of normal nuclei are required for myofiber survival. Have you looked at this question?

Dr. Blau: Actually, there is some evidence bearing on this from my lab and from others. One thing that we know is that there appears to be no selection at the level of the myoblast since by looking at mosaicism within muscle for X-expression, there is no evidence for selection against myoblasts containing the defective X chromosome. Based on the work of Terry Partridge and others who have looked at heterozygous muscle, for example in mdx mice, what is seen is that with progressive age, there are fewer and fewer patches of dystrophin absence. It suggests to me that the defective nuclei are somehow are being lost from the muscle fibers. It would be of interest to know how this occurs. It also appears to be true of Duchenne carriers.

Dr. Fischman: I would like to comment in regard to Helen Blau's observations on myosin distribution. In our own laboratory we have evidence for extensive myosin exchange between thick myofilaments. Admittedly our experiments differ from those of Dr. Blau's but indicate that thick filaments are not inert structures and that myosin and other myofibrillar proteins exchange and probably diffuse widely in the sarcoplasm. I believe that Dr. Blau's results are not incompatible with Fick's Law of diffusion with a standing wave of myosin emanating from the region around a given nucleus. I would not expect to see a change in myosin distribution as a function of time. The only way to test Dr. Blau's model, which I believe to be of great interest, is to inactivate a single nucleus and see if the distribution of myosin changed.

Dr. Stockdale: Dr. Fardeau, as I understand your experiments when you inject the satellite cells into your preparation you get regeneration of fibers which run in the longitudinal axis of the muscle. Could you give us a quantitative statement about how often that occurs and how often fibers run from one end to the other of the muscle?

Dr. Fardeau: The fibers are longitudinally oriented in the serial sections and exhibit the characteristics of normally proliferated myogenic cells.

Dr. Kaufman: I would like to ask Richard Bischoff whether in his elegant satellite cell preparations if the first labeled nuclei appear as doublets after fusion, or as Helen Blau suggested yesterday, that one of the two nuclei migrates to a distal site or gives rise to a residual population of satellite cells for future regeneration? I ask this question because it seems that from your data there is quite a synchronous activation of the satellite cells and possibly a similar synchrony of fusion.

Dr. Bischoff: I did not have time to say much about fusion but the proliferation of satellite cells is very synchronous, in fact, we often see all of the satellite cells in mitosis at the same time. The proliferating satellite cells do not fuse with the myofibers but with other satellite cells. They appear to be blocked from fusion with the myofibers. The myofibers undergo dedifferentiative changes in culture so that within four to five days they develop pseudopodial outgrowths and the proliferating satellite cells will fuse with these pseudopodia. Or if you add embryonic myoblasts or dissociated satellite cells from adult rat muscle that had been cultured in monolayer for a while, such cells also will not fuse with the mature muscle fibers but will fuse with the outgrowing sprout from the fibers. These sprouts appear to lack a basal lamina which may be involved in the fusion blockade.

Dr. Moxley: To follow up on Dr. Alan Peterson's comments about the inferences we make in cultured cell systems as compared to in vivo systems, I've been wondering if anyone knows of in vivo models to document the movements of nuclei within muscles. I am specifically trying to address a question that relates in part to Dr. Blau's presentation and to others about the migration of injected satellite cells within or outside of muscle fascicles. The other concern would be if you were lucky enough to get a fairly high number of fusion events within a single fiber, what assurance do we have that if a fiber were undergoing regeneration in one spot, that nuclei might have the opportunity to move intracellularly? Is there something different about mature muscle that has sarcomeres at an adult stage as opposed to myotubes in culture where intracellular, longitudinal elements may be less well developed?

Dr. Rubin: I can give a partial answer to your question. During normal embryonic development it has been known for some time and described by Alan Kelly that nuclear movement occurs in vivo. Since sole plate nuclei are more or less invariant, one must presume that they must get there by some migratory process. In adult muscle there is at least indirect evidence which suggests that the sole plate nuclei do not move. I do not know if anyone has carefully examined the in vivo movements of myonuclei away from the endplates.

SECTION 5

DEVELOPMENT OF MYOGENIC CELL CULTURES

IMPROVED MEDIA FOR RAPID CLONAL GROWTH

OF NORMAL HUMAN SKELETAL MUSCLE SATELLITE CELLS

Richard G. Ham, Judy A. St. Clair and Sarah D. Meyer

Department of Molecular, Cellular and Developmental Biology
Campus Box 347, University of Colorado
Boulder, CO 80309

INTRODUCTION AND HISTORICAL PERSPECTIVE

Many investigators who use cell culture as a research tool fail to give adequate attention to culture media. Virtually all "standard" nutrient media were published more than 20 years ago, and many widely used formulations are at least 30 years old. Thus, the culture medium is often the most obsolete and neglected component of the entire experimental system. This presentation describes recent improvements in cell culture media for human muscle satellite cells (HMSC) and seeks to encourage their use.

Mononucleate cells from skeletal muscle (myoblasts and satellite cells) are relatively easy to grow in classical culture systems based on the use of serum and chicken embryo extract (CEE). Chicken embryo myoblasts were one of the first types of normal cells to be both grown and induced to undergo differentiation in culture (reviewed by Konigsberg, 1963). However, prior to the studies summarized here, no nutrient medium has ever been designed specifically to be optimal for muscle cells from any species.

A variety of standard nutrient media have been used over the last 30 years for growth and differentiation of muscle cells (reviewed in Ham et al., 1988). However, medium F10 (Ham, 1963), which was originally developed for clonal growth of Chinese hamster ovary cells, has been the most frequent choice for clonal growth of muscle cells. F10 supplemented with serum and CEE was used by Hauschka and Konigsberg (1966) for chicken embryo myoblasts, and by Hauschka (1974) for fetal human myoblasts. The same formulation with conditioning was used by Blau and Webster (1981) for direct primary cloning of HMSC on collagen-coated surfaces, and it has been rather widely accepted as a "standard" muscle cell medium.

Although F10 works well with adequate amounts of serum and CEE, it was published 26 years ago, and lacks modern features such as trace element supplementation that are needed for serum-free growth. Other media commonly used for HMSC are also generally as old as F10 or older. In 1986, we began with support from MDA to develop media designed specifically for growth, differentiation, and maturation of HMSC with minimal levels of highly defined supplementation. These studies are still in progress, but major improvements have already been achieved in clonal growth of HMSC, including direct primary cloning without conditioned medium and serum-free growth with semi-defined supplements (Ham et al., 1988 and unpublished).

Nutrient Medium MCDB 120

The first step in the development of improved media for HMSC was an improved nutrient formulation optimized specifically for growth of HMSC with minimal supplementation. To avoid possible artifacts due to selective growth of contaminant cell types in experimental media, we used clonal cultures of HMSC prepared as described by Blau and Webster (1981) in all growth-response assays. An initial survey of standard and specialized media revealed that MCDB 131, which had been developed for human microvascular endothelial cells (Knedler and Ham, 1987), supported the best clonal growth of HMSC with minimal amounts of dialyzed serum and chicken embryo extract (or bovine pituitary extract). Medium F10 was almost equally effective, verifying its status as the "standard" medium of choice for HMSC. However, because MCDB 131 already had modern features that are generally needed for serum-free growth, we selected it as the starting point for developing an optimized medium for clonal growth of HMSC.

A process of determining the optimum concentration for clonal growth of each individual component of the nutrient medium, together with some further studies on nutrient interactions, resulted in the development of a new medium designated MCDB 120, which has consistently yielded better clonal growth than F10 or MCDB 131 with both conventional and serum-free supplementation (Ham et al., 1988). With appropriate supplementation, MCDB 120 can also be used effectively in experiments on differentiation and maturation of HMSC. We now use MCDB 120 in all routine studies on human muscle cells.

Serum-free Supplement SF

In studies that overlapped the final stages of development of MCDB 120, we replaced serum and CEE, which have traditionally been used as undefined supplements for growth of muscle cells, with a mixture of five defined and semi-defined supplements designated SF (Table 1). Clonal growth of HMSC in MCDB 120 with SF is generally slightly better than with serum and CEE. However, it is important to note that some components of SF are not fully defined, as discussed below.

Table 1. Serum-free supplement SF

| Component | Amount Added[a] | | Percent Growth |
	μg/ml	M/L	Without Component[b]
Bovine serum albumin	500	7.2E-6	5
Dexamethasone	0.39	1.0E-6	0.1
Epidermal Growth Factor	0.010	1.7E-9	4
Fetuin (Pederson)	500	1.1E-5	0
Insulin (bovine)	180	3.0E-5	40

[a]Both weight and molar concentrations are given for ease of comparison with other formulations.

[b] Total colony area per dish is expressed as percentage of the value obtained with complete SF. This experiment was done in an intermediate medium before the composition of MCDB 120 was finalized, but is representative of results in MCDB 120. See Ham et al. (1988) for details.

Development of SF occurred in three phases. In the first, CEE was replaced with bovine pituitary extract, and then with fibroblast growth factor (FGF), which is needed for clonal growth of HMSC with serum as the only other supplement. In the second phase, which was done with added FGF, the five components of SF collectively replaced the need for serum. In the third phase, the components of SF plus FGF were omitted individually to determine if each was needed in the presence of all of the others. FGF was found to have no effect beyond that of SF, and was omitted from the final formulation.

The absence of a requirement for FGF is puzzling in view of the key role that has been proposed for FGF in control of growth and differentiation in muscle cells from other species (Hauschka, this volume; Florini, 1987). FGF is not stable in the serum-free medium, and, in retrospect, it now appears that in the absence of serum, rapid degradation of the added FGF occurred, such that in order to obtain growth without serum, it was also necessary to bypass the need for FGF. This interpretation is strongly supported by the observation that growth without FGF in media containing serum plus SF is far better than growth in media with serum plus FGF (see "Doubly-supplemented Medium", later in this report). However, when FGF is stabilized with heparin, it slightly improves growth over that obtained with SF alone, and also partially replaces the need to include EGF in the SF formulation.

Clonal growth of HMSC in MCDB 120 plus SF is generally better than that obtained with serum and CEE (or serum and FGF). However, SF does not support growth in F10, clearly demonstrating the more stringent demands placed on the nutrient medium by serum-free growth. All components of SF except insulin are strictly needed for serum-free clonal growth of HMSC, and growth without insulin is less than half of control values (Table 1).

The ability of EGF to serve as a major mitogen for HMSC is contrary to most observations with muscle cells from other species (for references, see Ham et al., 1988; Florini, 1987). However, in our serum-free system, mouse epidermal growth factor is active at physiological concentrations, and we have recently obtained equivalent growth with recombinant human EGF (Meyer et al., 1989). Thus, all available evidence strongly indicates that the growth-promoting activity of EGF for HMSC that is observed in MCDB 120 plus the other four components of SF is due to EGF itself, and not caused by a contaminant in EGF prepared from natural sources.

EGF is the only component of SF whose role can be considered reasonably well defined. Serious questions still remain about the stimulatory activities of each of the the other four components. These are summarized briefly below and discussed in greater detail in Ham et al. (1988).

The synthetic glucocorticoid dexamethasone is only partly replaced by the natural glucocorticoid hydrocortisone, even after compensating for differences in biological potency. This suggests that dexamethasone may have a second effect that is not mediated by glucocorticoid receptors.

The amount of insulin that is needed for optimal growth of HMSC is far too large to be explained physiologically, even if it is cross-reacting with IGF-I receptors. Also, not all preparations are equally active. Thus, it appears likely that the growth-promoting activity may be a contaminant.

Serum albumin is well-known as a carrier of bound small molecules, and requirements for albumin usually prove to be for substances carried on the albumin. However, commercially-prepared "fatty acid-free" albumins appear to retain full potency, and at present we have no data on the nature of the growth-promoting activity other than the observation that the activities of albumins from different sources vary rather widely (Ham et al., 1988).

The fetuin preparation that we use is a relatively crude fraction of fetal bovine serum. The literature is filled with reports of diverse growth-promoting activities in fetuin preparations, but at this time we have no firm data on the possible roles that it may be playing in serum-free growth of HMSC.

Thus, although MCDB 120 plus SF can legitimately be called a "serum-free" medium, this formulation is still far from being "chemically defined". Also, at least one additional factor from serum is needed to achieve a maximum rate of growth, as described below.

Doubly-supplemented Medium

During studies seeking to understand why FGF was needed with serum but not with SF, we added FGF to a medium that contained both serum and SF. FGF had little effect, but we unexpectedly observed that the doubly-supplemented control medium supported far better clonal growth of HMSC than could be obtained either with SF alone or with serum plus FGF.

We still do not fully understand the biochemical basis for the improved growth in doubly-supplemented medium. However, it appears to be due to the presence in serum of one or more additional growth factors for HMSC beyond those in SF. The minimum requirements for the enhanced clonal growth are at least 5% dialyzed or whole serum plus EGF, dexamethasone, and insulin at the concentrations in SF. Serum provides adequate amounts of fetuin and albumin or equivalent activities. Dexamethasone and EGF both are critically important for clonal growth of HMSC in the doubly-supplemented medium, and growth is also reduced without high insulin. In this case also, recombinant human EGF is as effective as natural mouse EGF.

Thus, while MCDB 120 plus SF supports growth at least as good as in MCDB 120 with serum and CEE, growth in MCDB 120 doubly-supplemented with both serum and SF far exceeds growth with either type of supplementation alone. The doubly-supplemented medium is so superior that we now routinely use it (usually with 15% fetal bovine serum) for all of our HMSC stock cultures, and as a positive control on all growth assays. In addition, we now do direct primary cloning in the doubly-supplemented medium without conditioning or collagen-coating, as described below.

Double supplementation of F10 with both serum and SF also improves growth over that obtained in F10 with serum and CEE or with serum and FGF. However, growth in MCDB 120 is always better than growth in F10 with any set of supplements that is used.

DIRECT PRIMARY CLONING IN DOUBLY-SUPPLEMENTED MEDIUM

MCDB 120 doubly-supplemented with SF plus 15% fetal bovine serum can be used for direct primary cloning of HMSC without conditioning the medium or using collagen-coated culture dishes, both of which were needed with F10 plus serum and CEE (Blau and Webster, 1981). Satellite cells are released from biopsy discard tissue by trypsin-EDTA digestion, diluted in doubly-supplemented MCDB 120 to 1-5 clonable HMSC per ml (estimated from prior experience with similar donors). Uncoated 96 well tissue culture plates are then inoculated with 0.2 ml per well. Under these conditions the cells attach readily and grow well. The first division of satellite cells from young donors normally occurs within 3 days, and the clonal cultures are ready for transfer directly to 60 mm tissue culture petri dishes at 11-12 days. By about 20 days, the 60 mm dishes are approaching confluency and ready for harvesting and frozen storage of individual clones.

The number of clones obtained per milligram of tissue is roughly equivalent to that obtained with collagen-coating and conditioned medium. In addition, loss of proliferating cells by terminal differentiation is minimized in the doubly-supplemented medium, as discussed below. Clones containing approximately one million cells each are routinely ready to harvest after about 20 days. This corresponds to an average growth rate of about one doubling per day, which becomes even faster if a correction is made for the initial lag period of up to 3 days. Under optimal conditions, measured growth rates can be very rapid. We have obtained doubling times as short as 13-14 hours for clonal cultures seeded at relatively low densities in doubly-supplemented MCDB 120 at population doubling levels around 25-30 (after frozen storage and thawing).

The results described above are for cultures established from biopsy discard tissue obtained from young donors (around five years of age) whose muscle tissue has been found not to be diseased. Growth of cells from older donors is not as rapid, and total population doubling potential is reduced, often to around 30 doublings. We can readily obtain clonal cultures from adult (and even elderly) donors and grow them up sufficiently for frozen storage at about 20 population doublings. However, by the time the cells are thawed and placed back into active culture, they are beginning to show early signs of cellular senescence, and are no longer fully satisfactory for use in clonal growth assays.

We have also been able to initiate cultures from autopsy tissues with these methods. Muscle samples obtained by autopsy within 24 hours after death appear to be equivalent to biopsy specimens as sources of muscle satellite cells for culturing. We have also observed that biopsy or autopsy tissue fragments can be stored at refrigerator temperature for up to five days in cell culture saline solutions without any major loss of viability or colony-forming efficiency of the satellite cells.

CULTURES FROM VERY YOUNG DUCHENNE MUSCULAR DYSTROPHY PATIENTS

Direct primary cloning of HMSC from DMD patients in doubly-supplemented MCDB 120 without conditioning or collagen-coating also works very well. In two of the three clonings described below, direct comparisons verified that collagen-coating and conditioning provided no benefit over direct plating in unconditioned medium without collagen-coating. The small amount of tissue available from the third biopsy specimen precluded another comparison, but growth was excellent, as described below for the 5 month old donor.

Blau et al. (1983) reported that clonal cultures established from Duchenne muscular dystrophy (DMD) patients showed greatly reduced doubling potential and behaved like cultures from much older normal donors. Our experience with cultures from a DMD patient biopsied at 3 years of age was similar. The number of clonable satellite cells was low, and the clones that were obtained showed greatly reduced proliferative potential. However, we have also had the opportunity to establish cultures from two DMD patients who were biopsied unusually young at 7 months and 5 months because they had older brothers with DMD. In both cases, the number of clonable satellite cells was in the range expected for normal young muscle, and the clones that were obtained grew extremely well, and also differentiated well when switched to appropriate media. We have not completed detailed studies of growth rates and population doubling potential on these cultures. However, our preliminary data strongly support the theory that DMD patients have already exhausted most of the proliferative potential of their muscle satellite cells by the time that the major overt symptoms of the disease begin to be manifested.

The availability of clonal cultures of human DMD muscle satellite cells that multiply rapidly and undergo a reasonably large number of doublings will open many new opportunities for detailed in vitro studies on the pathology of DMD. We have not initiated cultures from fetal muscle. However, doubly-supplemented MCDB 120 worked well for direct cloning of autopsy tissue from a premature infant, and fetal muscle cultures (kindly provided by Dr. Helen Blau) grow well in it. Thus, we expect that cultures initiated from muscles of aborted DMD fetuses will also grow well in this medium.

DIFFERENTIATION

By performing all growth assays with clonal cultures of HMSC, we were able to focus entirely on obtaining optimal growth of muscle cells without concern about growth of fibroblasts or other contaminating cell types. Because cells are removed from the proliferative pool as they enter terminal differentiation, the process of optimizing growth also resulted in an unusually low level of differentiation. The net result was that the SF- and doubly-supplemented media supported dense growth with very little fusion. HMSC grown to confluency in these media exhibit an elongated fibroblast-like morphology, with the cells tightly packed together in parallel arrays, often with swirling patterns similar to those seen in cultures of normal fibroblasts. The only time that significant fusion is seen is when cultures are kept in the same medium with no replenishment for a long enough period to deplete differentiation-inhibiting activity.

When transferred to either Dulbecco's modified Eagle's medium (DME) or MCDB 120 with low insulin (10 µg/ml), dense cultures of HMSC undergo rapid fusion and biochemical differentiation. Serum-free supplement SF greatly reduces the amount of differentiation observed visually and measured by creatine kinase specific activity. Deletion of various components of SF has shown that the inhibitory activity is associated almost exclusively with the combination of fetuin, dexamethasone, and high insulin. Surprisingly, EGF and serum albumin, which are both needed for serum-free growth of HMSC, do not significantly inhibit their fusion or biochemical differentiation.

Furthermore, addition of EGF and albumin to the differentiation medium causes the differentiated cultures to be maintained in a healthier state for longer periods of time with substantially higher total protein and total creatine kinase levels and no significant change in creatine kinase specific activity. We now routinely do all of our differentiation experiments in MCDB 120 plus 10 µg/ml insulin, 10 ng/ml EGF, and 0.5 mg/ml bovine serum albumin. Preliminary data suggest that by keeping the myotubes alive and healthy longer, this formulation may also promote partial maturation (Ham and St. Clair, 1989). We have also obtained excellent differentiation of cultures from very young DMD patients under these same conditions.

Our current data reinforce questions being raised by others about the supposed dichotomy between growth and fusion and the role of growth factors as regulatory signals for muscle differentiation (Hauschka, this volume; Allen and Boxhorn, 1989). Crowded HMSC stop multiplying in our growth media without fusion. Whether or not they fuse and differentiate is determined by fetuin, dexamethasone, and high insulin, which have little or no effect on cellular proliferation in crowded cultures. Also, EGF and serum albumin, which are required for growth of HMSC in our serum-free medium, do not block their differentiation. Thus, there is far more to control over growth and differentiation of HMSC than a simple switch from one state to the other.

GROWTH OF CELLS FOR MYOBLAST TRANSFER THERAPY

Rapid growth of HMSC in doubly-supplemented MCDB 120 and the virtual elimination of cell loss due to terminal differentiation make this medium well suited for growing up large populations of cells for transplantation. HMSC from young donors can be expanded up to one million-fold within three weeks, with substantial proliferative potential left for further expansion within host muscles. The ability to initiate cultures from autopsy material also greatly expands available sources of cells for culturing.

Our current serum-free medium is not yet fully defined and growth rate and final cell density are both less than in the doubly-supplemented medium. We are working toward a fully defined medium for rapid growth, but for now, potential problems caused by serum are probably best dealt with by washing, by use of human serum, or by a final passage in serum-free medium.

Direct primary cloning provides a means for obtaining pure cultures of HMSC, thus eliminating other cell types that might increase the risk of rejection. However, pure cultures can probably be obtained more easily by using fluorescence activated cell sorting to isolate HMSC (Webster et al., 1988), followed by inoculation at low density into doubly-supplemented MCDB 120 for *in vitro* expansion of cell number.

ACKNOWLEDGMENTS

We thank Dr. Helen Blau and Dr. Cecelia Webster for their generous assistance and support during early phases of these studies. This research was supported by a grant from the Muscular Dystrophy Association.

REFERENCES

Allen, R. E., and Boxhorn, L. K., 1989, Regulation of skeletal muscle satellite cell proliferation and differentiation by transforming growth factor-beta, insulin-like growth factor I, and fibroblast growth factor. J. Cell. Physiol. 138:311.

Blau, H. M., and Webster, C., 1981, Isolation and characterization of human muscle cells. Proc. Natl. Acad. Sci. USA 78:5623.

Blau, H. M., Webster, C., and Pavlath, G. K., 1983, Defective myoblasts identified in Duchenne muscular dystrophy. Proc. Natl. Acad. Sci USA 80:4856.

Florini, J. R., 1987, Hormonal control of muscle growth. Musc. Nerve 10:577.

Ham, R. G., 1963, An improved nutrient solution for diploid Chinese hamster and human cell lines. Exp. Cell Res. 29:515.

Ham, R.G., and St. Clair, J.A., 1989, Differentiation and partial maturation of human skeletal muscle cells in serum-free culture. In Vitro Cell. Dev. Biol. 25:44A.

Ham, R. G., St. Clair, J. A., Webster, C., and Blau, H. M., 1988, Improved media for normal human muscle satellite cells: serum-free clonal growth and enhanced growth with low serum. In Vitro Cell. Dev. Biol. 24:833.

Hauschka, S. D., 1974, Clonal analysis of vertebrate myogenesis. II. Environmental influences on human muscle differentiation. Dev. Biol. 37:329.

Hauschka, S. D., and Konigsberg, I. R., 1966, The influence of collagen on the development of muscle clones. Proc. Natl. Acad. Sci. USA 55:119.

Knedler, A., and Ham, R. G., 1987, Optimized medium for clonal growth of human microvascular endothelial cells with minimal serum. In Vitro Cell. Dev. Biol. 23:481.

Konigsberg, I. R., 1963, Clonal analysis of myogenesis. Science 140:1273.

Meyer, S.D., St. Clair, J.A., and Ham, R.G., 1989, EGF is a mitogen for human skeletal muscle satellite cells (HMSC) but does not block their differentiation. J. Cell Biol. 109:305a.

Webster, C., Pavlath, G. K., Parks, D. R., Walsh, F. S., and Blau, H. M., 1988, Isolation of human myoblasts with the fluorescence-activated cell sorter. Exp. Cell. Res. 174:252.

RETROVIRAL LINEAGE MARKERS FOR ASSESSING MYOBLAST FATE IN VIVO

Helen M. Blau and Simon M. Hughes

Department of Pharmacology
Stanford University School of Medicine
Stanford, CA 94305

Using a retroviral marking system, we are monitoring normal muscle development in vivo. In contrast to other reports in this volume in which the fate of myoblasts that were injected into animals was followed, we have followed the fate of the animal's own myoblasts. Specifically, the question we are asking is: Do satellite cells migrate from one fiber to another? This, of course, is an important consideration in myoblast therapy, since the hope is that the injected myoblasts will be able to migrate and participate in myofiber formation at a distance.

Fully developed muscle fibers are separated from one another by a basal lamina composed of extracellular matrix components. We examined whether at this point in development, when fiber formation is complete and each fiber with its satellite myoblasts is ensheathed in its own basal lamina, myoblasts can participate in the growth of multiple fibers. This would require that myoblasts penetrate and cross basal lamina, a characteristic known to be shared only by tumor cells and lymphocytes.[1-3]

In these studies, retroviral constructs were injected into the hind limb of rats, the muscles serially sectioned, and assayed histochemically for β-galactosidase activity. For this purpose, we used defective retroviral constructs that following infection are only propagated to progeny cells by cell division.[4] The presence of these constructs was monitored by the expression of β-galactosidase which was targeted either to the nucleus or cytoplasm. An example of a single muscle fiber labeled in vivo with the cytoplasmic β-galactosidase construct is shown in Figure 1A and a clone of two labeled fibers in Figure 1B. In Figure 2 is an example of the nuclear β-galactosidase construct labeling multiple nuclei in a single fiber. If infected myoblasts divided and fused only with the muscle fiber to which they were closely juxtaposed, single isolated β-galactosidase labeled fibers would be expected. Instead, multiple labeled fibers in close proximity to one another were most frequently observed in these sections. Such clusters were well separated. As discussed in detail elsewhere,[5] the majority of clusters resulted from a single infection event. Briefly, when the two retroviral vectors encoding nuclear and cytoplasmic β-galactosidase were mixed and injected together, the frequency with which a cluster contained both markers was low. Thus, the majority of labeled clusters of multiple fibers occur by an infected myoblast dividing, and its progeny migrating across the basal lamina.

What can we conclude from these studies? First, myoblasts in normal development, in the absence of denervation, regeneration and degeneration or any perturbation of the animal's muscles (genetic or inflicted), are capable of crossing the basal lamina sheath and contributing to the growth of normal muscle. Second, a dual vector paradigm using two types of retrovirus targeted to different cellular compartments provides a useful method for assessing clonality and should have general application in studies using the retroviral marking system. Third, the retroviral vectors should have broad application in following the progeny of single myoblasts and their contribution to muscle development in vivo.

Why is this of interest to myoblast transplantation and cell therapy? First, if myoblasts can migrate across the basal lamina in the course of normal development in the intact muscle

Myoblast Transfer Therapy
Edited by R. Griggs and G. Karpati
Plenum Press, New York, 1990

in vivo, myoblast movement should be enhanced in degenerating muscle of mdx mice or Duchenne patients. Thus, novelly introduced myoblasts are likely to be able to contribute to the production of multiple myofibers. Moreover, that myoblasts migrate is of broad biological interest, because there are very few cell types that are capable of crossing the basal lamina. In this respect, it will be of interest to determine the relevant features that lymphocytes, tumor cells, and myoblasts have in common.

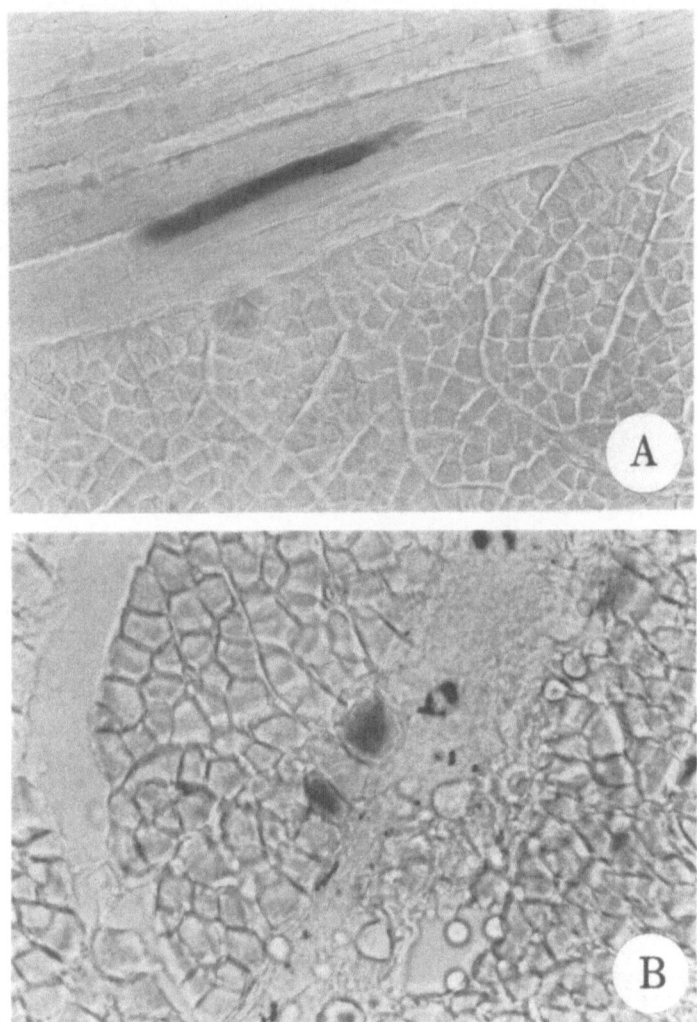

Figure 1. Retroviral marking of muscle fibers in vivo using a cytoplasmic β-galactosidase construct.

BAG, a retroviral vector which expresses β-galactosidase targeted to the cytoplasm, was injected into rat hindlimb muscle and analyzed histochemically on tissue sections. Labelled muscle fibers occurred either alone (A, longitudinal section) or in small clusters (B, transverse section).

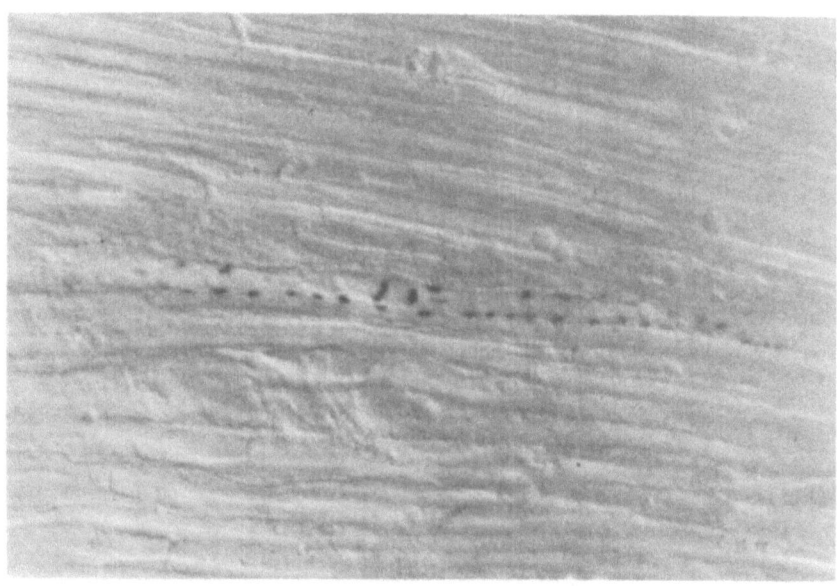

Figure 2. Retroviral marking of rat muscle nuclei in vivo.

A retroviral vector which specifically targets β-galactosidase expression to the nucleus was injected into rat hindlimb muscle in order to infect muscle cells in vivo. Histochemical staining of transverse sections for β-galactosidase indicates many nuclei are labeled within a single fiber. Since retroviruses infect only dividing cells, the expression of the β-galactosidase results from fusion of infected mitotic myoblasts or their progeny. Within a multinucleated muscle cell, β-galactosidase can diffuse and label nuclei in addition to those that encode it.

References

1. Anderson, A.O. and N.D. Anderson, Lymphocyte emigration from high endothelial venues in rat lymph nodes, Immunology 31:731 (1977).
2. Nakajima, M., T. Irimura, F.D. Ni, F.N. Di, and G.L. Nicolson, Heparan sulfate degradation: Relation to tumor invasive and metastatic properties of mouse B16 melanoma sublines, Science, 220:611 (1983).
3. Warfel, K.A. and M.T. Hull, Migration of lymphocytes through the cutaneous basal lamina in normal skin: an ultra structural study, Anat. Rec., 208: 349 (1984).
4. Varmus, H.E., T. Padgett, S. Heasley, G. Simon and J.M. Bishop, Cellular functions are required for the synthesis and integration of avian sarcoma, virus-specific DNA. Cell, 11:307 (1977).
5. Hughes, S.M. and H.M. Blau, Myoblasts migrate across basal lamina during skeletal muscle development, submitted to Nature (1989).

Discussion of Dr. Blau's paper

Unidentified Participant: We've had the opportunity to do two much younger patients. In addition to seeing much more proliferation capacity, we also saw a much higher fraction of satellite cells. The number of cells that we obtained from five-month-old and eight-month old patients approached that from normal tissue. This could mean that if we can get the cells early enough, there may be some hope of engineering.

MYOGENIC CONVERSION OF HUMAN NON-MUSCLE CELLS FOR THE

DIAGNOSIS AND THERAPY OF NEUROMUSCULAR DISEASES

A.F. Miranda, T. Mongini[*], E. Bonilla,
A.D. Miller[**] and W.E. Wright[+]

Columbia University, College of Physicians
and Surgeons, New York, NY
[*]University of Turin, Clinical Neurologica II
Turin, Italy
[**]Fred Hutchinson Cancer Research Center
Seattle, WA
[+]University of Texas Southwestern Medical School
Dallas, TX

INTRODUCTION

Control of tissue-specific gene expression in eukaryotes represents one of the central unanswered questions in developmental biology. Davis et al (1) isolated a muscle regulatory cDNA called MyoD from mouse cells: when transfected into fibroblasts or adipocytes, MyoD converts these cells to myoblasts, which, under appropriate conditions, can fuse and differentiate. Later, several additional genes, myd (2), myogenin (3) and Myf-5 (4), also capable of stably converting non-muscle cells to myoblasts were isolated. The availability of these probes will lead to a better understanding of the ordered regulation of gene expression in muscle and other differentiating tissues. These factors that can produce myogenic conversion may also have practical applications in the diagnosis or even the treatment of hereditary human myopathies, such as Duchenne Muscular dystrophy (DMD). This will be the theme of this presentation.

DIAGNOSIS OF DUCHENNE MUSCULAR DYSTROPHY (DMD)

Deletions, or, more rarely, duplications of the DMD gene have been identified in about 65% of patients: DMD can be diagnosed in these cases by Southern analysis, using DNA probes (5,6,7, 8 for review). An easier and more rapid method to recognize some DMD patients is based on the polymerase chain reaction (PCR) (9). Some DMD patients can also be identified by haplotyping, but this is not always feasible, because it requires tissues or cells from informative family members.

Immunocytochemistry and Western analysis of muscle biopsies using anti-dystrophin antibodies provides an alternative diagnostic procedure, showing lack of dystrophin in most or perhaps all DMD muscle biopsies (7,10,11,12). It has also been shown that dystrophin is expressed in normal cultured myotubes, but is lacking in myotubes from patients with DMD (13,14,15). While DNA analysis has been used to identify DMD mutations antenatally, dystrophin analysis is not useful in this situation because dystrophin is undetectable by immunocytochemistry in amniocytes, blood cells or chorionic villi. We, therefore, transfected 10T1/2 mouse cells, human fibroblasts, and human amniocytes with MyoD and Myogenin by Ca-phosphate precipitation (16), to see whether dystrophin would be expressed in the converted muscle cells. Using immunocytochemistry with anti-dystrophin antibodies produced by Hoffman et al (17), we found that dystrophin was, in fact expressed by differentiating convertants, together with myosin heavy chain (MHC). Although mouse cells could be transformed with great efficiency, only few converted human cells were observed. This is not surprising because, human primary cells are notoriously difficult to transform using Ca-phosphate techniques (18). However, using a MyoD-retrovirus engineered by Weintraub et al (19) we were able to generate stable myogenic cells from normal human amniocytes and fibroblasts: many of the infected cells fused, differentiated and expressed muscle-specific markers, such as creatine kinase, MHC and dystrophin (20). We are now trying to establish if dystrophin immunocytochemistry of MyoD-converted amniocytes and fibroblasts can be applied to the prenatal and postnatal identification of DMD. In addition, studies of clonal MyoD-converted fibroblasts may also be useful for detection of DMD carriers as previously demonstrated in clonal muscle cultures (21,22).

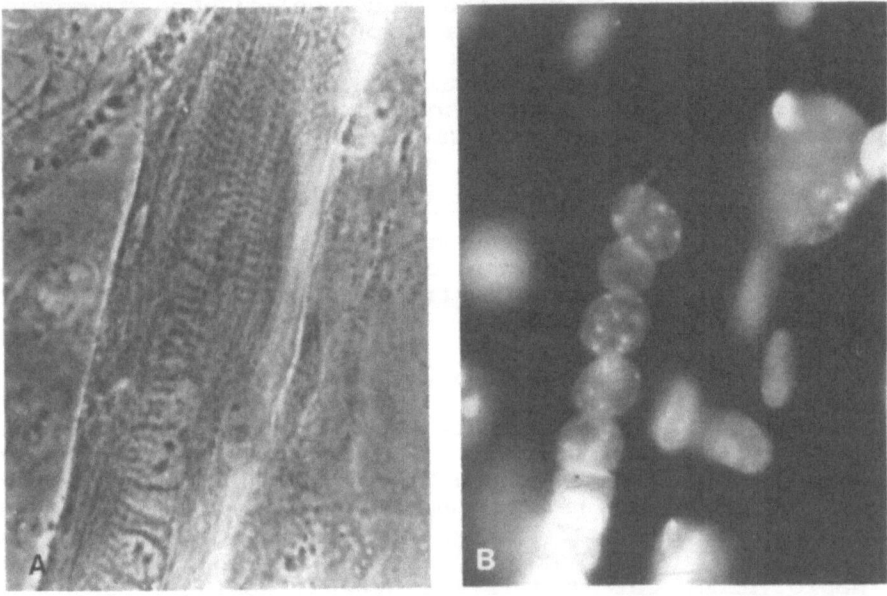

Fig. 1. (a) Phase micrograph of a hybrid human-mouse myotube, derived from MyoD-converted 10T½ myoblasts and native human myoblasts in a 10 day-old culture, showing prominent cross-striations.(b)The same myotube stained with the fluorescent dye Hoechst 33258, showing speckled mouse nuclei and uniformly stained human nuclei.

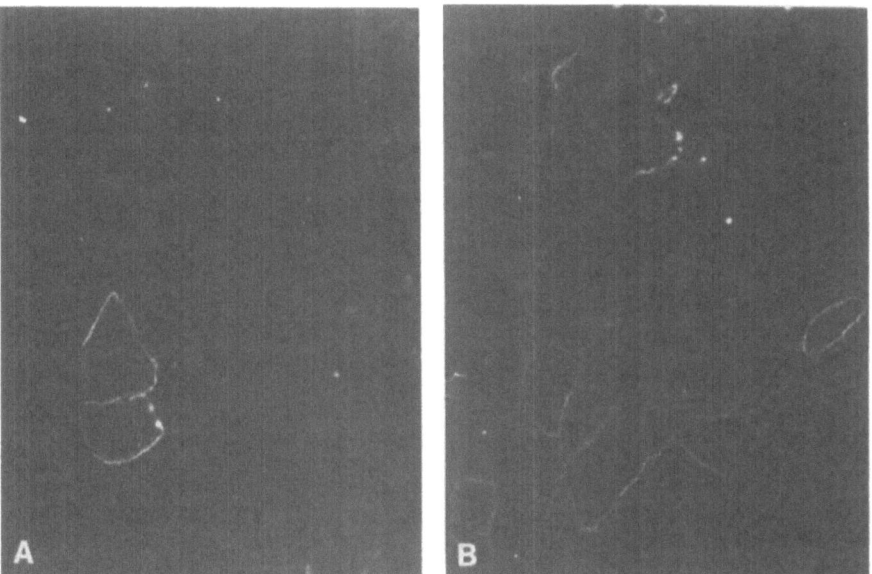

Fig. 2. Fluorescence immunocytochemistry on cryostat cross-sections of mdx mouse muscle stained with antidystrophin antibody: (a)Two muscle fibers show partial dystrophin staining at the surface, 9 days after injection with native untreated 10T½ cells. (b) Several muscle fibers show uniform or partial staining of the surfaces, 9 days after injection with MyoD-converted 10T½ cells.

MYOD-CONVERTED CELLS IN MYOBLAST TRANSFER THERAPY

Even if all the many hurdles in myoblast transfer therapy can be solved, a major obstacle is obtaining sufficient cells for injection. One important source would be human myoblasts isolated by electronic sorting (23). Since primary fibroblasts can be propagated easily, MyoD-converted fibroblasts represent another potential source of cells for injection.

We transformed mouse 10T1/2 cells with the MyoD-retrovirus and, after neomycin selection in medium containing G418, we co-cultured the transformants with cloned, native human myoblasts. After two days, the mixed cultures were fed with mitogen-poor medium. We found that most myotubes contained both 10T1/2 mouse nuclei and human nuclei (Fig. 1), indicating that MyoD-transformed cells had fused efficiently with native myoblasts, even across species lines. At nine days, the hybrid myotubes developed cross-striations and sometimes pulsated spontaneously.

It has already been shown that mouse or human myoblasts injected intramuscularly into dystrophin-deficient mdx mice produce hybrid muscle fibers that synthesize dystrophin (24,25). We have now observed that a single injection of MyoD-converted 10T1/2 cells in the anterior tibial muscle of the mdx mouse produces scattered dystrophin-positive fibers throughout the entire muscle (Fig. 2). Surprisingly, however, the contralateral anterior tibial muscle injected with non-transformed 10T1/2 cells also showed isolated muscle fibers expressing dystrophin at the sarcolemma (Fig. 2). This phenomenon could be explained in at least four different ways: (1) fusion of native mouse 10T1/2 cells with mdx muscle cells; (2) "spontaneous" myogenic differentiation of 10T1/2 stem cells, induced by their location in an in vivo muscle environment; (3) re-expression of a "fetal" isodystrophin of mdx; (4) fusion of MyoD-converted 10T1/2 myoblasts carried through the bloodstream.

We are now using specific nuclear markers to determine the efficiency of hybrid myotube formation in injected muscles. We are also determining whether some of the injected 10T1/2 myoblasts remain unfused (in satellite cell position) as a source of stem cells for future muscle regeneration.

MyoD-converted cells may also be used to treat human hereditary metabolic myopathies due to mutations of muscle-specific enzymes, such as phosphorylase and phospho-fructokinase (26 for review). In addition, once we can use genetic engineering to introduce normal functional genes in primary cells from patients with hereditary myopathies, then MyoD-converted fibroblasts from the same patient could be used in myoblast transfer therapy. This would be particularly useful in muscular dystrophy because native DMD myoblasts often senesce prematurely, while DMD fibroblasts seem to be unaffected (27).

REFERENCES

1. R. L. Davis, H. Weintraub, and A. B. Lassar, Expression of a single transfected cDNA converts fibroblasts to myoblasts, Cell 51:987 (1987).

2. D. F. Pinney, S. H. Pearson-White, S. F. Konieczny, K. E. Latham, and C. P. Emerson Jr., Myogenic lineage determination and differentiation: Evidence for a regulatory gene pathway, Cell 53:781 (1988).

3. W. E. Wright, D. A. Sassoon, and V.K. Lin, Myogenin, a factor regulating myogenesis, has a domain homologous to MyoD, Cell 56:607 (1989).

4. T. Braun, G. Buschhausen-Denker, E. Bober, E. Tannich and H. H. Arnold, A novel human muscle factor related to but distinct from MyoD1 induces myogenic conversion in 10T½ fibroblasts, EMBO J. 8:701 (1989).

5. B. T. Darras, P. Blattner, J. F. Harper, A. J. Spiro, S. Alter, and U. Francke, Intragenic deletions in 21 Duchenne muscular dystrophy (DMD)/Becker muscular dystrophy (BMD) families studied with the dystrophin cDNA: location of breakpoints on HindIII and BglII exon-containing fragment maps, meiotic and mitotic origin of the mutations, Am. J. Hum. Genet. 43:620 (1988).

6. B. T. Darras, M. Koenig, L.M. Kunkel, and U. Francke, Direct method for prenatal diagnosis and carrier detection in Duchenne-Becker muscular dystrophy using the entire dystrophin cDNA, Am. J. Med. Genet. 29:713 (1988).

7. E. P. Hoffman and L. M. Kunkel, Dystrophin abnormalities in Duchenne/ Becker muscular dystrophy, Neuron 2:1019 (1989).

8. S. M. Forrest, G. S. Cross, A. Speer, D. Gardner-Medwin, J. Burn, and K.E. Davies, Preferential deletion of exons in Duchenne and Becker muscular dystrophies, Nature 329:638 (1987).

9. J. S. Chamberlain, R. A. Gibbs, J. E. Ranier, P. N. Nguyen, and C. T. Caskey, Deletion screening of the Duchenne muscular dystrophy locus via multiplex DNA amplification, Nucleic Acid Res. 16:11141 (1988).

10. E. Bonilla, G. Samitt, A. F. Miranda, A. P. Hays, G. Salviati, S. DiMauro, L. M. Kunkel, E. P. Hoffman, and L.P. Rowland, Duchenne muscular dystrophy: deficiency of dystrophin at the muscle cell surface, Cell 54:447 (1988).

11. H. Sugita, K. Arahata, T. Ishiguro, Y. Suhara, T. Tsukahara, S. Ishiura, C. Eguchi, I. Nonaka, and E. Ozawa, Negative immunostaining of Duchenne muscular dystrophy and mdx muscle surface membrane with antibody against synthetic peptide fragment predicted from DMD cDNA, Proc. Jap. Acad. (B) 64:210 (1988).

12. E. E. Zubrzycka-Gaarn, D. E. Bulman, G. Karpati, A. H. M. Burghes, B. Belfall, H. J. Klamut, J. Talbot, R. S. Hodges, P. N. Ray, and R. G. Worton, The Duchenne muscular dystrophy gene product is localized in the sarcolemma of human skeletal muscle, Nature 333:466 (1988).

13. M. S. Ecob-Prince, M. A. Hill, and A. E. Brown, Localization of dystrophin in cultures of human muscles, Muscle and Nerve 12: 594 (1989).

14. A. A. Lev, C. C. Feener, L. M. Kunkel, and R. H. Brown, Expression of the Duchenne's muscular dystrophy gene in cultured muscle cells, J. Biol. Chem. 262:15817 (1987).

15. A. F Miranda, E. Bonilla, G. Martucci, C. T. Moraes, A. P. Hays, and S. DiMauro, Immunocytochemical study of dystrophin in muscle cultures from patients with Duchenne muscular dystrophy and unaffected control patients, Am J. Pathol. 132:410 (1988).

16. M. Wigler, A. Pellicer, S. Silverstein, R. Axel, G. Urlaub and L. Chasin, DNA-mediated transfer of the adenine phosphoribosyltransferase locus into mammalian cells, Proc. Natl. Acad. Sci. USA 76:1373 (1979).

17. E. P. Hoffman, R. H. Brown, and L. M. Kunkel, Dystrophin: The protein product of the Duchenne muscular dystrophy locus, Cell 51:919 (1987a).

18. T. D. Palmer, R. A. Hock, W. R. A. Osborne, and A. D. Miller, Efficient retrovirus-mediated transfer and expression of a human adenosine deaminase gene in diploid skin fibroblasts from an adenosine deaminase-deficient human, Proc. Natl. Acad. Sci. USA 84: 1055 (1987).

19. H. Weintraub, S. J. Tapscott, R. L. Davis, M. J. Thayer, M. A. Adam, A. B. Lassar, and A. D. Miller, Activation of muscle-specific genes in pigment, nerve, fat, liver and fibroblast cell lines by forced expression of MyoD, Proc. Natl. Acad. Sci. USA. 86:5434 (1989).

20. T. Mongini, A. F. Miranda, E. Bonilla, A. D. Miller, W. E. Wright, and H. Weintraub, Manuscript in preparation.

21. O. Hurko, E. P. Hoffman, L. McKee, D. R. Johns, and L. M. Kunkel, Dystrophin analysis in clonal myoblasts derived from a Duchenne muscular dystrophy carrier, Am. J. Hum. Genet. 44:820 (1989).

22. A. F. Miranda, U. Francke, E. Bonilla, G. Martucci, B. Schmidt, G. Salviati, and M. Rubin, Dystrophin immunocytochemistry in muscle culture: detection of a carrier of Duchenne muscular dystrophy, Am. J. Med. Genet. 32:268 (1989).

23. C. Webster, G. K. Pavlath, D. R. Parks, F. S. Walsh, and H. M. Blau, Isolation of human myoblasts with the fluorescence-activated cell sorter, Expl. Cell Res. 174:252 (1988).

24. G. Karpati, Y. Pouliot, E. Zubrzycka-Gaarn, S. Carpenter, P. N. Ray, R. G. Worton, and P. Holland, Dystrophin is expressed in mdx skeletal muscle fibers after normal myoblast implantation, Am. J. Pathol. 135:27 (1989).

25. T. A. Partridge, J. E. Morgan, G. R. Coulton, E. P. Hoffman and L. M. Kunkel, Conversion of mdx myofibres from dystrophin-negative to -positive by injection of normal myoblasts, Nature 337:176 (1989).

26. A. F. Miranda and T. Mongini, Diseased muscle in tissue culture, in: "Myology", A. G. Engel and B. Q. Banker, eds., McGraw Hill Book Comp., New York (1986).

27. H. M. Blau, C. Webster, and G. K. Pavlath, Defective myoblasts identified in Duchenne Muscular Dystrophy, Proc. Natl. Acad. Sci. USA. 80:4856 (1983).

GENERAL DISCUSSION

DEVELOPMENT OF MYOGENIC CELL CULTURES

Dr. Strohman: Dr. Ham, the most obvious question to me is: what is the significant advantage of using "super" medium as opposed to 10% or 15% fetal calf serum? It looks to me, at least the later growth kinetics, that Drs. Blau and Holland were getting almost as good growth. But maybe you can correct me on that.

Dr. Ham: I think it's true; once you've got a good monolayer of culture, they probably will grow almost as well. The big advantage of "super" medium is that it works equally well, right down to the clonal level. The doubly supplemented medium, I think, would permit, for example, taking fluorescence-sorted cells, and plating them at a very low density, maybe 100 cells in a culture flask, and having them take right off and grow. In addition, perhaps, big roller bottles with a few hundred or a few thousand FACS-isolated cells could provide a tremendous yield very rapidly.

I think once they're really going well, and are able to do their own medium conditioning, it does not make that much difference because Dr. Blau has gotten essentially the same results with conditioned medium.

Dr. Strohman: The "super" medium eliminates that lag period that both Holland and Blau were seeing. Is that true?

Dr. Ham: Dr. Blau might want to address that. I don't think we have any real advantages over her conditioned medium system, other than the fact that we don't have to make conditioned medium.

Dr. Wright: I wish to ask two questions. Dr. Ham introduced the use of 1% oxygen for cloning of human fibroblasts a while back, and it has been our experience that makes a big difference with cloning human myoblasts as well. I was wondering if he uses this system, or would he and Holland have any experience supporting our experience with that? Second question: Has anyone tried to apply the magnetic-activated cell sorting type of approach? Would it be a more convenient alternative to FACS sorting or primary cultures?

Dr. Ham: I will take on the low-oxygen question. Other than the fact that our laboratory is a mile-high altitude, we have not done any special low oxygen experiments. We automatically have reduced oxygen, and I think it does help. A long time ago, Wally McKeon did some experiments on fibroblasts, and even at Boulder's altitude, he could get some additional benefit by further reducing oxygen. But it was not enough to really work on. If you have a good medium with selenium, proper reducing agents, and so on, I do not think you need it.

Dr. Blau: I will speak to the second question. In our laboratory we are looking into the magnetic method for sorting, because it may be a much more readily applicable method for people who do not have the fluorescence-activated cell sorter available.

Dr. Wright: What are you using? MCBD-104 and 202 commercial preparations? I cannot attest that even in those media, the 1% oxygen at more or less sea-level makes a big difference in the growth, clonal growth of human myoblasts.

Dr. Stockdale: It is my understanding what Dr. Strohman is saying is that the "super" medium decreases cell cycle time, or does it make more cell cycle, or does it in any way lengthen the number of doublings that a particular cell can go through? Which, or all of those things?

Dr. Ham: We have not really analyzed these points properly. I think, first of all, it reduces the lag in the very early stages of getting clones started. Secondly, it does very strongly depress differentiation, so that essentially all of the cells stay in the cycle. I am not sure whether we get any real increase in the number of cycles they will do. For the normal cells, our experience seems to roughly match Dr. Blau's data, although we have not done those detailed studies. I think for individuals, say, in their twenties, thirties or forties, and even one that we cultured at age seventy, we are getting enough doublings to get up to a population doubling 20 to 25, freeze them and thaw those out and culture them a few more times. I think we are probably getting 30 or 35 doublings, and I think it roughly matches what Dr. Blau showed.

Dr. Stockdale: But you must be getting a change at cell cycle time, if you get these gigantic colonies relative to time. I mean, when you say these are great, big, all of them, you are really saying that the cell cycle time is shorter in those colonies than it would otherwise be.

Dr. Ham: I am sure the average cell cycle time is shorter because they are growing better. They are bigger colonies if you look at them early, not just the final stage. With established colonies, however, I do not think our times are better than what Dr. Blau had with conditioned medium, and I do not think they are better than most people get with a monolayer type culture. There is enough conditioning there.

Dr. Blau: I can see two real advantages to Dr. Ham's medium. One is getting around the conditioned medium. The other would be, if it's possible to grow the cells without serum, that would be advantageous in terms of myoblast transplantation. What I would like to ask you is, to what extent do you think is feasible for generating a large number of cells? Your "super" medium has serum in it. How good is the medium without the serum?

Dr. Ham: We can get most of the same results in the medium that we call serum-free, except that it is slower. We have not done the really long-term studies, to make sure we do not have reduced population doubling. I would, however, point out that there is a lot of fetuin and a lot of albumin, which collectively, must add up to about 90% of the protein of fetal bovine serum. So after another year or two with the new program we have going, we shall know what those components are doing, and keep only the active components.

Dr. Hauschka: I think we need to clarify something here, so people don't go out with a misconception. First, Dr. Ham and I have been in communication about his medium for a couple of years, and he kindly provided us with techniques for making it several years ago. So, in the experiments that I showed you, comparing the FGF-dependent, and independent cells of humans, those were all done in this MCDB nutrient medium of Dr. Ham, and an important point that we think we have found, in terms of the FGF requirement of these cells, and I want to reiterate that some myoblasts don't require FGF at all, I showed that were FGF-independent ones. But for the dependent ones, the only way that you can see that requirement is to add FGF almost continuously during the culture system. In the experiments that Dr. Ham showed you, he added the components at the beginning of the experiment, and he did not keep re-adding FGF. We have shown that FGF is lost from the medium of even clonal cultures with a very rapid half-life. You lose more than half of it every twelve hours. In our typical cultures, we are adding it every twelve hours and blocking the cultures. So when you add FGF at even very low levels, to Ham's MCDB medium, then you can get expression of not only the FGF-independent cells, but the FGF-dependent ones as well. I would say that at the moment, that would be a "super" medium also.

Dr. Emerson: Two questions for Dr. Blau. Has any work been done on Becker's dystrophy as far as looking at the lifetime of the myoblasts derived from them? Has the lifetime of potential recipient cells for transfection, for instance, fat cells, or muscle fibroblasts, been looked at in Duchenne patients?

Dr. Blau: My last slide, which I did not present, showed some myoblast proliferative capacity in polymyositis, dermatomyositis, and Becker's dystrophy. The proliferative capacity of myoblasts, say, from a Becker's child aged fifteen, was like that of a two-year-old Duchenne. So, it was what you would expect. It was diminished relative to normal, but it was certainly better than Duchenne, showing a sort of aging process, but not quite as rapid. Concerning the second question, the fibroblasts do not seem to be impaired; they do seem to have a normal proliferative capacity from Duchenne.

Dr. Emerson: And that's the only cell that's been looked at?

Dr. Blau: That's the only ones we've looked at.

Dr. Smith: Do you think that the reduced number of doublings in the Duchenne cells is due to muscle, or a membrane fragility, and also in relation to gene transfer, could the introduction of an intact, functional Duchenne's gene cause those doubling times to increase again?

Dr. Blau: My feeling from our studies with the doubling heterozygous carriers is that it is not related to a primary defect in those cells; it relates to growth. Especially looking at the age-related decline in the cell doublings, I think that it reflects a regenerative capacity, that these cells have undergone so many rounds of cell doublings that they've used up their proliferative capacity. Because if you get them from very young children, as Dr. Ham said, as well as we have seen that, too, from two-year-olds, you can get clones with 40 doublings which can give rise to billions of cells. It seems to be related to age, and the more regeneration that has occurred the more they have used up their proliferative capacity.

As for the normal cells, you can transfect them. We've done that with various muscle-constructs. So that is possible.

Dr. Grounds: I have two short questions, one for Dr. Blau and the other for Dr. Strohman. I am interested in why the mdx mice have this very successful muscle regeneration compared with Duchenne patients and the dystrophin-deficient dogs. I wonder if Dr. Blau would be interested to look what the aging effect was in mice, whether it was equivalent to that seen in Duchenne because we would not expect it if it was an aging phenomenon effectively. With respect to Dr. Strohman, I wonder whether he had observed in Duchenne patients, the high level of FGF around the muscle fibers.

Dr. Blau: With respect to your first question, we have not thus far worked on the mdx mouse, but we are just beginning to do studies. So we will address that question.

Dr. Strohman: The difference in the dystrophic phenotype between the mouse and humans is a key point. Right now, what we are seeing is what everybody knows: the mouse can evidently successfully regenerate fibers at least once, and perhaps more than once without engaging in a lot of connective tissue cell hyperplasia. So we think that the key to that question will be found in the modulation of connective tissue cells. In humans, of course, you have connective tissue cell hyperplasia.

We have looked in Duchenne's and Becker's and we do see FGF stored in the extracellular matrix of both of these, but in Duchenne there seems to be somewhat more. Those are very preliminary studies and we need to do a lot more.

Dr. Cooper: Dr. Witkowski, I think you made a good point and, certainly the scale-up of culturing is going to require a good deal of basic research in terms of things like growth factors. So, in that regard, I'd like to ask a few questions about growth factors. Dr. Strohman, in terms of sequence analogies between things like EGF, IGF and FGF, are there problems with cross-reactivity that could interfere with immunocytochemistry?

Dr. Strohman: Not that I know of. Not between FGF or the ones that you mentioned. I think we are just scratching the surface with regard to growth factors that are present in the muscle. Clearly, IGF is thoroughly implicated in one of the very early reactions that a satellite cell has to a growth stimulus. But cross-reactivity I do not think is a problem; that we know about.

Dr. Cooper: Dr. Ham, when you added insulin to your medium you saw some profound effects. I was wondering if you considered the possibility that it was mediated through an IGF-1 type receptor.

Dr. Ham: Yes, we have considered that possibility. We ran one test that seemed to indicate that was not the case, but at that time, we did not have a good positive control to be sure our IGF-1 preparation was good. We will be repeating those studies, and studying insulin requirement extensively as part of our new NIH program.

Dr. Cooper: Dr. Hauschka, I really liked your two-receptor model, and I was curious as to whether you thought maybe Dr. Bischoff's muscle-secreted factor might be the ligand for the second receptor.

Dr. Hauschka: Yes.

Dr. Chamberlain: I would just like to follow up on one of the points that Dr. Witkowski was making. I would like to ask any of the members of the panel if they feel that they could grow out 10^{11} or 10^{12} myoblasts in their laboratories, and could that also be expanded to treat DMD patients?

Dr. Blau: First of all, I think that the extrapolation and numbers that we've come up with are reasonable in theory. It's possible to generate the number of necessary myoblasts for transfer into a DMD patient from a small piece of tissue. In practice, of course, this is a very enormous undertaking. I think it is not one that is readily undertaken in the research laboratory, and I think it would be a mistake to do it in one's research laboratory where one's growing cells of all different kinds, and where one does not have the facilities for toxicity testing. In fact, I would like to make a strong point that was raised yesterday, also by my colleague, Dr. Larry Steinman: We both feel very strongly that such cells should only be grown under proper facilities with toxicity testing, so that there are no infectious agents, no toxic agents and tests are also done for tumogenicity; i.e., the cells are handled under controlled conditions. There are companies that are outfitted for just this kind of culture system that are scaled for growing human cells for use, for instance, cells in wound healing. These are precedents for using human cells in human beings that have been grown in culture. There are facilities that are dedicated to just that kind of activity. That is where I think these myoblasts should be grown and tested.

I also would like to emphasize the point that Dr. Witkowski made, namely, that a lot of work should be done in the mouse models first, and that we need to learn a good deal more about the immune system, and what those problems are, and how to circumvent them. We need to learn a lot about injection regimens, about satellite cell migration and to what extent cells can gain access to distant sites, and whether they restore function. A lot of work should be done in the animal before we go to clinical trials.

Dr. Stedman: As Dr. Witkowski elegantly points out, we're arriving at this number bottleneck which, right now, as Dr. Blau also emphasizes, is going to involve centralized facilities for culturing, if culturing alone is the answer to the number bottleneck. But I would like to just make a fairly strong statement at this point for the potential application if we're going to move away from living related donors toward multi-owner donors in the general age range, so that you can obtain 40 generations worth of subsequent division per satellite cell. There is a potential in something in the neighborhood of 25 to 30 kilograms of tissue potential available. If one can establish means for the very efficient sterile further manipulation of the cells and that is something that I am hoping to bring up in greater detail later because it relates directly to the activities of a number of the existing transplant surgical services in place at a number of hospitals in the country.

Dr. Strohman: Well, Hansell, since you brought it up, we discussed this a little bit last night that since we are probably going to have to immune-suppress the recipients anyway.

Dr. Van Ommen: I had a comment actually on the presentation of Dr. Miranda and this is a question. How do you see the future of converting nonmyoblasts into myoblasts for neonatal diagnosis? Because the way I see this is that the antenatal diagnosis is actually quite perfect using DNA techniques. We've done about 70 prenatal diagnoses on boys since the program in Leiden started and we had no difficulty in finding the abyss chromosome. I would like to add a word of caution, given our own results in detecting dystrophin in developing myogenic cells or myotubes that one actually can get to wrong conclusions by the absence of dystrophin in DMD cells.

Dr. Miranda: I have not seen any problems in finding any dystrophin in DMD cells. I do not think that the dystrophin analysis is going to be much better than other analyses, much simpler and much easier and much more accurate.

SECTION 6

MONITORING   CLINICAL   SUCCESS:
PHENOTYPIC   TRANSFORMATION

GENETIC AND BIOCHEMICAL DETERMINATIONS IN THE PRE-TRANSPLANT
WORKUP AND IN THE POST-TRANSPLANT ASSESSMENT PERIOD

Ronald G. Worton, Dennis E. Bulman,
Elizabeth E. Zubrzycka-Gaarn and Peter N. Ray

Genetics Department and Research Institute,
Hospital for Sick Children, and Department of
Medical Genetics, University of Toronto
555 University Avenue, Toronto, Canada M5G 1X8

INTRODUCTION

The topic that we have been asked to address is the pre-transplant workup, including selection of patients to receive myoblasts. Since the choice of patients for study and the nature of the pre-transplant workup depends critically on the planned post-transplant studies, this aspect of myoblast transplantation is included in the presentation.

Before introducing this topic, however, it may be appropriate to introduce a note of caution, by stating that in our research group we are not yet convinced that the field has progressed to the point where we are ready to begin myoblast transfer into boys with Duchenne or Becker muscular dystrophy. If, at this meeting we are able to reach a consensus that transplants should begin, then any myoblast transfer being planned must be considered to be in the realm of human experimentation, and not in the realm of disease modification or therapy.

We would like to make the additional point that there are a great many parameters to be worked out in developing myoblast transplantation as an approach to the treatment of muscular dystrophy. It will be difficult, if not impossible, to vary these parameters in a systematic fashion in boys undergoing experimental injections. Yet this is what must be done if we are to determine the optimum time of injection, the best route and frequency of injection, the optimum cell number to be injected, and so on. Therefore, we would like to make a plea at the outset for the continuation of animal experimentation with a priority equal to or greater than the priority given to the human experiments. Both the mouse and the dog models are useful in this regard and should be studied in depth for answers to the questions that are best answered in each species.

GENETIC AND BIOCHEMICAL STUDIES

In this presentation we suggest a series of pre- and post-transplantation studies that rely extensively on our knowledge of the DMD gene and of its protein product, dystrophin. Of course, the idea of myoblast transplantation and the early mouse experiments to test it (Watt et al., 1984; Law et al., 1988) pre-date the discovery of the DMD gene and we have heard arguments that the procedure could have been tried on patients without any prior knowledge of the DMD gene or its product. We do not subscribe to this point of view. In our view the discovery of the DMD gene and its product (see Worton and Thompson, 1988 for review) are crucial developments that bear directly on the current discussions of myoblast transfer. These discoveries are considered to be relevant to myoblast transfer for two main reasons. First, the new level of understanding brought about by these discoveries has indicated that the defect is related to the absence or the alteration of a very high molecular weight cytoskeletal protein (Koenig et al., 1988; Hoffman et al., 1988) that is localized at the sarcolemmal membrane of muscle fibers (Zubrzycka-Gaarn et al., 1988). Furthermore, the protein is encoded by an unusually large gene that will not be easy to manipulate (Koenig et al., 1987). Experience suggests that such a defect will be extremely difficult to correct by pharmacological means or by direct protein replacement therapy, and gene therapy for most genetic diseases, particularly this one, is still many years down the road. This forces us to consider myoblast transfer as the best current alternative for introducing new gene product into the muscle cells in the hope that enough dystrophin introduced into enough muscles of the body will result in a slowdown or arrest in the progression of the disease.

The second reason for considering the genetic and biochemical work as significant to myoblast transplantation is that in any attempt to correct a genetic defect through gene replacement or cellular transplantation (a form of gene replacement) one needs to be able to measure an effect of the procedure in order to optimize it. The measured end-points must be directly related to the correction of the basic defect; they must be quantifiable and the measurements must be reproducible. The ability to monitor the DMD gene and its product provide us with such a set of assays and these should form the basic underpinnings for the development of an optimal therapy based on myoblast transplantation. Recent experiments in the mdx mouse typify this approach (Partridge et al., 1989; Karpati et al., 1989).

Pre-transplant Work-up

In the early phases of myoblast transplantation it will be imperative to have a complete medical and scientific profile on the boys involved in the experiment. From the laboratory perspective these should be boys who have a well defined lesion in the DMD gene, and who have been characterized for the quantity and quality of dystrophin in skeletal muscle (Bulman et al., submitted). By choosing only boys with a defined lesion in the DMD gene, it will be possible to avoid the inclusion of boys with another disease such as limb-girdle MD as this would be an undesirable and unnecessary complication in the study. Well characterized lesions in the gene include deletions and duplications that occur in 60% and 6% of affected boys, respectively (Worton and Thompson, 1988). By choosing boys with a deletion or a duplication that gives rise to an identifiable novel junction fragment of DNA, it will make it possible to monitor the relative proportion of defective and normal DNA in the post-transplant biopsies as described below. By choosing boys who have little or no dystrophin by western blot analysis and no measurable immunostaining with anti-dystrophin Ab on muscle to estimate the amount and the tissue distribution of dystrophin produced by the donor nuclei.

On this basis we would consider it essential to include a complete DMD gene analysis and a complete dystrophin analysis as part of the pre-transplant workup. We suggest that the boys chosen for study should have a defined deletion or duplication in the DMD gene and, if possible, an identifiable novel DNA fragment from the deletion (or duplication) junction. Also essential is a level of immunoreactive dystrophin that is below the limits of detection on western blots or detectable dystrophin that is altered in size so that it will be clearly distinguishable from donor dystrophin following transplantation.

We have been asked to comment on the age of patients to be chosen for study. This is not an area in which we profess to have any special insight and we hope that this meeting will provide an opportunity for considerable discussion on the subject. However, perhaps we can begin the discussion by suggesting that the age of child chosen will depend to a great extent on the nature of the study. Intuitively one would think that if myoblast injections are to have maximum benefit for the recipient (in a therapy mode) they should be given at an early age, perhaps shortly after birth in the pre-clinical phase. However, in the context of human experimentation to determine safety, the feasibility and the efficacy of myoblast injections, perhaps an age of 5-7 is appropriate since certain muscles are undergoing extensive degeneration and at this stage one may have the best opportunity to observe an effective reversal of the degenerative process. Finally, if the experiment is designed only to test the safety and not the efficacy of the procedure, then an older child of 18-20 could be considered.

Post-transplant Monitoring

In addition to monitoring patients for signs of immune rejection, muscle bulk, muscle strength, and other features to be discussed later, it is our view that it is imperative to monitor the transplant for the successful introduction of normal gene copies into the muscle and the successful expression of those genes in terms of dystrophin production. This type of measurement has been considered to be unnecessary by some individuals and indeed it may be of limited importance if the injections are an unqualified success in terms of improved muscle performance (i.e., If it works, who cares how it works?). However, the significance of such measurements is very great indeed if the physiological response to the injections is less than anticipated. In such an event it will be essential to have enough information to determine the reason for the failure or for the poor response, and the knowledge of the number of gene copies introduced and the amount of dystrophin being produced is crucial. Without this knowledge it would be impossible to know what parameters to adjust in order to achieve a significant improvement, and a great many children may be subjected to useless experiments, adjusting parameters that need not be adjusted.

For these reasons it is our view that transplantation parameters should, in the first instance, be adjusted to give maximum sustained dystrophin levels. This will require that post-transplant biopsies be taken at intervals of perhaps 3-6 months. A small amount of tissue should be assayed for the presence of normal DMD gene, another portion for the amount of dystrophin and a third portion should be sectioned for analysis of dystrophin distribution.

Assuming that the transplanted patient has a duplication or deletion with a recognizable junction fragment, DNA from a very small amount of the biopsy should be subjected to PCR (polymerase chain reaction) analysis (Chamberlain et al., 1989) with two sets of amplification primers. One set chosen to amplify a segment of DNA that spans the junction and another set chosen to amplify the normal sequence that was destroyed by the deletion or duplication, would allow one to estimate the relative abundance of the two sequences in the post-transplant muscle. In the absence of information about the specific mutation in the patient's DMD gene, a known polymorphic difference between the donor and host DNA could be used for the same purpose. The proportion of amplified normal sequence relative to junction sequence would provide an estimate of the proportion of genetic material of donor origin. It would not, however, distinguish whether these gene copies were present in myonuclei, satellite cells or other cell types of donor origin.

The second assay is a quantitative western blot analysis for dystrophin. Unlike our initial anti-dystrophin antibodies that were very good at staining native dystrophin in a tissue section but not very good at revealing denatured dystrophin on a western blot (Zubrzycka-Gaarn et al., 1988), we have now prepared new antisera that give excellent quantitative results on western blots of normal muscle. Two of our best antibodies in this regard include one directed against amino acids 67 to 667 near the N-terminus of the protein (antibody 9219) and another directed at the last 17 amino acids at the C-terminal end of the protein (antibody 1461). The former was prepared in sheep using a TrpE fusion protein as antigen (Bulman et al., 1990) whereas the latter utilized a synthetic peptide conjugated to BSA to elicit an Ab response in rabbit (Zubrzycka-Gaarn et al., 1990). Both antisera have a high titre, and are highly specific for dystrophin.

Figure 1 shows the staining intensity of a western blot loaded with differing amounts of normal muscle. The dilution curve resulting from laser densitometry of the blot shows that the quantitative assay is linear up to 50 ug of protein and that the Ab is capable of detecting dystrophin in 2.5 ug of normal muscle. Since 50 ug is the normal amount applied to a western blot the Ab should be able to quantitate levels of dystrophin down to about 5% of normal. This would be more than adequate for quantitative estimates of dystrophin following myoblast transplantation.

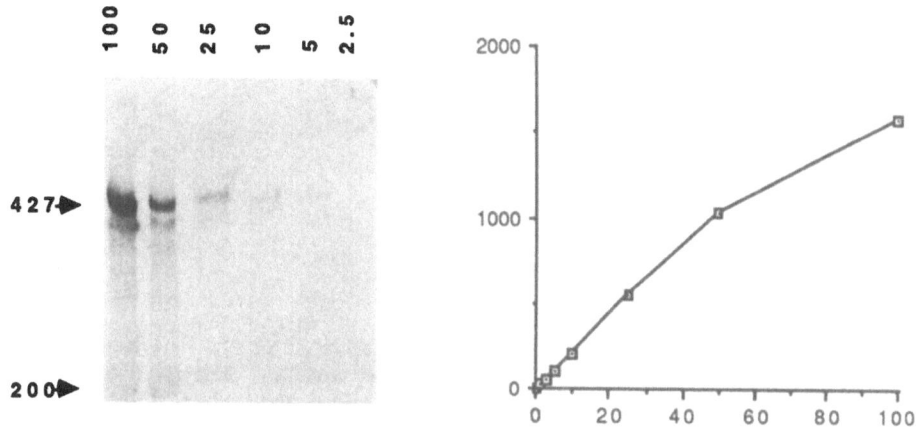

Fig. 1. Left: Western blot of normal muscle immunostained with N-terminal Ab 9219. The amount of muscle homogenate (ug) in each lane is indicated at the top of the figure. Right: Dilution curve obtained by laser densitometry of the western blot shown.

Figure 2 shows how these antibodies may be combined to detect altered dystrophin in BMD and DMD patients. Two Becker patients (lanes 2 and 3) have dystrophin of reduced size due to a deletion that removes several exons but leaves the reading frame of the message intact. Their dystrophin is detected with both N-terminus and C-terminus Ab (lanes 2 and 3) indicating that the two ends of the protein are intact but the size is reduced as a consequence of the deletion. In contrast, an intermediate and a DMD patient with residual dystrophin of reduced size (lanes 5 and 7) show no staining with C terminus Ab, consistent with a deletion that shifts the reading frame of the message and gives rise to a somewhat unstable protein, truncated by a stop codon in the altered reading frame following the deleted segment.

The third important assay is the distribution of dystrophin in sections of transplanted muscle as revealed by immunocytochemistry. This will not only confirm that dystrophin is being made in the muscle; it will also allow an evaluation of its distribution among the sarcolemma of the fibers of the injected muscle. A clustering of all the dystrophin in a few fibers that happened to be along the track of the injection needle would presumably be less beneficial than a more uniform distribution over the muscle. We need to have much more information on the ability of injected myoblasts to move through the muscle from the site of deposit and we need to know more about the ability of dystrophin to diffuse along the fiber away from the donor nucleus that is producing it. Early experiments with the mdx mouse injected with normal mouse myoblasts (Partridge et al., 1989) or human myoblasts (Karpati et al., 1989) appear promising, but there is much more experimentation to be done.

In conclusion, it is our view that a detailed genetic and biochemical analysis of patient muscle prior to myoblast transplantation is an integral part of the procedure. A patient should not be subjected to the transplantation

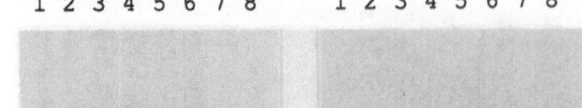

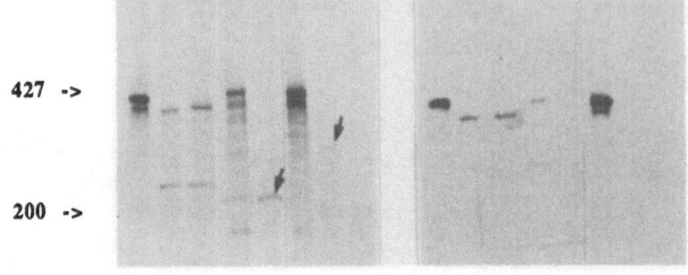

Fig. 2. Western blots of muscle stained with N-terminus Ab 9219 (left) and C-terminus Ab 1461 (right). Samples are from normal muscle (lane 1), non-DMD disease controls (lanes 4 and 6), BMD (lanes 2 and 3), DMD (lanes 7 and 8) and intermediate (lane 5). Altered size dystrophin in the intermediate and one of the DMD cases is indicated by an arrow. The dystrophin bands of reduced molecular weight are clearly visible in the BMD cases; the lower band just above 200 kD is thought to be a breakdown product of the altered dystrophin.

procedure without first having a detailed knowledge of the nature of his mutation and of the amount and the muscle fiber distribution of his dystrophin. Since the procedure is experimental, it is unfortunately necessary to experiment. It will be necessary to subject different cohorts of individuals to different "treatment" regimens and it will be necessary to follow the injections with one or more biopsies to monitor the success in terms of dystrophin replacement. Without these detailed measurements to gain valuable information, we would find it hard to justify the physical and the psychological trauma that the patient is bound to suffer in being subjected to the procedure. And that brings us back full circle to animal experiments. In order to minimize the number of young boys who will be asked to participate in this type of research, it is essential that we gain as much of the information as possible from studies of mice and dogs.

REFERENCES

1. Bulman, D.E., Murphy, E.G., Zubrzycka-Gaarn, E.E., Worton, R.G. and Ray, P.N. 1990. Differentiation of Duchenne and Becker muscular dystrophy phenotypes with amino- and carboxy-terminal antisera specific for dystrophin.

2. Chamberlain, J.S., Gibbs, R., Ranier, J.E., Nguyen, P.N. and Caskey, C.T. 1988. Deletion screening of the Duchenne muscular dystrophy locus via multiplex DNA amplification. Nucl. Acids Res., 1: 11141.

3. Hoffman, E.P., Fischbeck, K.H., Brown, R.H., Johnson, M., Medori, R. et al. 1988. Dystrophin quality and quantity determines the clinical severity of Duchenne/Becker Muscular Dystrophies. New Eng. J. Med., 318: 1363.

4. Karpati, G., Pouliot, Y., Zubrzycka-Gaarn, E.E., Carpenter, S., Ray, P.N., Worton, R.G. and Holland, P. 1989. Rapid communication, dystrophin is expressed in mdx skeletal muscle fibers after normal myoblast implantation. Am. J. Path., 135: 27.

5. Koenig, M., Hoffman, E.P., Bertelson, C.J., Monaco, A.P., Feener, C. and Kunkel, L.M. 1987. Complete cloning of the Duchenne muscular dystrophy (DMD) cDNA and preliminary genomic organization of the DMD gene in normal and affected individuals. Cell, 50: 509.

6. Koenig, M., Monaco, A.P. and Kunkel, L.M. 1988. The complete sequence of dystrophin predicts of rod-shaped cytoskeletal protein. Cell, 53: 219.

7. Law, P.K., Goodwing, T.G. and Wang, M.G. 1988. Normal myoblast injections provide genetic treatment for murine dystrophy, Muscle & Nerve, 11: 525.

8. Partridge, T.A., Morgan, J.E., Coulton, G.R., Hoffman, E.P. and Kunkel, L.M. 1989. Conversion of mdx myofibres from dystrophin-negative to -positive by injection of normal myoblasts. Nature, 337: 176.

9. Watt, D.J., Morgan, J.E. and Partridge, T.A. 1984. Use of mononuclear precursor cells to insert allogeneic genes into growing mouse muscles. Muscle & Nerve, 7: 741.

10. Worton, R.G. and Thompson, M.W. 1988. Genetics of Duchenne muscular dystrophy. Annu. Rev. Genet., 22: 601.

11. Zubrzycka-Gaarn, E.E., Bulman, D.E., Karpati, G., Burghes, A.H.M., Belfall, B. et al. 1988. The Duchenne muscular dystrophy gene is localized in the sarcolemma of human skeletal muscle fibers. Nature, 333: 466.

12. Zubrzycka-Gaarn, E.E., Hutter, O.F., Karpati, G., Bulman, D.E., Klamut, H.J., Hodges, R.S., Worton, R.G. and Ray, P.N. 1989. Dystrophin is tightly associated with the sarcolemma of mammalian skeletal muscle fibres. Exp. Cell Res., (in press).

Discussion of Dr. Worton's paper

Dr. Mendell: Does anybody have any particular comments related specifically to Dr. Worton's discussion about dystrophin monitoring?

Dr. Engel: Just a small one. I wouldn't give up on the idea of trying to treat patients with Becker dystrophy just because they are not expected to improve. If the patient actually gets an improvement of strength then that would be a positive clinical result, regardless of what the molecular genetics shows us.

Dr. Worton: The reason I shied away from that issue is certainly an arguable point. If you start with the young

Becker patient, who is predicted to have slower progression on the basis of his mutation, it may be difficult to quantitate the improvement.

Dr. Mendell: Thank you, Ron. The next group of speakers will address clinical tests of efficacy. The speakers selected have particular experience in this area. We will begin with Dr. Ted Munsat who will discuss quantitative muscle testing. We will ask him to provide leadership on how to measure strength following myoblast transfers.

USE OF QUANTITATIVE MYOMETRY IN THE EVALUATION

OF MYOBLAST TRANSFER THERAPY

Theodore L. Munsat

Tufts-New England Medical Center
Boston, MA

Because time is so limited, I've decided to present my conclusions to you first and then afterwards to present as much supporting data as time allows.

My conclusions, which are really in the form of suggestions, are as follows. I would suggest that maximum voluntary isometric contraction (MVIC) of muscle is the most direct, the most convenient, the most valid and the most reproducible way of testing muscle strength and should form the heart of any outcome assessment protocol.

Although the initial transplant studies will undoubtedly involve only single muscles, I would suggest that we should analyze deterioration curves for all body muscles, not only the ones being initially transplanted, as these data will provide very important natural history information. Also, the patient who has a successful transplant for a single muscle, will then become a very appropriate candidate for subsequent transplants to other muscle groups. In addition, assessing other muscles will allow us to evaluate both short and long term effects in muscle groups distant from those transplanted. I think this is quite important.

I would suggest that deterioration rates be plotted for a minimum of six months. Patients who have more rapid courses of deterioration should be selected for trials, as this has a salutory impact on sample size. For the initial transplant studies, six patients are sufficient if those patients are chosen with deterioration rates as similar to each other as possible, preferentially at the mean or greater. I would favor selecting patients in the age group five to nine.

The established and existing ways of measuring strength by manual muscle testing are probably inadequate, not only for single muscle transplants, but certainly for the future, when hopefully more widespread transplantation will be possible. Classical manual muscle testing such as the MRC scale which is a five-point scale, provides extremely ordinal data, so that a single point on the MRC scale may account for as much as 95% of muscle strength.

Expressing this another way, manual muscle testing is very insensitive to minor and even sometimes quite marked changes in muscle strength. When one plots manual muscle grades against strength for the same muscle group, you find that even though the manual muscle grade is unchanged, strength is showing gradual and linear deterioration.

The same applies to functional scales which may remain flat, even though a composite score for strength show progressive, linear decline.

When using timed functions and functional scales, one has to be aware of the fact that a "crescendo" effect can occur; a situation where the loss of only a few motor units can result in a dramatic reduction of functional scale score because those same motor units may be involved in several similar functions. For example, walking, climbing stairs, getting up from a stooped position all overlap. We, therefore, have utilized a rather simple strain gauge tensiometer with a fixed table for measuring muscle strength. The patient is placed in predetermined positions, the best of two maximum voluntary isometric contractions is obtained and the information is stored.

Contrary to what one might think, the psychologic aspects of MVIC are not meaningful and on test re-test both intra and inter rater reliability is very good. We observe percentage change on test re-test, depending on the muscle, in the range of six to ten percent. Really not too bad. This range is for adults and in children the numbers are a bit higher. In addition to the measurement instrument there are data transformations that need to be carried out. We transfer the raw data, be it in kilograms, or liters, for pulmonary function or what have you, to a standard reference base which is the mean of all muscular dystrophy patients for that particular function, so that the "Z" score, the transformed score, is actually expressed in standard deviation units. A plus one score means that the raw score is plus one standard deviation above the mean for all patients.

There is a certain advantage in creating Megascores once one has transformed the data. Megascores represent whole limbs or any other combination of muscles. One can then very easily plot scores against time, Megascores or individual scores. One can then determine how linear the rate of deterioration is. If you can establish linearity and the variance of that linearity, you have an extremely powerful statistical tool by which you can evaluate the efficacy of a therapeutic intervention in as small a number as even five or six patients.

The CIDD data of Duchenne muscular dystrophy obtained with manual muscle testing is very interesting. Between the ages of two and a half and eight, the curve is rather flat. This represents the trade-off between the deterioration of the disease and the normal growth effects on muscle strength. But soon, at about age six, the battle is lost and you begin to see a rather linear decline of function until the age eleven or twelve. Then patients for the most part stabilize or "bottom out". And, in fact, at this stage which corresponds with wheelchair use, there is remarkably little functioning muscle anywhere.

It is interesting that the actual strength deterioration period is only about five or six years, which is quite similar to what happens in ALS; except that in ALS, the patients die instead of bottoming out.

It is important to select patients who are positioned early in this deterioration curve and not to evaluate them at the point where the curve is flattened or where it has bottomed out. We have asked the question: How many monthly quantitative strength measurements do we need to do in order to establish the degree of linearity and to predict the future? It turns out to be six to nine. Of course, the more the better, but six to nine examinations will do.

Lastly, I want to make one comment about the sample size formula, which is critical to proper trial design. If you reduce sigma, which is the patient variance at entry to the study, it has a remarkable effect on "n". In other words, if patients are more similar to each other, our ability to detect change increases. This is an extremely important factor. Because of the twenty fold variation in deterioration rates in Duchenne patients, it becomes essential to analyze pre-transplantation deterioration plots, so that one can select the smallest number of patients which will give the greatest amount of information.

Discussion of Dr. Munsat's paper

Dr. Mendell: Ted, I would like to ask you to further comment on specific criteria for patient selection for myoblast transfers. During the afternoon we will try to develop a profile for patients to be selected for myoblast transfer trials.

Dr. Worton suggested selection to be confined to patients with known DNA deletions and in whom dystrophin has been shown to be deficient. You have recommended that 5 - 7 year old patients may represent the best candidates. Do you also recommend that boys be selected in part on the basis of their ability to cooperate for quantitative muscle testing?

Dr. Munsat: Yes, that is a very important admission criterion. What we actually do is we bring the parents and the child into the laboratory for two or three visits prior to serious testing efforts. This provides the opportunity to meet the physical therapists who are actually doing the testing and they get to understand the equipment so that it is not threatening. This is the learning phase and will be very important particularly in the first patients that undergo myoblast transfer.

Dr. Mendell: Okay, we will go on to the next speaker, Dr. Michael Brooke who has provided leadership for the CIDD group in monitoring patients with Duchenne dystrophy.

ISSUES IN MYOBLAST TRANSFER

Michael H. Brooke

University of Alberta
Edmonton, Alberta, Canada

The work which forms the basis for my comments was done by the clinical evaluators in the CIDD group (1). I shall present only the conclusions, not the studies themselves.

Firstly, I would like to stress that there is only one time to design a clinical trial of a drug, or any other therapeutic intervention, and that is before the drug has been discovered. Once it has been discovered, one has no time to lay the groundwork for such trials and there is an irresistible urge for investigators to begin random trials none of which are very carefully controlled.

Secondly, I would entirely agree with the comment that was made yesterday that the diagnosis of Duchenne muscular dystrophy in this day and age depends upon the demonstration of dystrophin deficiency or aberration and everything else is secondary.

But the treatment response that we all want has nothing to do with the dystrophin. The patient is ultimately interested in survival and strength and function. Therefore, we have to test this in any clinical trial.

There will be two types of clinical trials involving myoblast transfer. The first phase will involve a very few patients and perhaps a single muscle in each and will determine whether myoblast transfer is feasible; that it is safe, practical and only secondarily that it may increase muscle strength. The techniques in a Phase I trial will be quite different, in terms of evaluating the response, than the techniques used in a Phase II trial. In Phase II, one is testing the treatment, if not in the real world, then at least in a large enough population of patients over a longer period of time to obtain information in regard to the patients' function and longevity.

Measurements themselves have no inherent value. They only have value when applied to a particular clinical situation. In Phase I and Phase II trials, it is likely that the methods of evaluating muscle strength and function will be different.

Myoblast Transfer Therapy
Edited by R. Griggs and G. Karpati
Plenum Press, New York, 1990

Muscle strength has been evaluated in many different ways in boys with DMD. The first and the most widely used has been the manual muscle test and it is inherently unpopular. It is "obvious" that the manual muscle test is subjective; the data are not interval data; it is skewed to one side and has few inherently desirable features. And yet when you apply it to Duchenne muscular dystrophy something strange happens. If you look at a fairly large number of muscles it behaves as interval data with linear changes over time, particularly between the ages of five and fifteen. It has an acceptable variability and is sensitive enough to demonstrate changes in the course of the illness within four weeks at a level of significance of $p < 0.0001$. It does, in fact, work.

Manual muscle testing, however, is useless if one is testing a single muscle because that muscle stays so long in one particular grade. I agree with Dr. Munsat that fixed myometry is a very good system and probably the best one for quantitating the strength of single muscles.

Another popular variation of myometry is the hand held myometer. Several years ago in preparation for the CIDD project and again more recently we compared hand held myometers to manual muscle testing. We found no advantage of hand held myometry over manual muscle testing. Indeed, it may be considered simply as an extension of manual muscle testing. One is not only quantitating the patient's strength and resistance with a hand held myometer; you are also quantitating the evaluator's strength and resistance.

It may be helpful to those who are inexperienced in manual muscle testing and it may be a little more useful when you are looking at a single muscle but I do not think anybody who has used hand held myometry really espouses that particular method and certainly interclass correlation and intraclass correlation has not been any higher than with manual muscle testing.

Isokinetic dynamyometry measures muscle strength when the joint is extended or flexed at a fixed rate (speed). This rate can be varied and strength correlated with different speeds. For example, the subject is asked to straighten his leg out as fast and as hard as he can. The machine takes care of the speed of movement. No matter how hard or how little the patient pushes, the machine will move at the same speed. Its appeal lies in the fact that it is a physiological type of contraction.

One problem with the isokinetic dynamyometry is that the measurements depend to some extent upon the weight of the limb. A muscle which is trying to move a very heavy limb is going to do less well than a muscle which is moving a limb which is thin. Some machines will compensate for the weight of the limb and eliminate the problem. Isokinetic dynamyometry can only be used on a few muscles but in those muscles it provides measurements with high inter- and intrarater reliability and low variability.

In our evaluation of isokinetic dynamyometry, we found that a quarter of the patients with DMD could not use the machine because they could not move the limb against gravity and they were unable to exert enough torque to produce reliable reading. However, for the others it is an excellent evaluation tool.

Another problem in evaluating strength involves the patient's cooperation. We found it difficult to test children reliably under the age of five years.

There are some exceptions. We have had reliable testing as early as three years but it takes a really intelligent three year old to be able to do that.

In conclusion, I would agree with Dr. Munsat that in testing single muscles, fixed myometry is the preferable method. I would disagree a little in other aspects because in spite of all the theoretical arguments against manual muscle testing, they remain theoretical. When applied to a population of patients with Duchenne dystrophy, manual muscle testing seems to work quite well.

Discussion of Dr. Brooke's paper

Dr. Mendell: Mike, just to clarify one anticipated question before opening this up for general discussion. Would you agree with Dr. Munsat that manual muscle testing may not have a place in monitoring the success of myoblast transfer?

Dr. Brooke: I would absolutely agree that manual muscle testing has no place in Phase I of myoblast transfer studies looking at efficacy in a single muscle.

Dr. Mendell: I just have one other question. Do you think that the CIDD data on manual muscle testing would provide a more linear slope of deterioration if only patients with dystrophin absence were plotted rather than including milder phenotypes? Would you speculate on that point?

Dr. Brooke: As I have gotten older, I have learned never to speculate.

Dr. Mendell: Okay, let's go on then. The next speaker is Dr. Berch Griggs.

QUANTITATION OF MUSCLE MASS AND MUSCLE PROTEIN SYNTHESIS RATE:

DOCUMENTING A RESPONSE TO MYOBLAST TRANSFER

Robert C. Griggs

University of Rochester
School of Medicine and Dentistry
Department of Neurology
601 Elmwood Avenue - Box 673
Rochester, NY 14642

The goal of treatment in muscular dystrophy is to provide a favorable effect on the natural history of muscle strength. Such an effect can be produced by improvement in strength, by arrest of progression, or by slowing in the rate of progression. All laboratory measurements which assess muscle histology or electrophysiology, muscle mass, or measure muscle components in serum (e.g. creatine kinase) are indirect and of value in a therapeutic trial only to the extent that they either predict or substantiate a clinical response. If myoblast transfer treatment is to be used, it will be studied initially in one muscle or at most, a small number of individual muscles. There is currently no documented way of establishing a change in the natural history of strength for individual muscles in Duchenne dystrophy. Indirect, laboratory measures of "success" of myoblast transfer treatment applied to individual muscles are therefore of potential importance since they may: (1) predict the ultimate clinical success of the treatment; (2) establish a mechanism of action for myoblast transfer; (3) provide clues as to other approaches which could be combined with myoblast transfer to improve results.

ASSESSMENT OF MUSCLE MASS

Whole Body Measurements

The two laboratory procedures most useful for the assessment of body muscle mass are the 24-hour creatinine excretion and the total body potassium. Creatinine excretion has been found to be a reliable and reproducible indicator of muscle mass by numerous investigators (1,2). However, for creatinine excretion measurements to be reliable, care in data collection is essential. There is considerable variation in creatinine excretion throughout a 24-hour period. Therefore, 3 consecutive 24-hour urine collections are essential. Patients must also be on a flesh-free diet during the collections since approximately 7% of urinary creatinine is of dietary origin if patients are consuming flesh (3). Using this method, creatinine excretion data has a coefficient variation of 9% in normals and in patients with muscular dystrophy (2).

Total body potassium is also useful in determining muscle mass and is markedly reduced in patients with Duchenne dystrophy and other

Myoblast Transfer Therapy
Edited by R. Griggs and G. Karpati
Plenum Press, New York, 1990

TABLE 1. Clinical and Laboratory Measurements in Prednisone Study Groups
After Six Months of Treatment (modified from reference 6)

Measurement	Placebo	Prednisone		P values			
		0.75 mg/kg	1.5 mg/kg	Placebo vs 0.75 and 1.5 mg/kg	Placebo vs 0.75 mg/kg	Placebo vs 1.5 mg/kg	0.75 kg vs 1.5 mg/kg
	mean* (patients completing testing†)						
Muscle strength score	5.80 (35)	6.23(30)	6.25(30)	0.0001	0.0001	0.001	0.84
Forced vital capacity (liters)	1.52 (34)	1.68(29)	1.66(28)	0.0001	0.0004	0.002	0.63
Creatinine excretion	190(35)	248(31)	261(33)	0.0001	0.0001	0.0001	0.36
Log_e creatine kinase	8.17(32)	7.91(29)	7.87(30)	0.03	0.07	0.04	0.79

* Values represent the least-squares means derived from the average of the first two and last
  two visits

† The following patients were dropped from the study: one patient taking placebo, two taking
  0.75 mg of prednisone per kilogram, and one taking 1.5 mg per kilogram (never tested).

neuromuscular disease (2,4). Since only 50% of total body potassium is
in muscle, the rate of decline in totaly body potassium measurement slows
as disease progresses in Duchenne dystrophy (2).

Usefulness of Whole Body Muscle Mass Estimates in Therapeutic Trials

Studies in Duchenne dystrophy have shown that both creatinine
excretion and total body potassium are useful for confirming and
predicting an increase in muscle mass during the increase in strength
occurring with corticosteroids (5,6). Table 1 shows the increase in
creatinine excretion documented during the course of a six month trial of
prednisone treatment in Duchenne dystrophy (6). Similarly predictive
responses to treatment have been observed in other disorders (7).
However, a significant increase in muscle mass has been demonstrated by
creatinine excretion and total body potassium during a therapeutic trial
with anabolic steroids without there being a salutory affect on strength
(8,9). Thus, an increase in muscle mass does not necessarily imply that
the goal of increasing muscle strength has been achieved.

Regional Muscle Mass

There are four widely available techniques which assess muscle mass
in the extremities. The techniques are listed in order of both
increasing cost and accuracy: (1) volumetry; (2) ultrasonography; (3)
computed tomography (CT); (4) magnetic resonance imaging (MRI).
Ultrasonography is useful for detecting fat and connective tissue
replacement of muscle and is therefore more helpful than simple volumetry
but neither are as reliable as cross-sectional images obtained by CT or
MRI. Extensive published experience with CT has shown its accuracy in
differentiating dystrophic from inflammatory muscle disease and has also
demonstrated changes in the context of therapeutic trials. The
beneficial effect that exercise produced in corticosteroid atrophy was
readily demonstrated by CT (10) (Figure 1).

Magnetic resonance imaging is particularly suitable for studies of
children since it avoids the radiation exposure incurred with repeated CT

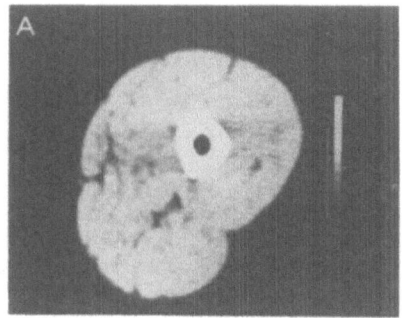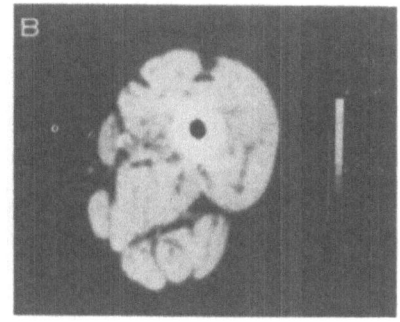

Fig. 1. A, CT of the left thigh of a normal subject. B, CT of the left
thigh of a patient treated with glucocorticoids. This patient
was matched with the subject in A. Note the mottled appearance
of the muscle and the smaller midthigh muscle area.

measurements (Figure 2). Cross-sectional area can be determined in
multiple tissue "slices" and reliably quantitated with a simple computer
program (11). MRI has been shown to be useful in distinguishing muscle
from subcutaneous tissues, from vascular and neural structures, from bone
and most importantly, from infiltration of muscle with fat and connective
tissue. The ability of MRI to determine the local effect of injected

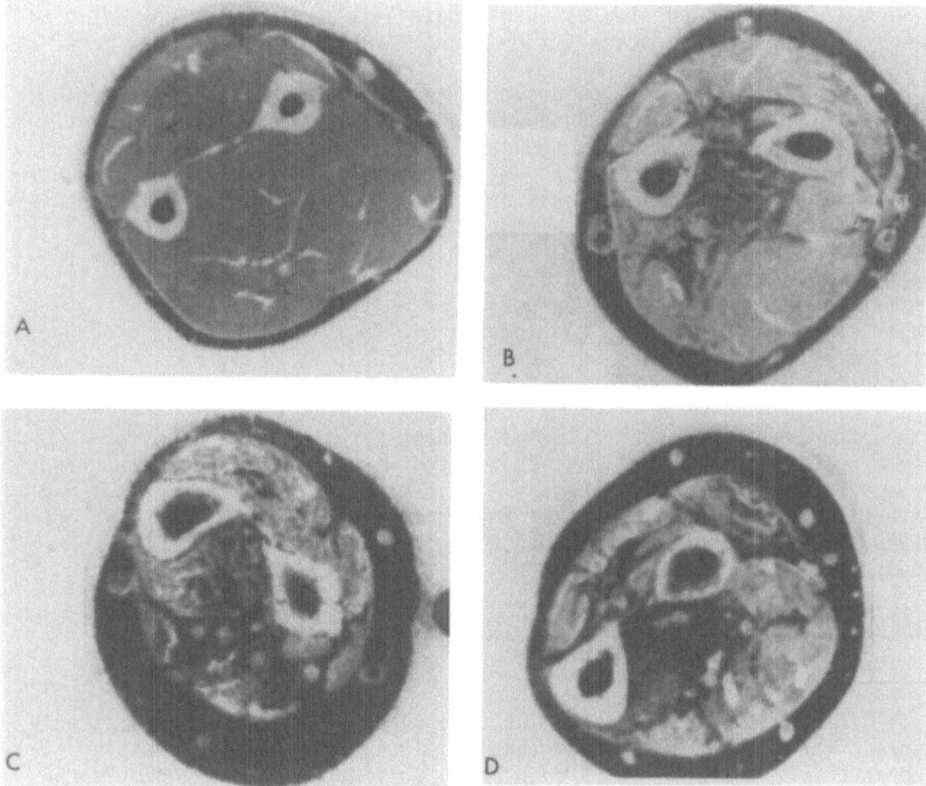

Fig. 2. MRI scans of the forearm of a normal subject (A) and in 3
patients with myotonic dystrophy (B-C). Individual muscles
and replacement of muscles by fat are readily apparent.

myoblasts will depend upon its ability to distinguish between true muscle and the increase in muscle size due to inflammation, fluid accumulation, or reactive changes in injected tissues.

MUSCLE PROTEIN SYNTHESIS

Techniques are now available for the safe, in vivo determination of rates of muscle protein synthesis and degradation in individual tissues of humans (12,13). Stable isotope-labelled amino acids can be infused for brief periods of time to label the precursor pool. Subsequent study of tissue incorporation of isotope permits calculation of the rate of muscle protein synthesis. Since stable isotopes are nonradioactive, the techniques for determining the rate of incorporation into tissue require the use of mass spectrometry. However, numerous laboratories throughout the world are now capable of these studies. A major advantage of stable isotope studies is the ability to investigate metabolism in normal humans including both normal and diseased children (14). Analysis can be performed on tissue samples of less than 10 mg size.

In Duchenne dystrophy there is evidence that disease progression may reflect "failed regeneration" with a decrease in muscle protein synthesis as the ability of satellite cells to regenerate and repair diseased muscle is exceeded (15). There is evidence that muscle protein synthesis is decreased in later stages of Duchenne dystrophy once weakness has appeared (16); it remains unclear whether muscle protein synthesis is increased in early stages of the disease and then fails to keep pace with destruction as muscle satellite cells reach the limit of their ability to regenerate.

The rationale for myoblast transfer therapy is the replacement of muscle cells lacking dystrophin with cultured myoblasts from individuals with normal dystrophin-producing capacities. One of the practical limitations of myoblast transfer will be the limitation in the absolute number of doublings of myoblasts obtained from histocompatibility antigen-matched donors. Furthermore, the strategy of using genetically-engineered dystrophin replacement with the patient's own myoblasts has been proposed in order to avoid issues of tissue rejection. In either instance, the potential senescence of transferred myoblasts will be a major issue with regard to the long-term effectiveness of the procedure. Studies of the synthetic and regenerative capacity of myoblasts may help define the potential for ultimate success of transfer therapy.

Proposals for study of myoblast transfer include a plan for histological study of the muscles treated with transferred cells. In vivo studies of protein synthesis can be combined with such histologic investigations to provide direct measurements of rates of synthesis of mixed skeletal muscle protein or of individual sarcoplasmic and contractile component proteins (12).

Usefulness of Studies of Muscle Protein Synthesis During Therapeutic

Trials

In vivo studies of muscle protein synthesis have shown remarkably dynamic changes in rates of synthesis. Thus, feeding (14), fasting (14), and inactivity (17) have all been shown to induce rapid changes in the rate of muscle protein synthesis. Similarly, anabolic hormone administration has been shown to increase muscle protein synthesis rate (18). It should be possible to obtain similar data with regard to the effect of treatment of individual muscles with myoblast transfer.

SUMMARY

Initial trials of myoblast transfer into individual muscles should include studies of regional muscle mass. Magnetic resonance imaging is a safe and accurate technique that should be studied in the context of these trials. Creatinine excretion should also be measured in order to provide baseline information for future studies of myoblast injections into a larger number of muscles. In vivo assessment of muscle protein synthesis can be combined with histologic study to determine if the treatment alters the rate of protein synthesis.

REFERENCES

1. Heymsfield, S. B., Arteaga, C., McManus, B. S., Smith, J., Moffitt, S., 1983, Measurements of muscle mass in humans: validity of the 24-hour urinary creatinine method, Am J Clin Nutr, 37:478.
2. Griggs, R. C., Forbes, G., Moxley, R. T., Herr, B. E., 1983, The assessment of muscle mass in progressive neuromuscular disease, Neurol, 33:158.
3. Crim, M. C., Calloway, D. H., Margen, S., 1975, Creatine metabolism in men: urinary creatine and creatinine excretions with creatine feeding, J Nutr, 105:428.
4. Forbes, G. B., Hursh, J. B., 1963, Age and sex tends in LBM calculated from K measurements: with a note on the theoretical basis for the procedure, Ann NY Acad Sci, 110:255.
5. Brooke, M. H., Fenichel, G. M., Griggs, R. C., et al, 1987, Clinical investigation of Duchenne muscular dystrophy: interesting results in a trial of prednisone, Arch Neurol, 44:812.
6. Mendell, J. R., Moxley, R. T., Griggs, R. C., et al, 1989, Randomized, double-blind six-month trial of prednisone in Duchenne's muscular dystrophy, N Eng J Med, 320:1592.
7. Griggs, R. C., Pandya, S., Moxley, R. T., Forbes, G., et al, 1981, Treatment of myopathic carnitine deficiency - quantitation of response to prednisone and carnitine, Trans Am Neurol Assoc, 106:199.
8. Griggs, R. C., Kingston, W. J., Herr, B. E., Forbes, G., Moxley, R. T., 1985, Effect of testosterone administration on total body potassium and creatinine excretion in myotonic dystrophy, Neurol, 35:1035.
9. Griggs, R. C., Pandya, S., Florence, J. M., Brooke, M. H., et al, 1989, Randomized controlled trial of testosterone in myotonic dystrophy. Neurol, 39:219.
10. Horber, F. F., Scheidegger, J. R., et al, 1985, Evidence that prednisone-induced myopathy is reversed by physical training, J Clin Endocrinol Metab, 61:83.
11. Jozefowicz, R. J., Ceckler, T. L., Flood, D. G., Herr, B. E., Hawthorne, B. W., Griggs, R. C., 1989, Assessment of regional muscle volume utilizing forearm magnetic resonance imaging, Neurol, 39(Suppl 1):134.
12. Halliday, D., McKeran, R. O., 1975, Measurement of muscle protein synthesis rate from serial muscle biopsies and total body protein turnover in many by continuous intravenous infusion of L-[alpha-^{15}N]lysine, Clin Sci Molec Med, 49:581.
13. Nair, K. S., Halliday, D., Griggs, R. C., 1988, Leucine incorporation into mixed skeletal muscle protein in man - its relationship to whole body leucine kinetics and muscle fiber composition, Am J Physiol, 254:E208.
14. Rennie, M. J., Edwards, R. H. T., Halliday, D., et al, 1982, Muscle protein synthesis in man: the effects of feeding and fasting, Clin Sci, 63:519.

15. Moxley, R. T., 1983, Metabolic studies in muscular dystrophy and their relationship to clinical pathophysiology: A focus upon myotonic dystrophy and Duchenne muscular dystrophy, <u>Seminars Neurol</u>, 3:308.
16. Griggs, R. C., Rennie, M. J., 1983, Muscle wasting in muscular dystrophy: decreased protein synthesis or increased degradation? <u>Ann Neurol</u>, 13:125.
17. Shengraw, R. E., Stuart, C. A., Prince, M. J., Peters, E. J., Wolfe, R. R. 1988, Insulin responsiveness of protein metabolism in vivo following bedrest in humans, <u>Am J Physiol</u>, 255:E548.
18. Griggs, R. C., Halliday, D., Kingston, W., Moxley, R. T., 1986, Effects of testosterone on muscle protein synthesis in myotonic dystrophy. <u>Ann Neurol</u>, 20:590.

Discussion of Dr. Griggs' paper

Dr. Mendell: One point of clarification. Do you think that MRI has a greater power to distinguish fat, connective tissue and vascular volume of muscle than CT scanning?

Dr. Griggs: I really cannot comment on that. I just don't know. We have not done comparative studies on MRI and CT.

Dr. Mendell: Perhaps someone in the audience can address that issue during the general discussion. Let me introduce Dr. Peter Law, who has obviously had a great deal of experience in monitoring success in experimental animals undergoing myoblast transfer studies.

PLAUSIBLE STRUCTURAL/FUNCTIONAL/BEHAVIORAL/BIOCHEMICAL

TRANSFORMATIONS FOLLOWING MYOBLAST TRANSFER THERAPY

Peter K. Law, Tena G. Goodwin, H.-J. Li
and Ming Chen

Departments of Neurology and Physiology/Biophysics
University of Tennessee, Memphis, TN

CLINICAL SUCCESS MEANS MUSCLE FUNCTION IMPROVEMENT

As of April 7, the Institutional Review Board has approved our proposal on human trial with myoblast transfer therapy.

Six immunosuppressed Duchenne muscular dystrophy (DMD) boys between the ages of 6 to 10 years will receive myoblast injections in the right extensor digitorum brevis (EDB) muscle which dorssiflexes the toes. By this age, significant muscle growth, degeneration and regeneration have occurred[1]. Satellite cells should be abound. The EDB is chosen because of its early involvement in dystrophy[2,3], its easy access for myoblast injection, and its distinct innervation as the last muscle by the common peroneal nerve such that its mechano-physiologic properties can be objectively measured in vivo without movement artifacts. Its small size necessitates fewer injections of myoblasts to effect improved muscle function. The left EDB will not be injected and thus will serve as a control.

Immediately prior to myoblast injection and three months after, the mechanophysiologic properties of the left and right EDBs will be measured in vivo with a force transducer connected via a stiff wire looped around the big toe. Isometric twitch and tetanus tensions (at 30 stimuli/sec) will be measured in response to supramaximal peroneal nerve stimulation at the ankle. Comparison of the mechanophysiology between the right and the left EDBs three months after myoblast injection, and between the same EDBs before and after myoblast injection will determine if the procedure exerts any beneficial effect on the dystrophic muscle. Clinical success is indicated by increase in twitch and tetanus tensions in the right EDB as compared to decrease with the left EDB with time.

OTHER MONITORS

A review of the pathogenesis using the C57BL/6J-dy^{2J}dy^{2J} mouse[4] as an animal model for hereditary muscle degeneration

Myoblast Transfer Therapy
Edited by R. Griggs and G. Karpati
Plenum Press, New York, 1990

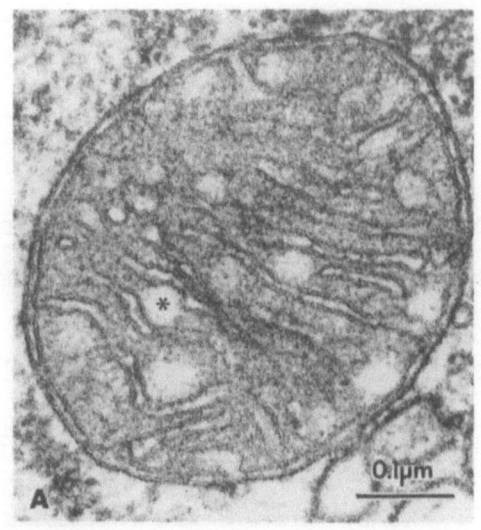

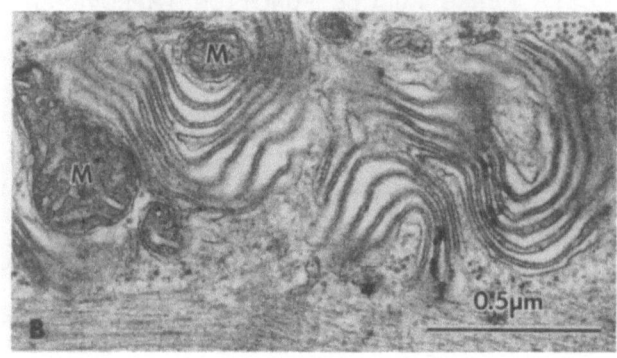

Fig. 1. (A) Mitochondria with areas of low density (*) in the matrix space. These mitochondria resemble those exposed to high free calcium. (B) Degenerative mitochondria. (From Ref. 15)

will reveal the spectrum of parameters that can be used to monitor clinical success or phenotypic transformation. Our laboratory has concentrated on attempts to improve muscle genetics, structure, function, and the behavior and life-span of the dystrophic mouse[5-10].

PATHOGENESIS OF MURINE DYSTROPHY

In 1972, Law and Atwood reported the first direct evidence of dystrophic muscle cell membrane abnormality. Using "cable" property measurements, it was shown that the cell membrane of the dystrophic mouse soleus was leaky to sodium and potassium ions. The leakage of sodium ions resulted in the lowering of the sodium equilibrium potential such that abortive spikes of reduced amplitudes and rates of rise were generated. Such abortive spikes would not overshoot the resting membrane potential. They propagated close to the endplate region but would not propagate the whole length of the muscle fiber[12]. They generated localized contractions that pulled on the

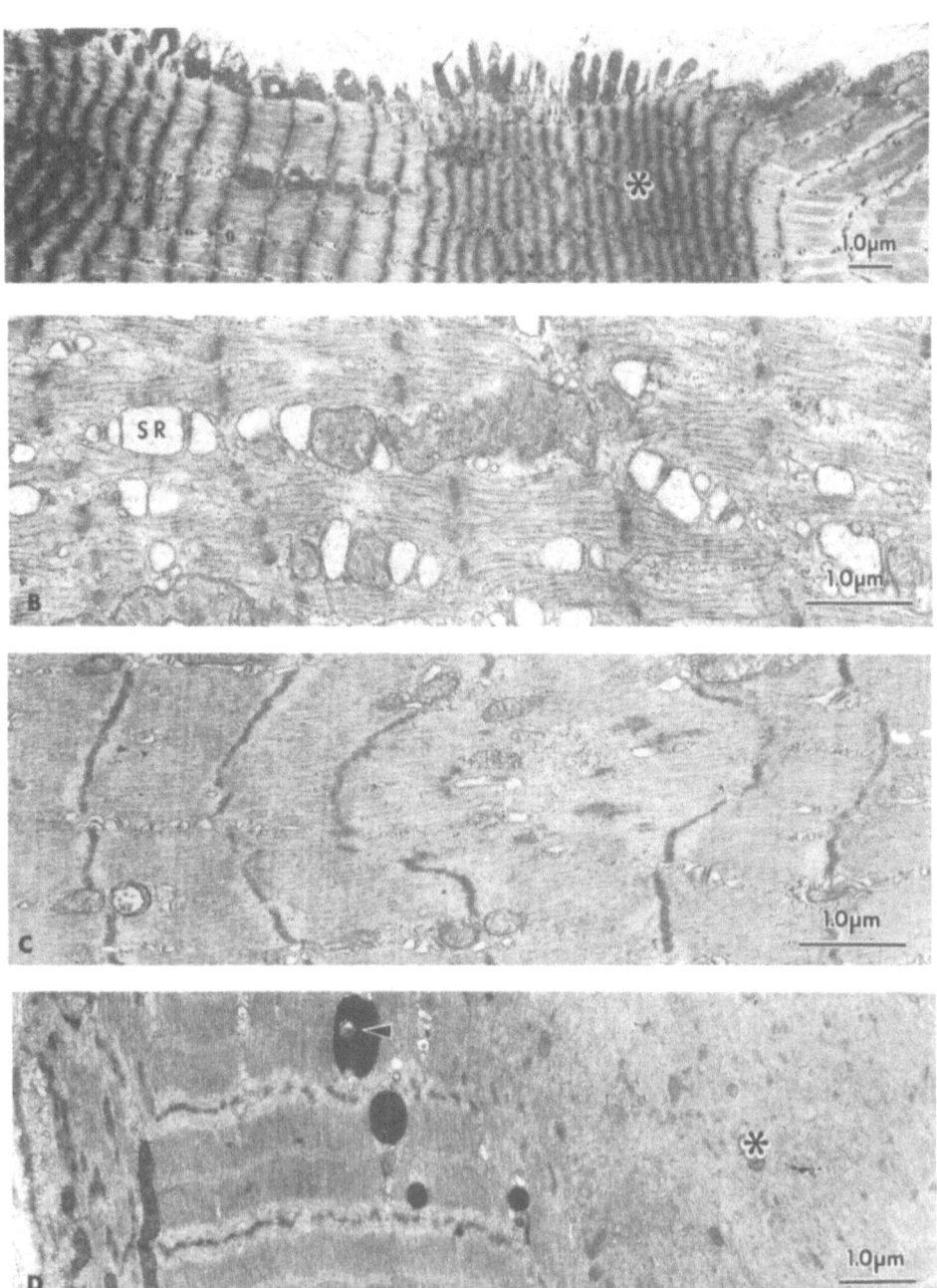

Figure 2. (A) Localized hypercontraction. Highly contracted
regions (*) were observed in some myofibrils composed of
about 20 adjacent sarcomeres. The I-band was absent and
the Z-line was broadened. These regions always contained
abnormal mitochondria (B) Swollen sarcoplasmic reticulum
(SR) could be observed. (C) Sarcomeres adjacent to the
hypercontracted area showed myofibrillar disarrangement
and Z-line streaming. (D) Localized myofibrillar disarray
(*) and lysosome-like dense bodies(▲). (From Ref. 15)

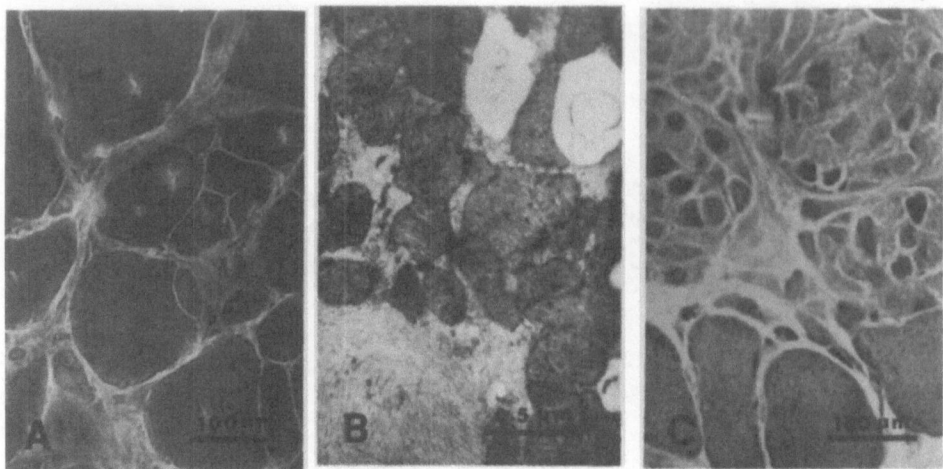

Fig. 3.   (A) Dystrophic characteristics included fiber splitting, central nucleation, variation in fiber shape and size, increase in intercellular connective tissues and infiltration of connective tissues into myofiber (arrow) due to breakdown of sarcolemma. Modified Gomeri Trichrome stain. (B) Electron micrograph at arrow showing the absence of sarcolemma. Collagen fibers are now adjacent to subsarcolemmal mitochondria. (C) Phagocytic necrosis showing replacement of contractile filaments by connective tissues, with debris to be removed by macrophages. Modified Gomeri Trichrome stain.

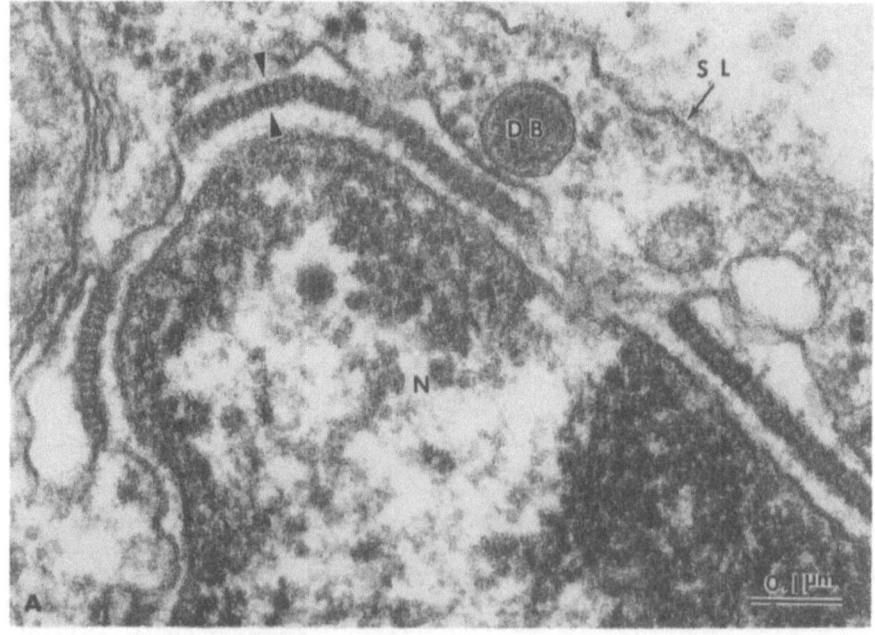

Fig. 4. (A) A discontinuous electron dense band of uniform width (240 Å) was found between the inner and outer membranes of the nuclear envelope. SL, sarcolemma; DB, dense body; N, nucleus.

244

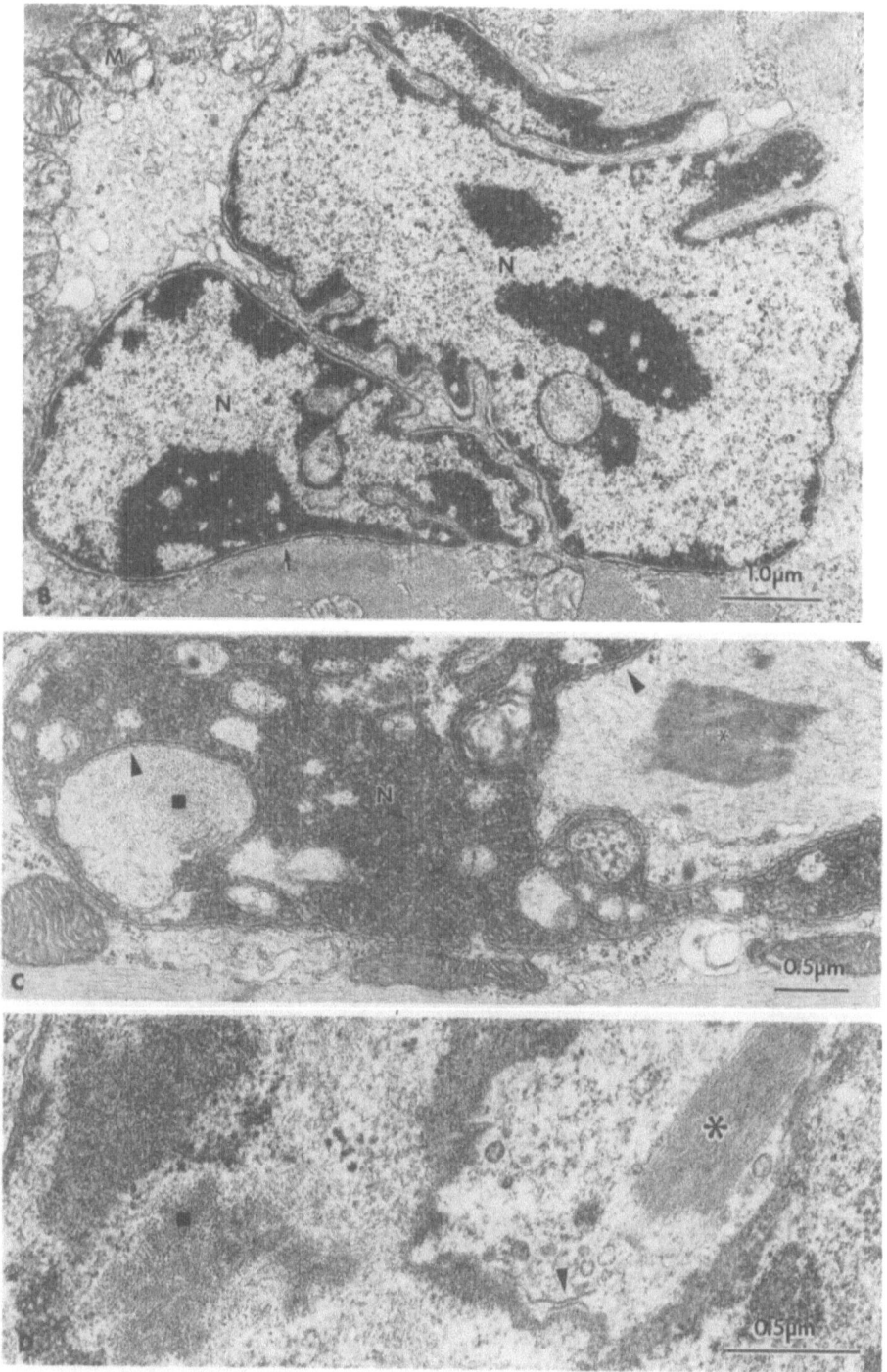

Fig. 4. (B) Cytoplasmic invaginations associated with abnormal nuclear membrane having the discontinuous band encirclement (↑). M, mitochondria. (C) Cross-sections of cytoplasmic invaginations showing Z-line fragment (*) and actin filaments (■) surrounded by double nuclear membrane (▲). (D) A double membrane fragment (▲), together with the direct apposition of Z-line fragment (*), actin filaments (■) and chromatin implied degeneration of the nuclear membrane. (From Ref. 15)

sarcomeres farther away from the endplates. These latter sarcomeres were not activated by the abortive spikes to contract[12,13]. One or more tears in the plasma membrane commonly known as the delta lesions[14] eventually occurred as a result of such pulling[12,13].

Through the delta lesions came the influx of calcium ions down the steep calcium concentration gradient. The excessive intracellular free calcium caused the local mitochondria to become abnormal, ruining the ATP generating system (Fig. 1). Sarcomeres close to the delta lesions generated localized hypercontraction (Fig. 2A) because of the absence of ATP and the presence of excessive calcium ions[13,15]. The latter also caused the swelling of the sarcoplasmic reticulum[15] (Fig.2B). Neighboring sarcomeres responded to the localized hypercontraction with myofibrillar disarrangement and Z-line streaming[15,16] (Fig. 2C). Eventually lysosomal activity sets in (Fig. 2D), causing muscle degeneration and weakening[15]. All of these ultrastructural parameters may be used to monitor phenotypic transformation following myoblast transfer therapy.

Histological abnormalities include central nucleation, muscle fiber splitting, variation in fiber shape and size, and infiltration of connective tissue through the sarcolemma into the muscle fiber. There is also increase in intercellular connective tissue. Phagocytic necrosis occurs as the fiber debris are removed by macrophages (Fig. 3). Such structural abnormalities should be less apparent in dystrophic muscles with successful myoblast transfer.

Speaking of abnormality of structural protein in the membrane, Figure 4A shows an electron-dense band (240 Å in

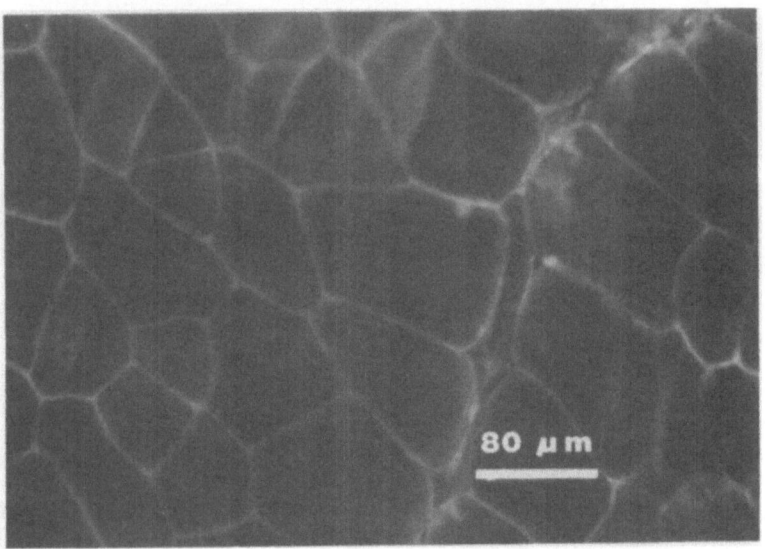

Fig. 5. Frozen cross-section of normal human muscle fibers stained immunocytochemically with anti-dystrophin antibody. Immunofluorescence was localized to the sarcolemma. Staining procedure followed that of Partridge et al[25].

width) between the inner and the outer membranes of the nuclear envelope[15]. Cytoplasmic invaginations were found in such nuclei indicating that the nuclear membrane was abnormal (Fig. 4B). Cross-sections of these invaginations revealed the double nuclear membrane at the early phase of degeneration (Fig. 4C). At later stage, only fragments of the nuclear membrane remained (Fig. 4D). Thus, membrane abnormalities was present not only in the sarcolemma but also in the nuclear membrane. This raises the question as to whether dystrophin[17-19] is present in the normal nuclear membrane and absent in the DMD nuclear membrane.

Fiber type differentiation is another useful parameter to monitor phenotype transformation after myoblast transfer. Fiber type differentiation is less defined in dystrophic muscles, possibly due to failure to fully mature[20]. Provided that it is properly reinnervated, the transformed dystrophic muscle injected with normal myoblasts should show a normal fiber type differentiation[8]. Furthermore, if donor myoblasts are surviving and developing in the host muscle, the latter should develop into a larger and heavier muscle.

The most important monitor is the objective measurements of muscle function. This involves supramaximal nerve stimulation to elicit maximal isometric twitch and tetanus tensions. Comparison will then be made between the myoblast-injected versus the contralateral uninjected and/or the sham-injected muscle. Comparison should also be made of the same muscles before versus after injection at various periods of time.

Phase 1 clinical trial should involve myoblast injection into one single muscle to establish the safety and efficacy of the procedure. Proven successful, Phase 1 will evolve into Phase 2 in which major muscle groups in the leg and the intercostal muscles will receive myoblast transfer. Isokinetic dynamometry will be useful to monitor both concentric and eccentric passive isokinetic contractions[21,22]. Measurements of forces at various velocities of contraction, and at various angles or ranges of movements can be obtained and compared between test and control muscle groups with time.

Improvement in function of major muscle groups in the legs should translate into behavioral improvement in gait, posture, and balance for the patient all of which can be measured objectively[23,24].

Finally, it is extremely important to monitor those biochemical changes that are associated with the genetic changes. Dystrophin is a very useful biochemical phenotype marker that is present in normal muscle cells (Fig. 5) but not in DMD muscle cells. Its presence in the myoblast-injected dystrophic muscle will indicate the survival, development, and functioning of donor myoblasts in the host muscle[25].

ACKNOWLEDGMENTS

The authors gratefully acknowledge Dr. Eric Hoffman for the supply of anti-dystrophin antibody, and Ms. Claudine O'Steen and Ms. Helen Ham for typing. This research was supported by USPHS Grant NS-20251 and NS-26185 from the National Institute of Neurological and Communicative Disorders and Stroke and by a grant from the Muscular Dystrophy Association, Inc.

REFERENCES

1.   J. Walton, "Disorders of Voluntary Muscle", Churchill Livingstone, New York (1981) p.487.

2.   R.K. Roelofs, G.S. deArengo, P.K. Law, D. Kinsman and J.H. Park Treatment of Duchenne muscular dystrophy and D. L-Penicillamine: Results of a double-blind trial. Arch. Neurol. 36:266 (1979).

3.   McComas A.J., M.E. Brandstater, A.R.M. Upton, J. Delbeke, C. deFaria and K. Toyonaga. Sick motor neurons and dystrophy: A reappraisal, in "Pathogenesis of Human Muscular Dystrophies".   L.P. Rowland ed. Excerpta Medica, Amsterdam (1977) p. 180.

4.   H. Meier, and J.L. Southard. Muscular dystrophy in the mouse caused by an allele at the dy-locus. Life Sci. 9:137 (1970).

5.   P.K. Law. Reduced regenerative capability of dystrophic mouse muscle. Exp. Neurol. 60:231 (1978).

6.   P.K. Law and J.L. Yap.   New muscle transplant method produces normal twitch tension in dystrophic muscle.   Muscle & Nerve 2:356 (1979).

7.   P.K. Law. Beneficial effects of transplanting normal limb-bud mesenchyme into dystrophic mouse muscle.   Muscle & Nerve 5:619 (1982).

8.   P.K. Law, T.G. Goodwin, and M.G. Wang.   Normal myoblast injections provide genetic treatment for murine dystrophy. Muscle & Nerve 11:525 (1988).

9.   P.K. Law. T.G. Goodwin, and H.J. Li.   Histoincompatible myoblast injection improves muscle structure and function of dystrophic mice. Transplant Proc. 20:1114 (1988).

10. P.K. Law, H-J. Li, T.G. Goodwin, G. Ajamoughli, X.Y. Zhang and M. Chen. Pathogenesis and treatment of hereditary muscular dystrophy. In press.

11. P.K. Law, and H.L. Atwood. Nonequivalence of surgical and natural denervation in dystrophic mouse muscles. Exp. Neurol. 34:300 (1972).

12. P.K. Law, H.L. Atwood, and A.J. McComas. Functional denervation in dystrophic mouse muscles. Exp. Neurol. 51:434 (1976).

13. P.K. Law. "Leaky" membrane causes weakness in dystrophic soleus fibers. Adv. Physiol. Sci. 5:213 (1980).

14. B. Mokri and A.G. Engel. Duchenne dystrophy: electron microscopic findings pointing to a basic or early abnormality in the plasma membrane of the muscle fiber. Neurology 25:1111 (1975).

15. P.K. Law, A. Saito, and S. Fleischer. Ultrastructural changes in muscle and motor end-plate of the dystrophic mouse. Exp. Neurol. 80:361 (1983).

16. A. Saito, P.K. Law. and S. Fleischer. Study of neurotrophism with ultrastructure of normal/dystrophic parabiotic mice. Muscle & Nerve 6:14 (1983).

17. E.P. Hoffman, R.H. Brown, L.M. Kunkel, Dystrophin: the protein product of the Duchenne muscular dystrophy locus. Cell 51:919 (1987).

18. K.P. Campbell and S.D. Kahl. Association of dystrophin and an integral membrane glycoprotein. Nature 338:259 (1989).

19. G. Salviati et al. Cell fractionation studies indicate that dystrophin is a protein of surface membranes of skeletal muscle. Biochem. J. 258-837 (1989).

20. E. Cosmos and J. Butler. Differentiation of muscle transplanted between normal and dystrophic chickens in "Research in Muscle Development and the Muscle Spindle" B.Q. Banker et al. ed. Excerpta Medica. Amsterdam, (1972) p. 149.

21. L. Gransberg and E. Knutsson. Determination of dynamic muscle strength in man with acceleration controlled isokinetic movements. Acta Physiol. Scand. 119:317 (1983).

22. E. Knutsson, and A. Martensson. Isokinetic measurements of muscle strength in hysterical paresis. EEG Clin. Neurophysiol. 61:370 (1985).

23. E. Knutsson and A. Martensson. Posture and gait in parkinsonian patients. In "Disorders of Posture and Gait." W. Bles and R.H. Brandt ed. Elsevier, Amsterdam, (1986) p.217.

24. J. Isaacson, L. Gransberg and E. Knutsson. Three-dimensional electrogoniometric gait recording. J. Biomechanics 19:627 (1986).

25. T.A. Partridge, J.E. Morgan, G.R. Coulton, E.P. Hoffman and L.M. Kunkel. Conversion of mdx myofibres from dystrophin-negative to -positive by injection of normal myoblasts. Nature 33:176 (1989).

Discussion of Dr. Law's paper

Dr. Mendell: Thank you very much, Peter. In general you seem to agree with the guidelines set forth by our previous speakers. I wonder if you could comment specifically on isokinetic monitoring of strength. Dr. Brooke's point was that maybe this would be very difficult for children.

Dr. Law: I think that computerized dynamyometry will allow one to accurately adjust for the strength of the patient.

I do not believe that there will be a problem with accurate monitoring.

Dr. Mendell: The next speaker is Dr. Michael Hudecki from the State University of New York in Buffalo. He has experience in clinical testing of the Mdx mouse and the dystrophic chicken.

MDX MOUSE AS THERAPEUTIC MODEL SYSTEM: DEVELOPMENT AND

IMPLEMENTATION OF PHENOTYPIC MONITORING

Michael S. Hudecki and Catherine M. Pollina

Department of Biological Sciences
State University of New York at Buffalo

INTRODUCTION

As Michael Brooke so aptly stated at the beginning of this session on phenotypic monitoring, it is imperative to have a quantifiable test system in place prior to assessing treatment entities for any of the muscular dystrophies (Brooke et al., 1981). In this manner, the various medical, experimental, and ethical considerations have been resolved in advance providing an unequivocal foundation for determining efficacy (regardless of the particular therapeutic approach under consideration). Similarly, in the preclinical study of myopathic animal models, it is equally pertinent to establish reliable phenotypic end-points before the implementation of a therapeutic study. Hence, efficacy or a lack thereof can be objectively and rationally determined against a backdrop of standardized markers of the disease.

For several years, our preclinical efforts have been directed at constructing and deploying a multi-component therapeutic test system using the autosomal recessive dystrophic chicken (Cosmos et al., 1980). Therapeutic approaches ranging from individual drug entities (e.g., Hudecki and Barnard, 1976; Hudecki et al., 1980, 1981) to exercise (Hudecki et al., 1978) and electrostimulation (Hudecki et al., 1985) have been evaluated against an array of quantifiable and standardized myopathic phenotypes including: weakness, stiffness, functional disability, plasma creatine kinase (CK) elevation, and muscle degeneration determined histomorphometrically and biochemically. On account of the prominent phenotypic commonality between the chicken and human myopathies (including the X-linked Duchenne disease), the avian disease has for many years been an invaluable animal model in which to systematically confirm as well as to explore drug efficacy.

A breakthrough of some proportion recently developed in the area of dystrophic chemotherapy when the X-linked dystrophic mouse (Bulfield et al., 1984) and dog (Cooper et al., 1988) were characterized and found to be nearly genetically homologous to the X-linked Duchenne form of

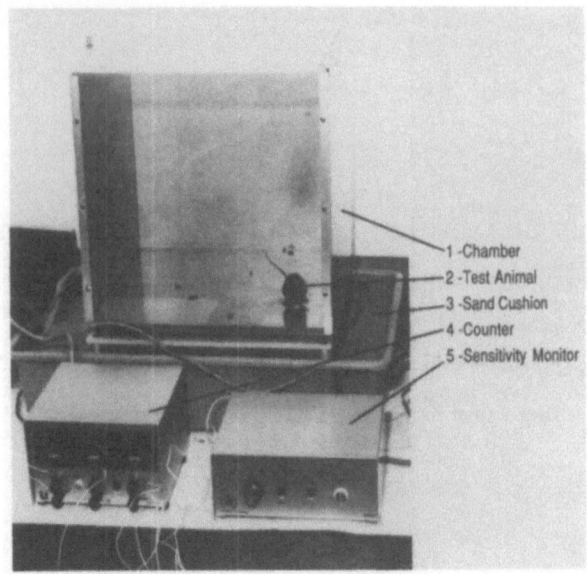

Fig. 1.   Photograph of activity chamber equipped with vibration-sensitive floor. Chamber is cushioned in sand to minimize extraneous vibrations.

muscular dystrophy.   Furthermore, dystrophin protein, the missing gene product in Duchenne dystrophy (Hoffman et al., 1987a) has also been shown to be absent in the mdx mouse (Hoffman et al., 1987b) and canine (Cooper et al., 1988) forms. In contrast, we in collaboration with Hoffman and Kunkel found dystrophin protein to be expressed at normal levels in various tissues of the dystrophic chicken, including the affected fast-twitch musculature (Hoffman et al, 1988). While the particular role of dystrophin in the myopathogenesis is as yet unknown, the lack of this protein in the X-linked animal and human diseases has led us to redirect our preclinical evaluations from the dystrophin-competent chicken model, and focus on the mdx mouse.

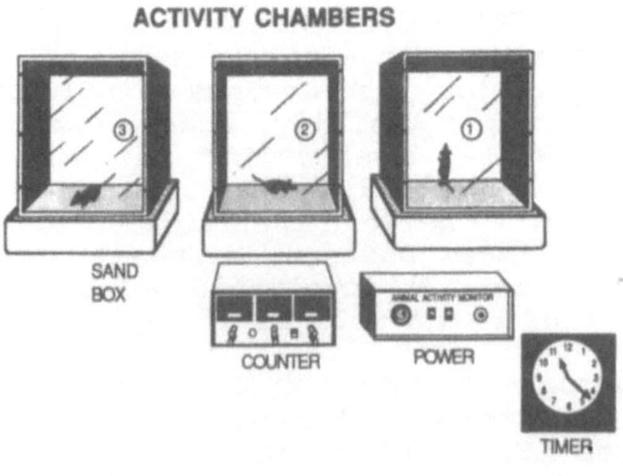

Fig. 2.   Illustration of three activity chambers connected to digital counter which record each physical movement as an "event".

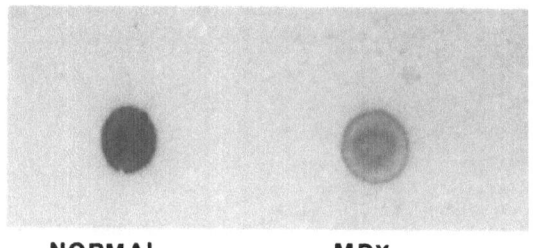

NORMAL   **MDX**

Fig. 3. Photograph of normal and mdx blood creatine kinase
activity reacted on filter paper and visualized under a
long-range UV light.

Consequently our present overall aim like in our prior
avian studies is to construct a multi-component, standardized
test system with the mdx mouse which will be responsive to
therapeutic intervention (including the subject of this
conference: myoblast transfer technology and its attendant
application).   Towards this goal, our initial quantitations
have included: non-invasive functional testing, endurance
measurements, blood creatine kinase activity, and selected
histomorphological and biochemical determinations.

METHODS

Functional Testing

Taking advantage of the natural curiosity and motor
behavior of mice, a small animal activity chamber (Lafayette
Instruments, Lafayette, Indiana) equipped with a 1x1 foot
vibration-sensitive floor (Figure 1) was utilized to record
spontaneous physical movements such as walking, preening,
stretching and scratching.   Each physical movement is
registered automatically by a digital counter (Figure 2) as an
"event".   Typically each mouse is tested for a continuous 20
minute period with the results expressed as EPM (events per
minute) or EPM/gm (EPM per gram body weight).

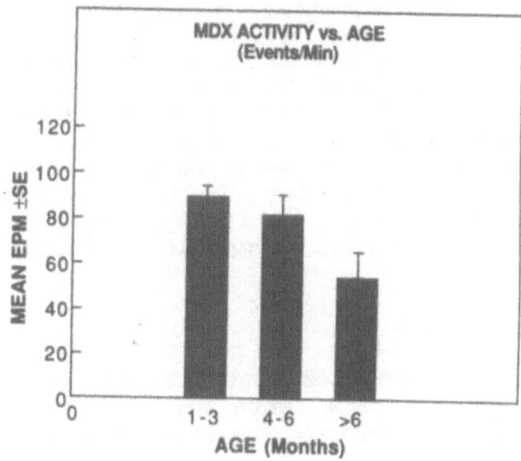

Fig. 4. Mean activity measurements of mdx mice versus age.
EPM = Events per minute +/- standard error.

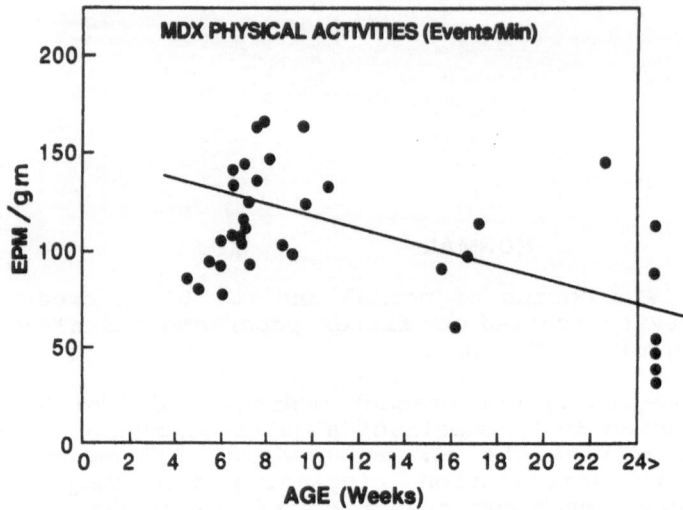

Fig. 5. Individual activity measurements of mdx mice versus age. EPM = Events per minute. Each point represents a single animal.

CK Activity

Blood levels of creatine kinase (CK) activity were determined either by a commercially available UV CK test kit procedure (Kit No. 46, Sigma Chemical Co., St. Louis, MO), or a fluorescence filter paper spot test (Figure 3) developed by Orfanos and Naylor (1984). The latter technique offers a major advantage peculiar to a small animal like the mouse; i.e., only a drop of whole blood is required which can be easily obtained at regular intervals by capillary tube from a tail vein puncture.

Muscle Histomorphology

Transverse sections of the gastrocnemius, extensor digitorum longus, tibialis and soleus muscles were excised, fixed and stained with hematoxylin and eosin, and trichrome as described before (Hudecki et al., 1985). Sections were analyzed by a computer-assisted image analysis system connected to a digitizing tablet (Houston HiPad, Houston, TX) whereupon the microscopic image field is projected via camera lucida for tabulation and/or measurement with a lighted cursor. Specific reference parameters which were assessed include: relative fiber necrosis, regeneration (viz., nuclear number and positioning within fibers), splitting, vacuolation, diameter, and infiltration of fibrous, lymphoid or adipose tissues.

RESULTS AND DISCUSSION

The spontaneous physical behavior of mdx mice appears to decrease with age (Figure 4) when results were obtained during single 20 minute test periods and were expressed as events per minute (EPM +/- SE).

Reduced activity with age was also observed when the test values of each animal were illustrated (Figure 5). The

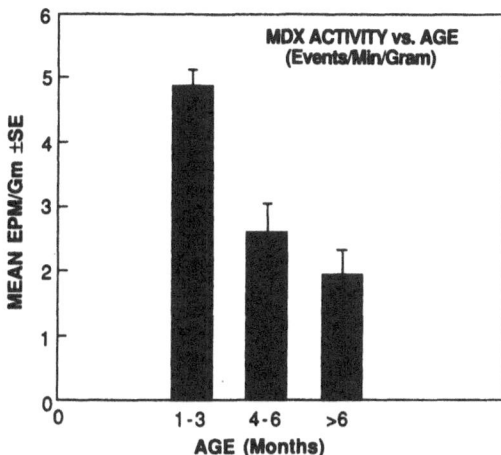

Fig. 6. Mean activity measurements of mdx mice versus age. EPM/gm = Events per minute per gram body weight +/- standard error.

tendency towards decreasing activity was accentuated and made more consistent when the same 20 minute test results were expressed on a per gram body weight basis (Figure 6).

With increasing age and accompanying body weight changes, the slope of the line of best fit increases (Figure 7) compared to values where body weight changes were taken into consideration (Figure 5).

In data not shown, neither the time or the frequency of testing (i.e., morning vs. afternoon, daily vs. weekly), nor the sex of the mdx animals had a significant effect on the

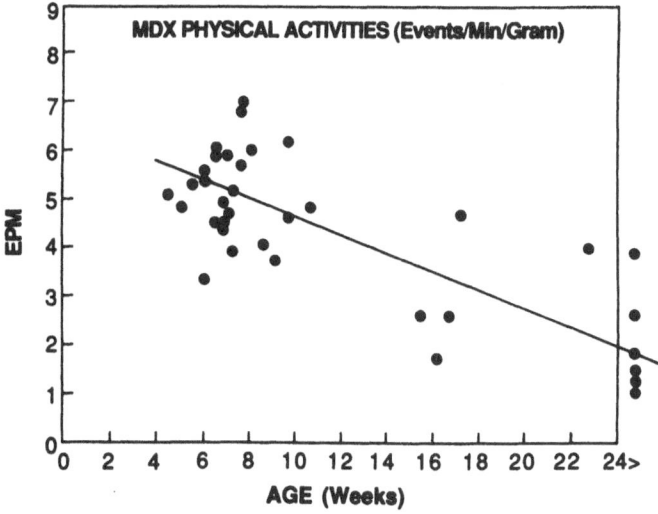

Fig. 7. Individual activity measurements of mdx mice versus age. EPM/gm = Events per minute per gram body weight. Each point represents a single animal.

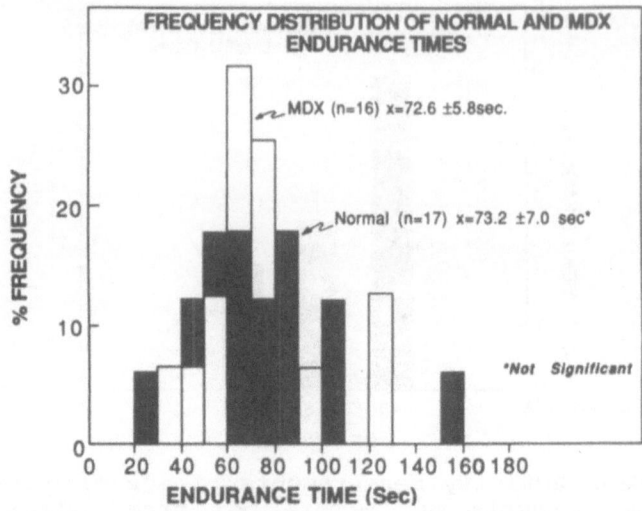

Fig. 8. Frequency distribution of endurance times determined on normal and mdx mice running continuously on a chamber treadmill at 9.5 cm/sec.

activity measurements illustrated in Figures 4-7. Furthermore, in all the one- to 14-month old mice tested, there were no significant differences in activities (expressed as EPM or EPM/gm) between the mdx mice and age-, sex-, and body weight-matched normal C57B1/10 mice. Hence mdx mice after one month appear to express "normal" spontaneous functioning determined within the limits and conditions of the test chambers.

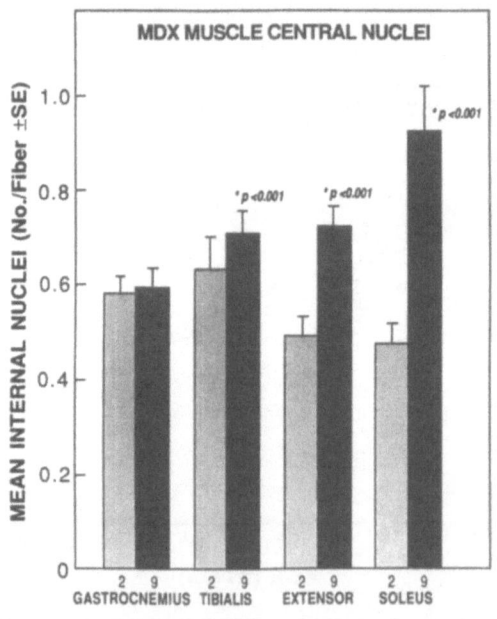

Fig. 9. Relative number of centrally positioned nuclei in mdx muscles at 2 and 9 months of age.

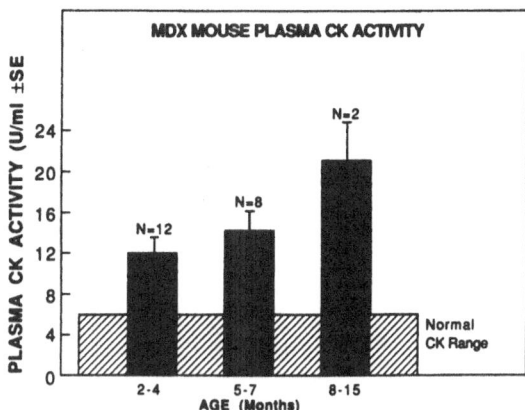

Fig. 10. Plasma creatine kinase activity of mdx mice versus age. Grey bar represents the upper range of normal enzyme activity throughout the same age period.

However, in order to determine if there were quantifiable differences in physical endurance between the mdx and C57Bl/10 normal mice, the animals were run continuously until exhaustion on a chambered treadmill set at 9.5 cm per second. As can be seen in the frequency distribution of endurance times (Figure 8), there were no significant differences in the clocked fatigue times between the mdx mice (72.6 +/- 5.8 seconds, N=16) and the normal times (73.2 +/- 7.0 seconds, N=17).

The expression of normal-like spontaneous functional behavior (viz., EPM or EPM/gm quantitated in activity chamber) or endurance (viz., time to exhaustion quantitated on treadmill) by the mdx mouse appears to be an attribute of its

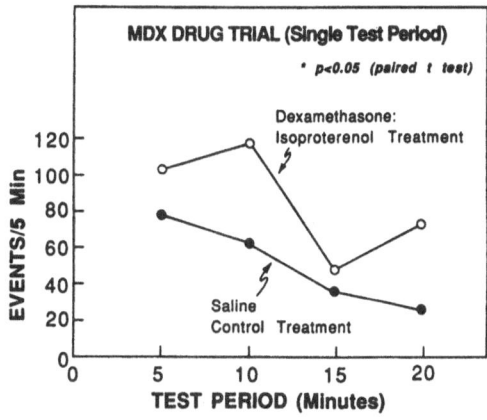

Fig. 11. Typical 20 minute test activity profile of mdx mice treated with dexamethasone and imipramine (0.02 and 2.0 mg per Kg respectively). Twelve month old mdx mice (N=3) were treated with either the drug combination or saline diluent (N=3) via a two week capacity implantable osmotic mini-pump over a 6 week period. Pumps were surgically positioned in the mid-dorsal cervical/intrascapular region of each mouse.

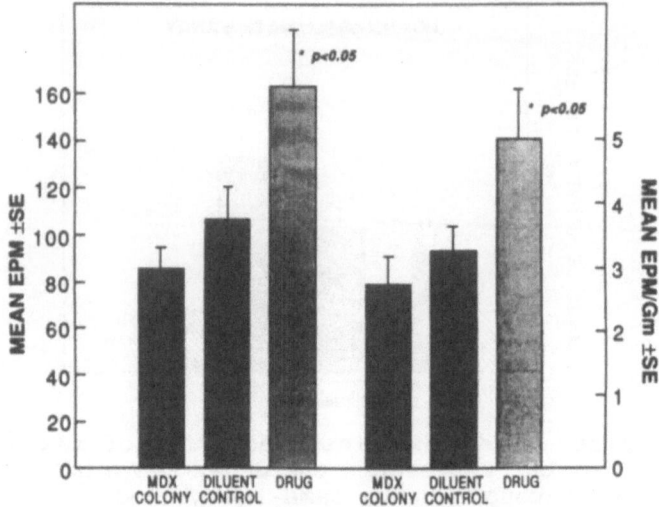

Fig. 12. Mean activity measurements of mdx mice treated (gray
bars) with dexamethasone/imipramine combination
(conditions as in Fig. 11). Solid bars refer to values
from saline treated mice, and from non-experimental mdx
colony mice. Mean EPM and EPM/gm =events per minute, and
events per minute per gram body weight +/- standard error,
respectively.

relative success at muscle fiber regeneration (Cullen and
Jaros, 1988). As seen in the centrally located myonuclei as
a histological index, prominent fiber regeneration is already
apparent in four hind-limb muscles at two months of age (Figure
9). Moreover, the regenerative process tends to increase
further with age as seen in the significantly higher values at
9 months (with exception of the gastrocnemius muscle). In
contrast, normal C57Bl/10 myonuclei as would be expected were
confined to the periphery of the fiber. In spite of the
absence of dystrophin, it appears that the combination of fiber
growth (quantitated increases in fiber diameter) and
regeneration (quantitated increments in central myonuclei)

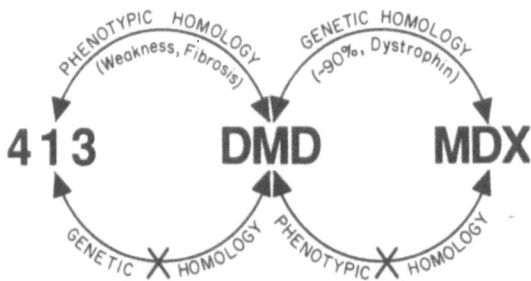

Fig. 13. Scheme representing the paradoxes in molecular
etiology confronting therapeutic investigations of
Duchenne muscular dystrophy (DMD) utilizing the X-linked
dystrophic mouse (mdx) and the autosomal recessive
dystrophic chicken (413) as preclinical animal model
systems.

258

provides the basis for the relatively normal functioning of the mdx mouse.

The progressive increase with age in plasma activity in mdx mice (Figure 10) appears to be related to the increased rate of myofibrillar protein turnover (Sano et al., 1988; Turner et al., 1988) concomitant with the extant regeneration rather than as a secondary expression of the dwindling fiber necrosis. On the other hand the early elevation in plasma CK is most likely a consequence of the early necrotic events (Sawada et a., 1986). With the filter paper CK assay (Orfanos and Naylor, 1984) which requires only a single drop of blood obtained by a relatively non-invasive tail puncture technique, it is anticipated that this enzyme value will be a useful adjunct to the functional testing of the mdx mice.

To determine if the physical behavior of mdx mice could be quantifiably altered by therapeutic intervention, one year old animals in a preliminary trial were treated with dexamethasone and imipramine administered by an implantable osmotic mini-pump (Alza Co., Palo Alto, CA) over a 6 week period. Concurrently the drug-treated and saline-treated control mice were tested for physical activity three times per week as described earlier. Figure 11 shows results from a typical 20 minute test period. Drug-treated mdx mice exhibited significant elevation in activity (p<0.05 paired t test) compared to the control values which tended to decline throughout the 20 minute test period.

Pooled data from the 6 week trial also showed a significant increment in the physical behavior of the treated mice (Figure 12). Furthermore, there appears to be a "diluent/implant pump" effect on physical activity since the diluent control values (i.e., mdx mice implanted with saline loaded pumps) were found to be higher than those obtained from the "colony" mice (i.e., mdx mice not implanted with pumps).

To date we have learned that mdx mice express normal-like spontaneous physical behavior and endurance, Hence a practical aim of therapeutic efforts involving this X-linked myopathy is to elicit above-normal functioning as seen in our initial trial with the dexamethasone/imipramine combination. In this connection, we plan to alter the conditions of our physical testing to enhance the overall sensitivity of the test.

Firstly, we plan to include challenges (e.g., ramp and ladder assemblies) within the activity chamber itself. In this manner the mice will be presented with optional barriers encouraging expression of a greater range of physical activities not presently allowed on the flat vibration-sensitive floor of the chamber (Figures 1 and 2). Secondly, we will incorporate a customized variable-speed, multi-laned treadmill (Law and Schafer, 1989) in our protocol in order to quantitate on a regular basis the physical endurance properties of the mice. With dual functional testing (i.e., activity chamber and treadmill), we expect to be able to quantitatively and selectively screen the potential efficacies of such disparate approaches to mdx therapies as drug perfusion (e.g., Christie, 1988) and myoblast transfer (Partridge et al., 1989; Karpati et al., 1989; Law et al., 1988). With regard to the former approach, we presently have a number of trials in progress consisting of various drug entities and the phenotypic monitoring of muscular functioning and endurance, plasma CK and myohistopathology. With regard to the latter approach, we will be collaborating with the principals of this conference in the assessment of the functional activity and endurance of mdx mice whose hind limb muscles have been bilaterally injected with dystrophin-competent normal myoblast cells. Furthermore, as suggested by the work of Morgan and Partridge (1989), we plan to functionally test mdx mice whose hind limb musculature have been irradiated at an early age in order to arrest the overt post-necrotic regeneration resulting from satellite cell proliferation and fusion. With this experimental approach, it is anticipated that a functionally and physically "weak" mdx mouse will develop which in turn may be a more relevant model system to test potential therapies for the progressive Duchenne weakness.

In therapeutic trials where the goal is to elicit positive change in muscle strength, function and histopathology, the paradox in expression between the animal and Duchenne diseases can confound therapeutic assessment (Figure 13). On the one hand, dystrophin deficiency leads to weakness and histofibrosis in the X-linked Duchenne myopathy and not in the murine mdx disorder. Alternatively the autosomal recessive defect expressed in the Line 413 chicken shows progressive weakness, disability, and connective tissue infiltration yet without an apparent deficit in dystrophin protein (Hoffman et al., 1988).

However, the mdx muscle shares with the chicken fundamental abnormalities in calcium transport and sequestration, and in calcium-mediated enzyme activities (Hudecki, et al., 1976; Turner et al., 1988). In this regard, we find that the chicken sarcoplasmic reticular calcium-activated ATPase is unresponsive to exogenous calmodulin (viz., "calmodulin resistance") while as expected the normal chicken muscle enzyme is calmodulin-stimulable (Galindo et al., 1988). Moreover, inositol 1,4,5-trisphosphate (IP_3) exclusively relieves the dystrophic enzyme inhibition. Similarly, we find that calmodulin-resistance of the calcium-activated ATPase, and the IP_3 effect on unblocking the enzyme inhibitor are also observed in mdx muscle (Zapalowski et al., 1989). Therefore, while uncertainties in phenotypic expression exist between the mouse and chicken models of Duchenne dystrophy, there are key biochemical commonalities among animal diseases which will serve to better characterize and experimentally treat the human myopathies.

ACKNOWLEDGMENTS

The authors are grateful to: James Pilc, Jean Hsaio, Sandy McAvoy, Dennis Bennett, Joseph Glavy, Jinacki Burroughs and Ave Maria Francis for their assistance in the mdx quantitations; Clyde Herreid and Julie Roberts for determining the endurance times; David Cooper for his suggestions in the development of the mdx test system; and Jim Stamos for his preparation of the illustrations and Dawn Styres for preparation of the manuscript. This work is supported by a grant from the Muscular Dystrophy Association's Task Force on Drug Development.

REFERENCES

Brooke, M. H., Griggs, R. C., Mendell, J. R., Fenichel, G. M., Shumate, J. B. and Pellegrino, R. J., 1981, Clinical trial in Duchenne dystrophy. I. The design of the protocol, Muscle and Nerve, 4: 186-197.

Bulfield, G., Siller, W. G., Wright, P. A. L. and Moore, K. J., 1984, X-chromosome-linked muscular dystrophy (mdx) in the mouse, Proceedings of the National Academy of Sciences USA, 81: 1189-1192.

Christie, K. N., 1988, Chymostatin has no apparent beneficial effect on muscular dystrophy in the mdx mouse, J. Neurol. Sci., 84: 341.

Cooper, B. J., Winand, M. J., Stedman, H., Valentine, B. A., Hoffman, E. P., Kunkel, L, M., Scott, M. O., Fishbeck, K. H., Kornegay, J. N., Avery, R. J., Williams, J. R., Schmechel, R. D. and Sylvester, J. E., 1988, The homologue of the duchenne locus is defective in X-linked muscular dystrophy of dogs, Nature, 334: 154-156.

Cosmos, E., Butler, J., Mazliah, J. and Allard, E. P., 1980, Animal models of muscle disease. Part 1. Avian dystrophy, Muscle and Nerve, 3: 427-435.

Cullen, M. J. and Jaros, E., 1988, Ultrastructure of the skeletal muscle in the X-chromosome linked dystrophic (mdx) mouse. Comparison with Duchenne muscular dystrophy, Acta Neuropathol., 77: 69-81.

Galindo J. G., Hudecki, M. S., Davis, F. B., Davis, P. J., Thacore, H. R., Pollina, C. M., Blas, S. D. and Schoenl M., 1988, Abnormal response to calmodulin in vitro of dystrophic chicken muscle membrane Ca^{2+}-ATPase activity, Biochemistry, 27: 7519-7524.

Hoffman, E. P., Brown, R. H. and Kunkel, L. M., 1987a, Dystrophin: The protein product of the Duchenne muscular dystrophy locus, Cell, 51: 919-928.

Hoffman, E. P., Hudecki, M. S., Rosenberg, P. A., Pollina, C. M. and Kunkel, L. M., 1988, Cell and fiber-type distribution of dystrophin, Neuron, 1: 411-420.

Hoffman, E. P., Monaco, A. P., Feener, C. C. and Kunkel, L. M., 1987b, Conservation of the Duchenne muscular dystrophy gene in mice and humans, Science, 238: 347-350.

Hudecki, M. S. and Barnard, E. A., 1976, Retardation of symptoms of dystrophy in genetically dystrophic chickens by chemotherapy, Res. Commun. Chem. Pathol. Pharmacol., 14: 167-176.

Hudecki, M. S., Caffiero, A. T., Gregorio, C. C. and Pollina, C. M., 1985, Effects of percutaneous electrical stimulation on functional ability, plasma creatine kinase and pectoralis musculature of normal and genetically dystrophic chickens, Exp. Neurol., 90: 53-72.

Hudecki, M. S., Kibler, P. K., Davis, P. J., Davis, F. B., Thacore, H. R., Pollina, C. M. and Blas, S. D., 1986, Abnormal gene expression of calmodulin in dystrophic chicken muscle, Res. Commun. Biochem. Biophys., 137: 507-512.

Hudecki, M. S., Pollina, C. M., Bhargava, A. K., Fitzpatrick, J. E., Privitera, C. A. and Schmidt, D., 1978, Effects of exercise on chickens with hereditary muscular dystrophy, Exp. Neurol., 61: 65-73.

Hudecki, M. S., Pollina, C. M., Bhargava, A. K., Hudecki, R. S., 1980, Screening of anti-seritoninergic drugs employing the genetically dystrophic chicken, Arch. Neurol., 73: 173-185.

Hudecki, M. S., Pollina C. M., Heffner, R. R. and Bhargava, A. K., 1981, Enhanced functional ability in drug-treated dystrophic chickens: Trial results with indomethacin, diphenylhydantoin and prednisolone, Exp. Neurol., 73: 173-185.

Karpati, G., Pouliot, Y., Carpenter, S. and Holland, P., 1989, Implantation of nondystrophic allogenic myoblasts into dystrophic muscles of mdx mice produces "mosaic" fibers of normal microscopic phenotype, in: "Cellular and Molecular Biology of Muscle Development", Alan R. Liss, Inc., New York.

Law, P. K., Goodwin, T. G. and Wang, M. G., 1988, Normal myoblast injections provide genetic treatment from murine dystrophy, Muscle and Nerve, 11: 525-533.

Law, P. K. and Schafer, B., 1989, Personal communication.

Morgan, J. E. and Partridge, T. A., 1989, Personal communication.

Orfanos, A. P. and Naylor, E. W., 1984, A rapid screening test for Duchenne muscular dystrophy using dried blood samples, Clin. Chim. Acta, 138: 267-274.

Partridge, T. A., Morgan, Coulton, G. R., Hoffman, E. P. and Kunkel, L. M., 1989, Conversion of mdx myofibers from dystrophin-negative to -positive by injection of normal myoblasts, Nature, 337: 176-179.

Sano, M., Wada, Y., Ii, K., Kominami, E., Katunuma, N. and Tsukagoshi, H., 1988, Immunolocalization of cathepsins B, H and L in skeletal muscle of X-linked muscular dystrophy (mdx) mouse, Acta Neuropathol., 75: 217-225.

Sawada, H., Tsuji, S., Kusumoto, S., Dai, Y. and Matsushita, H., 1986, Preclinical increase in activity of muscle microsomal trypsin-like protease in murine muscular dystrophy, C57Bl/10, mdx, FEBS Lett., 199: 193-197.

Turner, P. R., Westwood, T., Regan, C. M. and Steinhardt, R. A., 1988, Increased protein degradation results from elevated free calcium levels found in muscle from mdx mice, Nature, 335: 735-738.

Zapalowski, C., Hudecki, M. S., Davis, F. B., Davis, P. J., Pollina, C. M. and Blas, S. D., 1989, Characterization of a biochemical abnormality in the murine model (mdx) of muscular dystrophy, Clin. Res., 37: 464a.

Discussion of Dr. Hudecki's paper

Dr. Mendell: We will open the discussion for general participation at this time.

Dr. Russo: As a clinician it seems to me that, given our choice, we would prefer to study boys in the five to eight year group primarily because they would be more suitable for quantitative testing and also because they have a more linear downhill slope of change. On the other hand, the information from the cell biologists suggests that this may be the incorrect route. It might be better to perform myoblast transfers on boys two to three years of age because of greater satellite cell responsivity. I also think that whether a younger or older group is chosen, we need more objective measures of muscle function. Possibly something as simple as M-amplitude on maximum nerve stimulation which is totally involuntary and separate from patient effort.

Dr. Mendell: Since you did not direct that comment to anyone specifically, I will ask Dr. Munsat to respond.

Dr. Munsat: We are planning to incorporate maximum electrical stimulation of peripheral nerve and recording of twitch tension. I think it is a very important part of the evaluation procedure.

The question about which age to study is a very difficult one. I think it depends in great part on the goals and what outcome assessments you are looking at. For example, if you're merely interested in safety and dystrophin expression, that might suggest one group. If you are interested in more functional kinds of outcomes, then you might want to look at an older group. My guess is that eventually transfer will be studied in all age groups.

Unidentified participant: In reference to the point that you made that cell biologist would encourage myoblast transfers into younger patients, I am not sure that evidence has emerged from experimental studies that would indicate that older host muscle fibers would be more reluctant to accept transferred myoblasts.

Dr. Russo: Possibly I did misinterpret Dr. Blau's comments in the previous question of the responsivity of the satellite cells and their ability to go through multiple generations.

Unidentified participant: But the dividing cells are derived from the donor, not the patient.

Dr. Russo: No, I understand that.

Dr. Russman: Dr. Munsat, we have discussed this previously but maybe you can elaborate more about the quantitative testing in children. You suggested that the variability of repeated observations is greater and the question I have is how great is it? Is the quantitative analysis so sensitive that we need to observe the children over a longer period of time to determine if the strength is altered with the potential transplant?

Dr. Munsat: Yes, in adults the variance depending on the muscle group is six, eight, ten percent. It also varies with the stage of disease and intellectual capability. The variance in children again, depending on which group you are looking at, is another twenty-five to thirty percent greater than for adults. As you increase the measurement error of your testing equipment, as you increase the variability of the patient's status at time of entry, you increase the numbers required and the length of time for the observation.

Dr. Russman: Do you have any information as to what muscle groups you will be testing, and over how long a period of time, considering the variability of these observations? How long would it take us to learn something from the quantitative assessments assuming a transplant takes place?

Dr. Munsat: Well, I think we need to develop a quantitative data base on all available muscles in all Duchenne patients for as long as we can. I think we need to replicate the very excellent information in the CIDD data bank but do it with a testing technique that is a bit more sensitive. That is going to be a lengthy undertaking. I think if everybody gets to work on accumulating this data, maybe in three to five years we will have enough to establish natural history controls similar to the CIDD data base for manual muscle testing.

Dr. Brown: I wanted to pick up on a point that Dr. Blau made earlier and suggest to Dr. Mendell that possible guidelines be prepared for construction of pilot trials. This would include attention to the status of the donor and the status of the cells to be transplanted. I would suggest the following: any potential donor be serologically screened for infection both for HIV, CMV and various epitopes, not only at the time of donation of the cells but again at some subsequent time, perhaps a minimum of six months down the line.

We all know that HIV patients may be serologically negative for up to three or four years. Six months would be perhaps a minimum.

With respect to cell testing, it seems to me arguable that similar serological testing should be done on all myoblasts grown. In addition cultured myoblasts should be screened for infectivity. I think that EM should be carefully done on myoblasts to look for viral particles and finally, as Dr. Blau mentioned, the cells should be examined for tumorgenicity.

In another context, if one goes to private companies to look for mass culturing of myoblasts, it is undoubtedly the case that the FDA will get involved and will certainly demand such measures be taken.

My final question or point along these lines is to ask Dr. Law whether or not he has contacted the FDA and how they view the kinds of proposals he is engaged in?

Dr. Law: Thank you very much for your concern. All of the concerns you raised have been incorporated in the protocol but due to the time constraints, I was not able to discuss them.

As for thé FDA's posture, they are not going to regulate myoblast transfer because it is cell therapy. They are treating it just like an organ transplant.

As far as screening of myoblasts before injection, I think that all of this will be done in the clinical research center which has facilities to screen for viruses.

Dr. Spiro: With regard to the question of CT scan in muscle, we have had extensive experience in Duchenne and Becker dystrophy and spinal muscular atrophy.

For the myoblast transfer studies, I think the CT scan would lend itself very nicely.

One of the major advantages that I see using the CT scan is that you could probably extend the studies to a much younger population, as well as to an older patient group.

For example Becker's patients on CT scan have some muscles which appear very normal and other muscles which appear terribly involved. I think that type of patient would benefit from selective myoblast transplants.

Dr. Mendell: Thank you. Now we are going on to discuss the golden retriever animal model of dystrophin deficiency. We will hear presentations by Drs. Kornegay, Bartlett and Cooper before resuming floor discussion.

GOLDEN RETRIEVER MUSCULAR DYSTROPHY: MONITORING FOR SUCCESS

Joe N. Kornegay, Nicholas J.H. Sharp, Richard J.
Bartlett, Steven D. Van Camp, C. Tyler Burt, Wu
Yen Hung, Lester Kwock and Allen D. Roses

College of Veterinary Medicine, North Carolina State
University, 4700 Hillsborough St., Raleigh, NC
(Kornegay, Sharp, Van Camp), Duke University Medical
Center, Durham, NC (Bartlett, Roses), National
Institute of Environmental Health Sciences, Research
Triangle Park, NC (Burt), and Department of
Radiology, Memorial Hospital, University of North
Carolina, Chapel Hill, NC (Kwock)

INTRODUCTION

There are many phenotypic features that can be monitored
to, in the end, assess the success of any form of treatment for
muscular dystrophy (Table 1). Of course, in this setting, we
are speaking principally of myoblast transplantation.

GENERAL FEATURES

Strength certainly will be the final determinant that will
point to success, but there are several other features that
may, in some ways, be more objective. Some of these will be
discussed here with particular reference to golden retriever
muscular dystrophy (GRMD).

In referring to Table 1, it is important to recognize
echocardiography to emphasize that Duchenne muscular dystrophy
(DMD) and GRMD are systemic diseases that do not simply affect

Table 1. Selected Phenotypic Features to Monitor in
Evaluating Myoblast Transplantation

Strength
Body size, weight
Echocardiography
Creatine kinase
Electromyography
Nuclear magnetic resonance
Pathologic expression
Dystrophin

Myoblast Transfer Therapy
Edited by R. Griggs and G. Karpati
Plenum Press, New York, 1990

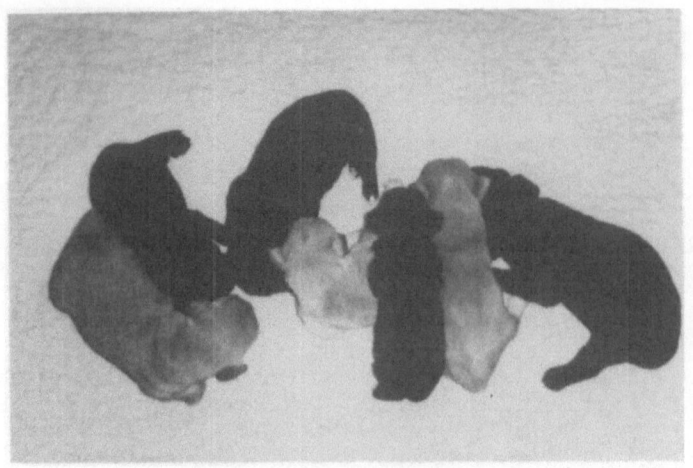

Fig. 1. Litter including some 10-day-old cross-bred dogs with GRMD. Two normal littermates flank five that are clinically affected. Note that the normal dogs are considerably larger.

skeletal muscle. Evaluating parameters for cardiac function and, potentially, even smooth muscle function will be important.

One of the features that could be relevant to assessing the effect of systemic, but not localized, therapy is dog size or weight. Cooper and his associates and our group have recognized that dogs with GRMD are stunted (Valentine et al., 1988; Kornegay et al., 1989). This is shown in Figure 1, where normal dogs can be clearly distinguished from stunted affected dogs.

CREATINE KINASE

Creatine kinase (CK) elevation is a dramatic phenotypic feature of the canine model (Kornegay, 1986; Valentine et al., 1988). Values in excess of 200,000 IU/l are noted at 1 to 2 days of age and plateau at around 10-20,000 IU/l. It will be difficult, of course, to assess localized transplantation by monitoring serum CK levels. However, some form of monitoring method, i.e., measuring values in blood from an isolated, treated muscle following repetitive nerve stimulation, may be of value.

ELECTROMYOGRAPHY

The golden retriever model may have particular relevance with regard to electromyographic (EMG) evaluation. There are certain EMG features that, while being recognized, are not a prominent feature in Duchenne patients. High frequency potentials, reaching up to 2,000 Hz on power spectral analysis, can be seen in GRMD dogs (Kornegay et al., 1988). Lessening of these potentials in the transplanted muscle could be an indicator of successful transplantation.

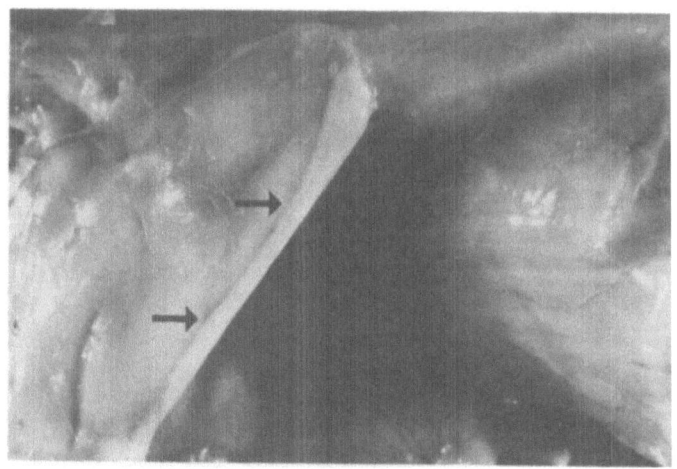

Fig. 2. Pelvic limb and caudal abdomen of a 1-day-old dog with
GRMD. The cranial head of the sartorius muscle (arrows)
is pale and easily demarcated from the adjacent muscles.
Marked mineralization of individual myofibers was noted
upon microscopic examination.

NUCLEAR MAGNETIC RESONANCE

Nuclear magnetic resonance (NMR) is a potential means of
evaluating the success of transplantation. Spectroscopy has
already been utilized in Duchenne patients; increased inorganic
phosphate relative to phosphocreatine is noted (Younkin et al.,
1987). We have found similar results in a limited number of
dogs with GRMD. Magnetic resonance images have also been
obtained from these dogs. Individual muscle bellies can be
easily demarcated in either sagittal or transverse planes.
Most muscles of affected dogs are atrophied. There is also
increased marbling, perhaps reflecting connective tissue and
fat deposition.

PATHOLOGY

The pathologic features of GRMD could be discussed at
length. Two features that are relatively specific for DMD have
also been noted in dogs with GRMD (Kornegay, 1988; Kornegay et
al., 1988). Unlike almost all other spontaneous canine
myopathies, mineralized myofibers are a prominent feature in
some affected dogs (Figure 2).

Myofiber mineralization seen on electron microscopic
examination is initially most pronounced in mitochondria
(Figure 3). Mineralization of myofibers is also a relatively
specific feature of DMD (Bodensteiner and Engel, 1978).
Reversal of mineralization, conceivably, could be an objective
means of recognizing successful treatment.

A particular pathologic feature that has long been
recognized in Duchenne patients is hypertrophy of the calf
muscles. Somewhat similar changes are seen in dogs with GRMD;
some proximal muscles at least appear to be hypertrophied
(Kornegay et al., 1988). We were initially unsure as to

Fig. 3. Electron photomicrograph of muscle from a 1-day-old
dog with GRMD. The myofiber at the bottom of the figure
is necrotic and has focal areas of increased electron
density corresponding to mineralization of individual
mitochondria.

whether this was true or pseudohypertrophy. Using morphometric
analysis, we have assessed mean myofiber diameter and, to
address a separate point, myofiber type distribution. There
seems to be true myofiber hypertrophy in certain muscles in
dogs with GRMD. On histographic evaluation of mean myofiber
diameters of these muscles, there is a shift to the right
(Kornegay et al., 1988).

Some investigators have shown that there is type 1
myofiber predominance in Duchenne patients (Dubowitz and
Brooke, 1973; Webster et al., 1988). We were initially
reluctant to suggest that the same was true of GRMD dogs
(Kornegay et al., 1988). However, more recently, our
morphometric data suggest that type 1 predominance may be
present in certain muscles of at least some dogs (Lane, 1988).
Since both mean myofiber diameter and myofiber type
distribution are quantifiable, monitoring changes subsequent
to treatment may be useful as an objective indicator of
successful treatment.

DYSTROPHIN

Dystrophin expression probably will be one of the most
definitive means of assessing successful treatment. As is true
of Duchenne patients and the mdx mouse, there is a paucity of
dystrophin in dogs with GRMD (Figure. 4). Interestingly, in
our initial studies using a nonaffinity purified antibody, we
noted scattered myofibers that were enveloped by what was
assumed to be dystrophin. However, when we dual labeled the
same sections using antibodies to erythrocyte-derived spectrin,
a similar, if not identical, pattern of staining was noted.
This suggested that the "dystrophin" staining was nonspecific,
i.e., perhaps reflecting cross reactivity of different
antibodies.

This raises the general question of putative protein up regulation. Clearly, if dystrophin expression is going to be a critical determinant of success, we must take into consideration other proteins that may be up regulated and potentially confused with dystrophin. In preliminary studies, we have evaluated expression of spectrin and actinin, which have homology with dystrophin. A pattern of subsarcolemmal expression similar to that of dystrophin was seen with spectrin. We were not able to definitely distinguish between the amount of spectrin expressed in normal versus GRMD dogs. In contrast, again based on only preliminary results, there appeared to be greater expression of α actinin in GRMD dogs. The staining was principally at the periphery of the myofiber and seemed to be confined principally to nuclei. In monitoring success, up-regulated α actinin or some other protein could be confused with dystrophin, if the antibodies used cross-react.

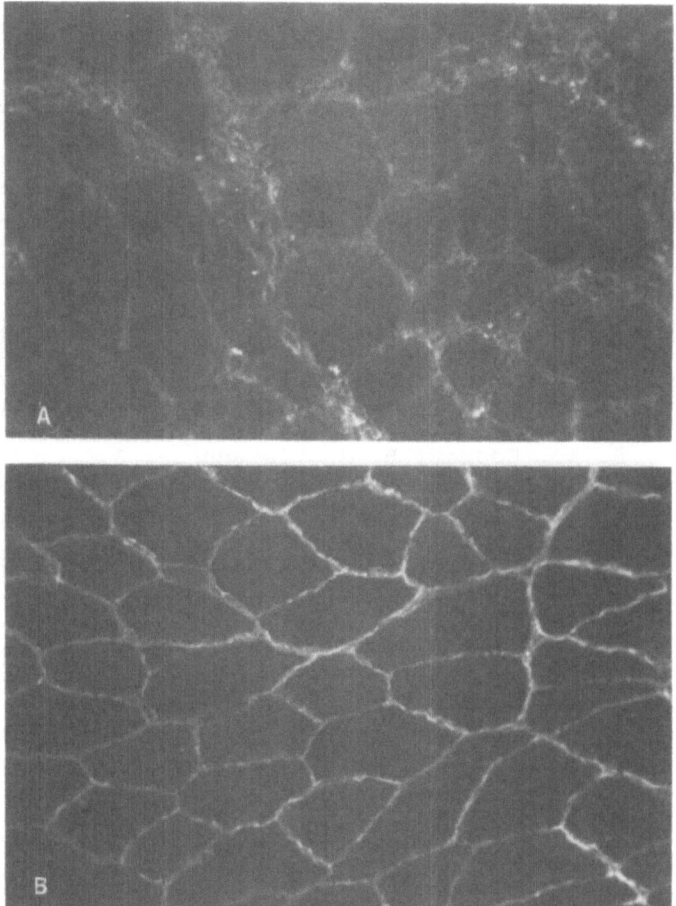

Fig. 4. Photomicrographs of muscle from a dog with GRMD (A) and a normal littermate (B). Note the typical subsarcolemmal pattern of dystrophin staining in the normal dog and the relative absence in the affected dog. (Antibody provided by Dr. Zubrzycka-Gaarn).

CONCLUSION

In conclusion, it should be emphasized that the golden retriever model is a progressive, X-linked form of muscular dystrophy that is phenotypically (Kornegay et al., 1988) and genotypically (Cooper et al., 1988) analogous to DMD. This model can be used to answer questions relevant to therapies that may not be appropriate for initial use in humans.

REFERENCES

Bodensteiner, J. B. and Engel, A. G., 1975, Intracellular calcium accumulation in Duchenne dystrophy and other myopathies: A study of 567,000 muscle fibers in 114 biopsies, Neurology, 28:439.

Cooper, B. J., Winand, N. J., Stedman, H., et al., 1988, The homologue of the Duchenne locus is defective in X-linked muscular dystrophy of dogs, Nature, 334:154.

Dubowitz, V. and Brooke, M. H., 1973, "Muscle Biopsy: A Modern Approach," W. B. Saunders Co., Philadelphia.

Kornegay, J. N., 1986, Golden retriever myopathy, in: "Current Veterinary Therapy IX," R. W. Kirk, ed., W. B. Saunders Co., Philadelphia.

Kornegay, J. N., 1988, Golden retriever muscular dystrophy: a model of Duchenne muscular dystrophy, Discuss. Neurosci., 5:118.

Kornegay, J. N., Tuler S. M., Miller, D. M., et al., 1988, Muscular dystrophy in a litter of golden retriever dogs, Muscle Nerve, 11:1056.

Kornegay, J. N., Sharp, N. J. H., Bartlett, R. J., et al., 1989, Phenotypic aspects of a canine model of Duchenne muscular dystrophy. Neurology, 39 (Suppl. 1):154.

Lane, S. B., 1988, Histochemical and morphometric analysis of normal and dystrophic myofibers in golden retriever dogs, in: "Proceedings of the Sixth Annual Veterinary Medical Forum, San Diego, 727.

Valentine, B. A., Cooper, B. J., deLahunta, A., et al., 1988, Canine X-linked muscular dystrophy. An animal model of Duchenne muscular dystrophy: clinical studies, J. Neurol. Sci., 88:69.

Webster, C., Silberstein, L., Hays, A. P., et al., 1988, Fast muscle fibers are preferentially affected in Duchenne muscular dystrophy, Cell, 52:503.

Younkin, D. P., Berman P., Sladky, J., et al., 1987, 31P NMR studies in Duchenne muscular dystrophy: age related changes. Neurology, 37:165.

MOLECULAR MARKERS FOR MYOBLAST TRANSPLANTATION IN GRMD

R.J. Bartlett[1,2], N.J.H. Sharp[1,2], W.-Y. Hung[1],
J.N. Kornegay[2] and A.D. Roses[1]

[1]Duke University Medical Center
Durham, NC and
[2]College of Veterinary Medicine
North Carolina State University
Raleigh, NC

Golden retriever muscular dystrophy (GRMD) has been proposed to be an animal model for Duchenne muscular dystrophy (DMD) (Kornegay et al., 1989). We have been studying GRMD to determine the underlying defect in this model using methodology and cDNA probes developed for the study of DMD (Koenig et al., 1987). Observed molecular differences between normal and affected animals may be useful for monitoring myoblast transfer in this model similar to what Ron Worton proposed as a means of defining tissue-specific markers for myoblast transplantation in DMD families (Worton reference this book). To illustrate the potential use of these markers of transplantation, an example of a human pedigree with a restriction fragment polymorphism which differs between individuals is illustrated in Figure 1. In this particular pedigree, there are two patients with deletions. Two potential markers would be useful for testing persistence of donor myoblasts after transplantation. The molecular markers in this case would be the presence of the deleted portion of the gene found in normal donor myoblasts, and if the donor was a non-carrier female, a gene dosage analysis which discriminates between the relative dosage of the X and Y-specific sequences would identify the donor cells.

Since transplantation may cause adverse immune responses in recipient individuals, donor and patient need to be histo-compatible. In a family such as this, the three sisters as well as the father would be potential candidates for myoblast donation once carrier status of the sisters was ruled out. In any case, demonstration of persistence of transplanted myoblasts using molecular markers would involve a re-biopsy of the transplanted region and testing for the presence of the donor-specific DNA evident when probed with the portion of the cDNA deleted in the patient.

In like manner, we have been examining the dog or canine dystrophin gene, to determine where we might find markers that

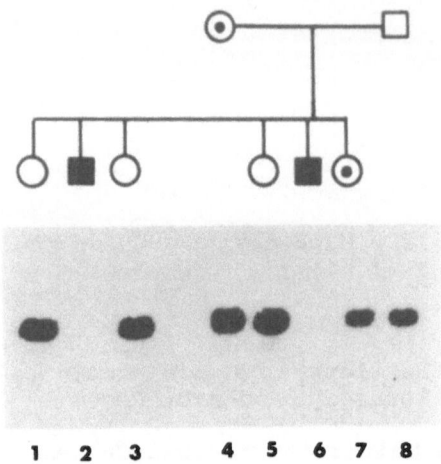

Fig 1. Human DMD Pedigree. DNA from the indicated individuals
was digested to completion with TaqI and blotted as before
(Bartlett et al., 1987). The probe used was 87-8 (Monaco
et al., 1986).

will differentiate normal and affected myoblasts (host vs.
donor) under conditions of experimental transplantation. An
example may be seen in Figure 2. A blot containing DNA from
normal and affected human and canine individuals was hybridized
to a small portion of the human cDNA probe designated 9-7
(ibid) specifically from the 5' end. There are five bands
observed in the lanes containing human DNA, suggesting five
exons. The same result is found when examining the lanes
containing canine DNA. Interestingly enough, you can see that
you find disease-specific polymorphisms that are present in the
dog (Figure 2A).

In like manner, polymorphisms are also seen at the 3' end
of the gene using the distal 1.0 kbp of the human probe (Figure
2B). With appropriate breeding selection of heterozygous
carrier females, we will have a number of useful molecular
markers for monitoring persistence of donor myoblasts in our
GRMD experiments. As a side note, a diagram of the structure
of the human dystrophin gene is shown in Figure 3. There are
believed to be a series of 65-75 exons in the human gene
(Koenig et al., 1987). When the entire human cDNA was used to
examine the canine gene, the canine gene was found to be
similar in structure: each probe recognized approximately the
same number of TaqI bands in human as well as canine DNA
(Figure 2). Thus, evolutionary conservation of this gene
includes not only sequence specificity as suggested by the
strong hybridization signals with the exon-specific cDNA
probes, but also a similar exon/intron structure as
demonstrated by the corresponding number of hybridizing bands
on Southern digests.

By changing our focus to the putative proximal 3' portion
of the canine gene, a different polymorphic pattern may be seen
when a portion of the human dystrophin cDNA from base-pairs
11,500 to 13,000[1] is used (Figure 4). Here the blot is

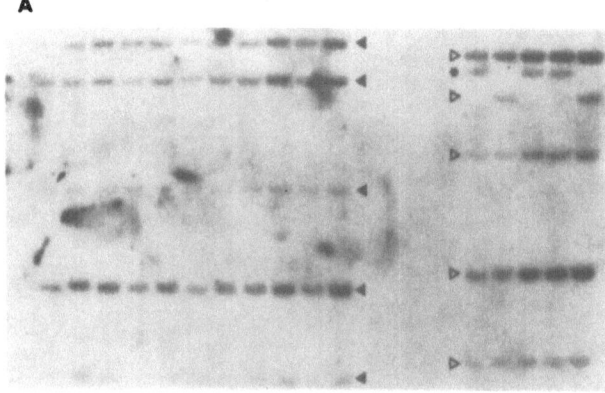

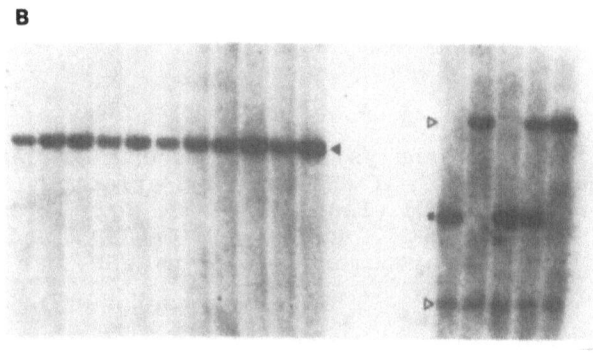

Fig. 2. Comparative Hybridization of Human and Canine Genomic
DNA. DNA from the human and canine individuals was
digested to completion and blotted as in Figure 1. The
blot was separately probed with human dystrophin cDNA
fragments: Panel A was probed with extreme 5' portion of
the cloned insert from 9-7 (Koenig et al., 1987); Panel
B was probed with the extreme 3' end of 63-1 (ibid).
Human samples are on the left, and canine samples are GRMD
male (1), normal male (2), carrier (3,4) and normal female
(5). Arrows indicate normal bands; asterisk indicates
abnormal bands.

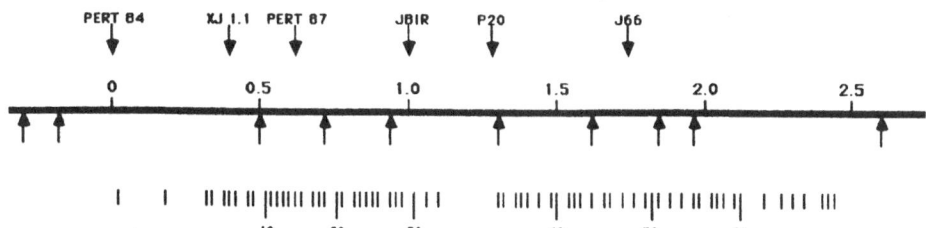

Fig. 3. A Diagrammatic Representation of the Human Dystrophin
Gene. The large size of the human dystrophin gene (Van
Ommen et al., 1986 and Burmeister et al., 1987) and exon
structure (Koenig et al., 1987; 1988) are merely
illustrated here. (This drawing is not to scale, nor is
the positioning of the exons intended to provide specific
locations.)

275

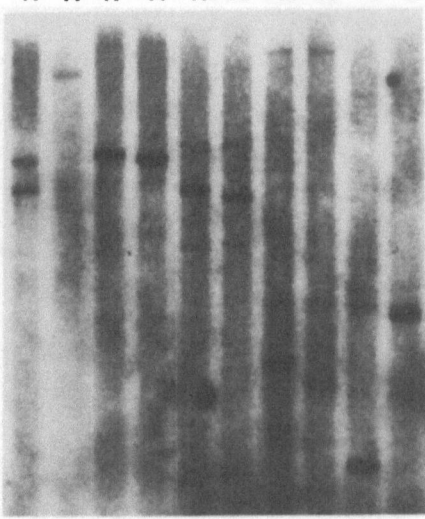

N A N A N A N A N A

1 2 3 4 5 6 7 8 9 10

Fig. 4. Multi-Enzyme Analysis of Normal and GRMD Affected Male
DNA. DNA from normal (N) and GRMD affected (A) males was
digested with BamHI (lanes 1 and 2), BglI (3 and 4), BglII
(5 and 6), HinCII (7 and 8) and MspI (9 and 10). Each
comparative digest demonstrates an RFLP.

formatted with an alternating set of normal and affected canine
samples digested with different restriction enzymes. This
region is extremely interesting because there is a polymorphism
for every enzyme examined thus far. In addition to those
indicated in Figure 4, there are other enzymes that have been
tested which demonstrate polymorphisms and these include
HinDIII, XbaI, SacI and PstI. In most cases, each pair of
enzyme digests suggests a small deletion of approximately 1200
to 1500 base pairs in the affected dogs. The exceptions are
shown here for both BamHI and MspI. The first pair, suggests
what looks like the loss of a restriction site, while the last
pair has obtained a new restriction site further downstream.
These polymorphisms segregate in litters of normal and affected
pups (Figure 5). These polymorphisms appear to be clearly
associated with the gene and probably very close to the GRMD
lesion. Going from left to right, we have the carrier mother
on the extreme left-hand side, and the affected father, on the
extreme right. In between, we have the offspring. The mother
is heterozygous for the probe bands with the lower band being
that associated with the normal allele, and the next lane shows
a normal male followed by an affected male, an affected female
(double intensity of hybridization), and so on, so you can see
that the specific polymorphism segregates with the disease.
Other enzymes listed above provide a similar pattern in this
pedigree as well as all of the other informative pedigrees.
This probe is within the 3' untranslated region of the
dystrophin gene transcript (Koenig et al., 1987). Since this
region does not code for the dystrophin protein, it must be
assumed that a specific defect associated with the disease must
produce a different etiology from that in the coding regions.
It may have something to do with a messenger RNA function other
than the reading frame of the protein itself. Perhaps this

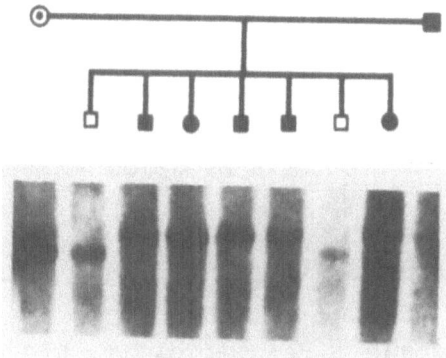

Fig. 5. Segregation of Disease-Specific Polymorphism in Canine Pedigree. DNA samples from the indicated animals were digested with HinDIII and probed with the same fragment of the human dystrophin cDNA as in Fig. 4.

would suggest something more to do with the stability of the messenger RNA.

What other methods of analysis, and particularly non-invasive methods, might be proposed to determine if the transplanted myoblasts are being included in newly regenerating muscle? Certainly, direct molecular analysis of injected myoblasts using RFLPs as described above, must require invasive techniques. These techniques will require a re-biopsy of in situ material from the transplantation muscle. What we'd like to do is use the dog model to develop something that would be non-invasive. Initial collaboration with Michel Fardeau and Hala Alameddine (Reference Alameddine and Fardeau this book) will assess the fluorescence-tagged myoblasts system in GRMD. While this method should prove to be a more sensitive method for detection of fusion than molecular markers, a biopsy will still be required. In contrast, we have begun to examine inclusion of a reporter gene in the donor myoblasts. This is similar to what Henry Klamut has done with the CAT gene (chloramphenicol transferase) fused to the human dystrophin gene. The reporter gene that we are proposing to use is the luciferase gene from the firefly fused to the human or canine dystrophin promotor. This reporter gene has been used as a chemo-luminescent tag which allows the experimental monitoring of photon production in the presence of ATP and luciferin (de Wet et al., 1987; Zhang et al, 1988). Thus, an in vivo assay would permit identification of fused donor myoblasts in regenerating tissue due to observation of the expression of a reporter gene product capable of producing light. It might be possible to use this reporter system in a donor myoblast injected into a topical muscle, at or very close to, the skin surface in a patient so that detection of the presence of 560 nm light would measure the progress of fusion without need for a biopsy. Joe Kornegay discussed the potential use of NMR to determine the clinical state of the muscle via analysis of phosphocreatine and inorganic phosphates (reference this book). It is hoped that either or a combination of both the chemo-luminescence and the NMR techniques may be developed into a possible system for non-invasive monitoring of fusion of donor myoblasts into newly regenerating muscle tissue.

REFERENCES

Bartlett, R.J., Pericak-Vance, M.A., Gilbert, J.R., Yamaoka, L.H., Herbstreith, M.H., Hung, W.-Y., Mohandas, T., Bruns, G., Laberge, C., Thibault, M.-C., Ross, D.A. and Roses, A.D. 1987. A new probe for the diagnosis of myotonic muscular dystrophy. Science, 235: 1648-1650.

Burmister, M. and Lehrach, H. 1986. Long-range restriction map around the Duchenne muscular dystrophy gene. Nature, 324: 582-585.

Koenig, M., Hoffman, E.P., Bertelson, C.J., Monaco, A.P., Feener, C. and Kunkel, L.M. 1987. Complete cloning of the Duchenne muscular dystrophy (DMD) cDNA and preliminary genomic organization of the DMD gene in normal and affected individuals. Cell, 50: 509-517.

Koenig, M., Monaco, A.P. and Kunkel, L.M. 1988. The complete sequence of dystrophin predicts a rod-shaped cytoskeletal protein. Cell, 53: 219-228.

Kornegay, J.N., Tuler, S.M., Miller, D.M. and Vann Camp, S.E. 1988. Muscular dystrophy in a litter of golden retriever dogs. Muscle and Nerve, 11: 1056-1064.

Monaco, A.P., Bertelson, C.J., Middlesworth, W., Colletti, C.-A., Aldridge, J., Fischbeck, K.H., Bartlett, R.J., Pericak-Vance, M.A., Roses, A.D. and Kunkel, L.M. 1985. Detection of deletions spanning the Duchenne muscular dystrophy locus using a tightly linked DNA segment. Nature, 316: 842-845.

Van Ommen, G.J.B., Verkerk, J.M.H., Hofker, M.H., Monaco, A.P., Kunkel, L.M., Ray, P., Worton, R., Wieringa, B., Bakker, E. and Pearson, P.L. 1986. A physical map of 4 million bp around the Duchenne muscular dystrophy gene on the human X-chromosome. Cell, 47: 499-504.

de Wet, J.R., Wood, K.V., DeLuca, M., Helinski, D.R. and Subramani, S. 1987. Fire fly luciferase gene: Structure and expression in mammalian cells. Molecular and Cellular Biology, 7: 725-737.

Zhang, J., Kornecki, E., Jackman, J. and Ehrlich, H. 1988. ATP secretion and extracellular protein phosphorylation by CNA neurons in primary cultures. Brain Research Bulletin, 21: 459-464.

THE ROLE OF THE *XMD* DOG IN THE ASSESSMENT OF MYOBLAST

TRANSFER THERAPY

Barry J. Cooper

Department of Pathology
NY State College of Veterinary Medicine
Cornell University, Ithaca, NY 14853

The success of myoblast transfer therapy can be assessed in a variety of ways. However, from the point of view of the utilization of experimental models in the development of this form of therapy, the methods used to determine its success depend on the particular biological questions that are being addressed.

The *xmd* dog lends itself to the study of biological phenomena important to the development of myoblast transfer for a number of reasons. There is compelling evidence that the canine disease involves a defect in the Duchenne locus, resulting in the absence of dystrophin,[1] and the clinical and pathologic expression of the disease is very similar to that of DMD in humans. In particular, the canine disease involves ongoing necrosis of muscle cells, infiltration by macrophages, and active regeneration (figure 1a). Normal myoblasts will therefore be injected into a necrotizing and regenerating tissue environment very similar to that encountered in young human patients. Furthermore, over the long term the *xmd* dog develops muscle fibrosis (figure 1b) and expresses clinical signs which would lend themselves to the assessment of the success of the therapy.

There are several biologic questions that need to be addressed in model systems in order to facilitate the development of this approach in man. For instance, what is the capacity of engrafted myoblasts to survive in an environment containing literally millions of macrophages that are presumably activated to phagocytose cellular debris? What is the ability of injected myoblasts to continue to function as satellite cells as well as to fuse with one another and with host muscle cells, and thus to continue to contribute to the repair of dystrophic muscle? How many myoblasts need to be injected? What is the capacity of injected myoblasts to migrate in the environment of active muscle necrosis and regeneration and, thus, how far apart do injections need to be made? What is the effect of cellular mosaicism on the synthesis and distribution of dystrophin, and to what degree can mosaic cells function and resist degeneration? Finally,

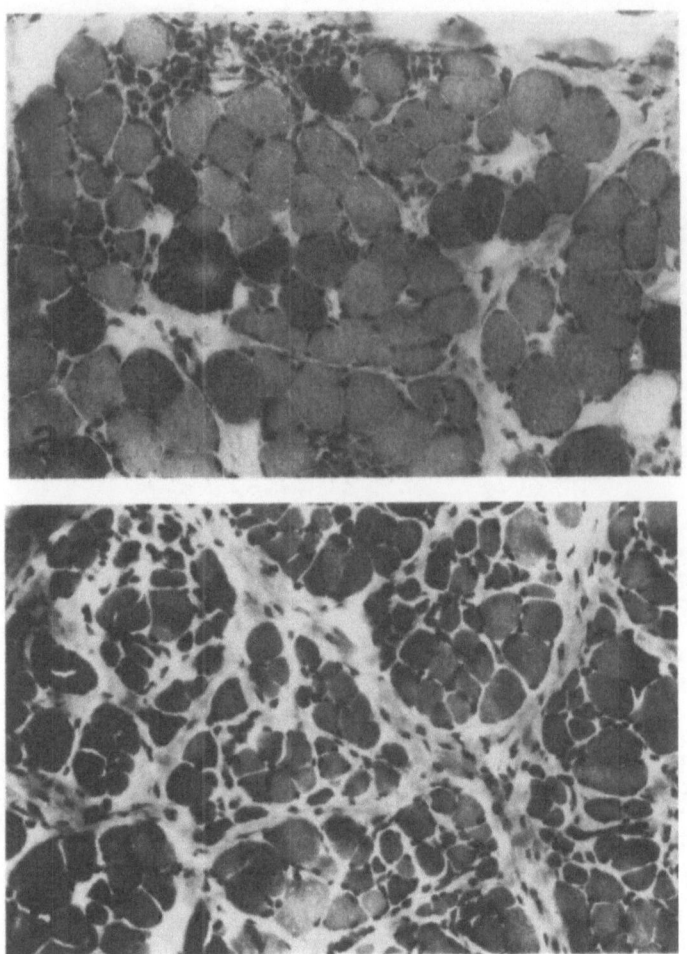

Figure 1. Muscle lesions from *xmd* dogs. a) early lesions are
characterized by the presence of darkly stained, acutely
injured fibers, histiocytic infiltrates, and clusters of
regenerating fibers. b) chronic lesions are characterized
by extensive fibrosis.

to what degree can successful engraftment with normal myoblasts
arrest the progression of dystrophic lesions and what clinical
benefits result? It is apparent that several of these
questions depend on the use of a model, such as the *xmd* dog,
in which the degeneration, regeneration, histiocytic
infiltration, and long term fibrosis characteristic of DMD are
recapitulated.

The ability of engrafted myoblasts to survive, to
proliferate locally, and to express dystrophin in dystrophic
host muscle can best be assessed by demonstrating the presence
of dystrophin in the treated muscle. This can be done by
western blotting and immunohistochemical staining. The
combination of these techniques should allow assessment of the
degree to which the protein is expressed and of its spatial
distribution. The ability of injected myoblasts to continue
to proliferate and contribute to long term repair of dystrophic

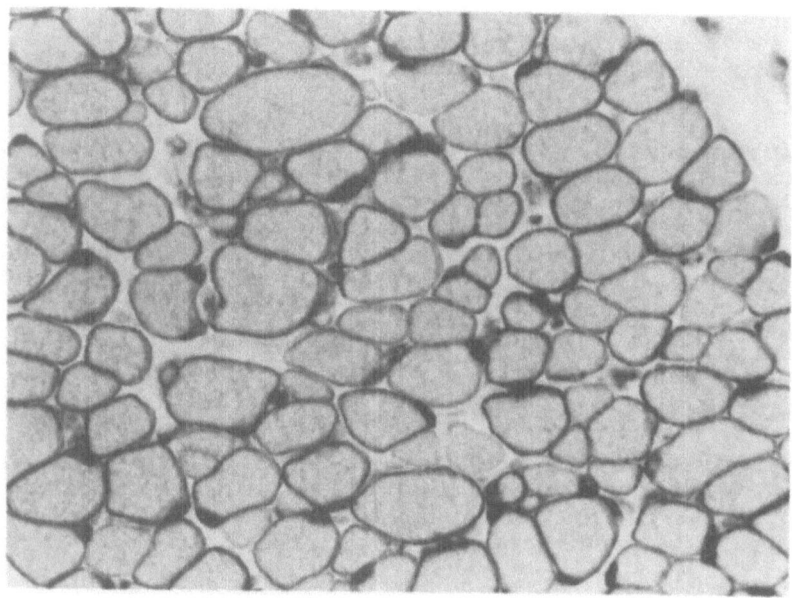

Figure 2. Muscle from an obligate carrier immunostained for dystrophin. Fibers lacking dystrophin are scattered throughout the muscle.

lesions can be assessed by studying changes in the expression of dystrophin over time. Thus, careful study of the expression of dystrophin in treated muscle would seem to offer the basic method for assessment of the fundamental questions posed above.

Assuming that grafted myoblasts can fuse with host muscle and express dystrophin, there are a number of criteria that can be addressed to assess the ability of the treatment to arrest the progression of the disease. Basic morphologic and morphometric techniques can be used to quantitate the number of necrotic cells, the number of calcium-positive cells, the size of muscle cells, the number of regenerating cells, and the degree to which fibrosis occurs. Again, these can be assessed at various times after the injection of normal myoblasts in order to study the long-term effects of treatment. The question of clinical improvement can probably be best addressed by measuring the strength of treated muscles. This will probably be done by measuring the force generated by stimulating treated muscles. To address the questions outlined here the engraftment of single, easily accessible muscles that lend themselves to the measurement of force generation would seem to be the best approach. Once the feasibility of the approach has been proven in this way, studies of "whole body" therapy can begin both in animal models and in human patients.

Another fundamental question that needs to be further studied is the way in which dystrophin is expressed and distributed in cells that are mosaic for nuclei competent or incompetent to direct the synthesis of dystrophin. We have begun to approach this question by studying obligate carrier females as models for cellular mosaicism. Briefly our results indicate that young carriers do express dystrophin in a mosaic pattern (figure 2), even in single muscle cells, and that this effect rapidly disappears as the animal matures. Whether recovery is due to incorporation of competent satellite cells or to intracellular diffusion of dystrophin requires further study.

In summary, the *xmd* dog is uniquely suited to exploratory studies of myoblast transfer therapy because the clinical and pathologic consequences of the absence of dystrophin so closely resemble those in man. The tissue environment into which normal myoblasts would be injected is similar to that of dystrophic human patients, and the progressive nature of the canine disease lends itself to study of the long-term benefits of the therapy.

ACKNOWLEDGEMENTS

The author wishes to thank Dr. Beth Valentine, Dr. Nena Winand and Ms. Eleene Gallagher for their role in the research that forms the basis for this discussion. This research was supported by a grant from the Muscular Dystrophy Association.

REFERENCE

1. B.J. Cooper, N.J. Winand, H. Stedman, B.A. Valentine, E.P. Hoffman, L.M. Kunkel, M. Oronzi Scott, K.H. Fischbeck, J.N. Kornegay, R.J. Avery, J.R. Williams, R.D. Schmickel, and J.E. Sylvester, The homologue of the Duchenne locus is defective in X-linked muscular dystrophy of dogs. Nature 334:154-156 (1988).

Discussion of Drs. Kornegay's, Bartlett's and Cooper's papers

Dr. Mendell: Thank you. We would welcome any questions related to the dog model.

Dr. Whalen: Dr. Cooper, the idea that you were advocating is that certain nuclei, the competent nuclei in the mosaic fibers of carriers, make dystrophin in their domain which is eventually mobilized into areas of dystrophin incompetent nuclei. Is that the idea?

Dr. Cooper: My idea is there must be a nuclear domain effect.

Dr. Whalen: Well, have you looked for that using dystrophin antibody staining in longitudinal sections?

Dr. Cooper: Yes, we have looked in serial sections and there are individual fibers that are mosaic as George Karpati has shown also in the Mdx mouse. This indicates that there is a nuclear domain effect along the length of muscle fibers.

My idea is that initially, dystrophin will localize near the competent nuclei. I also suspect that with time dystrophin may be able to be mobilized away from that site of synthesis. We need to look at that since it is important to know.

Dr. Whalen: The important information is that you have actually seen mosaic fibers.

Dr. Karpati: Barry, can I comment on the dystrophin-negative fibers in your heterozygote females, which seem to disappear with age. And as you pointed out, similar studies have been done by our group in the mdx mouse, and I guess also by Simon Watkins and Lou Kunkel. I think that approximately 30% of dystrophin-negative fiber segments that one finds in a very young animal are reduced to a very small percentage in older animals.

A finding demonstrated in the heart gives us an important clue as to what is taking place. In the heart of young mdx animals, there is approximately 30% of dystrophin-negative fiber segments on any given cross-section. That is in a ten-day-old animal. If you look at a 60-day-old animal, the percentage is practically the same. Our interpretation is that the reason the percentage remains static in the heart, as opposed to skeletal muscle, relates to lack of satellite cells in cardiac muscle.

Dr. Cooper: I would just like to comment that in the heart there is no mosaicism. The heart has single muscle cells with single nuclei. It lacks satellite cells, but it also lacks the opportunity for the protein to diffuse along the length of a single cell.

It is interesting, though, that in the dog, where cardiomyopathy is an absolutely consistent feature of this disease, carrier females, which lack dystrophin in 50% of the mass of their myocardium, do not develop any clinical evidence of cardiomyopathy. It would be very interesting to understand why.

Dr. Engel: The comments about the calcium in muscle fibers in your dog model reminds me of work we did with Roger Kula some years ago demonstrating calcium in the muscle fibers of Duchenne dystrophy by light and electron microscopy. With regard to this discussion it might be interesting to follow the calcium in the muscle by diphosphenate scanning. This could be added to the armamentarium of following the patients.

Dr. Miller: I have a question for Dr. Kornegay with respect to the monitoring of EMG. Is it possible that the high-frequency discharges observed in the dog have no functional correlate. Might it not be more important, in terms of EMG, to look at the compound muscle action potential amplitude as was suggested earlier? Most importantly, do you plan to measure muscular force? Also with regard to spectroscopy, could the altered inorganic phosphorus to phosphocreatine ratio reflect disuse and not be specifically related to the underlying disease?

Dr. Kornegay: Well, I think, in fact, they are non-specific features. We know that the pseudomyotonia is a non-specific feature. We also know that the spectroscopy findings of increased inorganic phosphorus relative to phosphocreatine is non-specific. The question is though whether this type of information could be used to assess treatment. I think if, in fact, we could show improvement in any of these parameters, that would be an indication that we were achieving success. In regard to the M-wave, this could represent a quantifiable measurement. I think evaluating repetitive nerve stimulation and assessing creatine kinase represent other good features to utilize.

Unidentified participant: Not disregarding the urgency of curing this disease in humans, I, as a cell biologist, find this dog model really compelling. It just seems so valuable to study this model in detail for a couple of years before rushing ahead to an incredibly difficult technical and ethical situation in the humans. I get the sense, in talking to different people around the room, that lots of us sort of feel that way, and I just want to urge tremendous support of this mode. Maybe we should be a little more cautious about rushing ahead with human trials in the next half-year or so.

Dr. Mendell: Does anybody have any preliminary data on myoblast transfer in the dog that they can briefly summarize?

Dr. Cooper: We have begun to do these experiments. The preliminary data was not presented because of time constraints. In these studies we chose to immunosuppress with a fairly modest dose of cyclosporin-A. We used very young animals, beginning at two weeks of age. This drastically altered the phenotype of the lesion, with much more mineralization. Frank, true mineralization, and formation of huge numbers of giant cells in the muscle, which we do not normally see. We have got to go back and re-evaluate whether we are going to immunosuppress these animals, or whether we are going to try to use purified, non-allergenic muscle cell populations.

Dr. Partridge: This is just to comment further on the mosaicism in the hearts of the mdx mouse. We confirmed the same sort of mosaicism in the female heterozygotes. In the heart, there is a sharp demarcation between positive and negative cells that runs right along the intercollated discs. This means that there is really no sharing of dystrophin along these fibers beyond the intercollated discs. That really bodes very badly for the use of any sort of analogous treatment in the heart condition.

Dr. Kornegay: We also have preliminary data on myoblast transplantation in dogs. Our results are perhaps somewhat different from Dr. Cooper's but in the end the final denominator is the same. We have injected one dog locally and two dogs systemically, and our dogs were immunosuppressed with cyclosporin. We did not encounter the same problems that Dr. Cooper has alluded to with regard to the propensity for seeing greater fiber mineralization, but we have not been able thus far to definitely prove that there actually has been dystrophin expression.

That is really the reason that I was trying to make points with regard to the definitive, fairly objective markers to be used. Whether it is degree of fiber mineralization, the actual morphometry of the lesion, or whether it is the degree of dystrophin expression, all are necessary in truly evaluating successful treatment.

THE DILEMMA OF MANIFESTING CARRIERS IN THE CONTEXT OF

MYOBLAST TRANSPLANTATION

Louis M. Kunkel

Division of Genetics and Howard Hughes
Medical Institute
The Children's Hospital
Boston, MA

What I would like to do for the end of this session is add a little caveat to myoblast transplantation. Basically, the caveat comes from work we have been doing on understanding the role of dystrophin in normal muscle function. When Eric Hoffman first came to me concerning collaboration with Terry Partridge on cell transplantation, I was a bit skeptical about the feasibility of the procedure. But as the experiments progressed, they looked more and more promising. The nagging question in both my mind and others was the occurrence of manifesting carriers. Having two cell populations (those capable of producing dystrophin and those incapable) one would expect that a carrier, much like a recipient of a transplant, would be expected to compensate for the abnormal cells and with time have only dystrophin producing cells. One way to address the issue would be to follow the fate of the two cell populations during embryonic development in an obligate carrier. These experiments would only be possible in an animal model, so we set up a collaboration with Simon Watkins to study cross-sections of muscle taken from obligate mdx carrier mice at different times during normal development. As might be predicted, early in development there were obvious dystrophin negative fibers among dystrophin positive ones. The number of negative fibers decreased with time such that by 60 days after birth, the vast majority of fibers were dystrophin positive.

However, the question still remained as to why there were manifesting carriers in the human population. Indeed, a very nice paper from Sugita's group in Japan demonstrated that some of these females could have as many as 30% of fibers dystrophin negative. This could even be in older women who had measurable symptoms of muscle weakness. So, why these females were unable to use dystrophin competent cells to compensate for the cells incapable of producing dystrophin remains a mystery. To further complicate the picture, while I was a visiting lecturer in Dallas, Jay Cook presented a wheelchair-bound female whom he wanted me to comment on. The woman was in her late teens and could have been easily mistaken for a Duchenne male of similar age. DNA deletion analysis had been performed on the woman and her completely normal identical twin sister. Both

Myoblast Transfer Therapy
Edited by R. Griggs and G. Karpati
Plenum Press, New York, 1990

were obligate carriers by having a detectable deletion on one of their X chromosomes. Why were the women so disparate in their phenotypic presentations? The first way to address the question was to have a muscle biopsy of the affected twin tested for dystrophin. When Eric Hoffman did the test, it was found that the affected twin had somewhere between 5 and 10% of normal levels of normally sized dystrophin. To determine if all fibers of muscle had uniform low levels or whether just a few fibers were expressing normal levels of dystrophin, cross-sections of her muscle were stained by Simon Watkins with anti-dystrophin antibodies and the staining of muscle observed by immunofluorescence. What Simon found was that there were small islands of dystrophin positive fibers in a sea of negative necrotic fibers. A plausible explanation for the results found in this twin was that there was a disproportionate inactivation of the normal X chromosome in this one twin relative to her sister. The observation of monozygotic twins with one expressing an X-linked disorder and the other not is not uncommon to the literature. Inactivation has always been used as an explanation of the observed phenotype. Our case was a good example of the phenomenon. The question, much like that seen in manifesting carriers, was why this woman's normal myoblasts were unable to correct the problem with developmental time. Taken in context with the myoblast transplantation proposed to be used with boys with DMD, what would be the prospects for correction be when we couldn't expect to achieve the levels of dystrophin positive myoblasts found in this woman by the transplants? In other words, if nature can't do it in the case of this twin girl and a manifesting carrier, how can we expect to have success with normal cells transplanted into a severely dystrophic boy?

The real problem may lie in our meager knowledge of normal dystrophin function in the tissues which normally express the protein. The most notable lack of knowledge is that of dystrophin's function in the nervous system. Could the unfortunate non-random inactivation have occurred in the nervous system? This might cause most nerves to be incapable of producing dystrophin in this tissue. With the inability of nerve to regenerate, this might be contributing to the phenotype observed in this female and other less severely affected manifesting females? This would imply that dystrophin expression in nerve might play a major role in the disease and that transplantation of myoblasts into Duchenne muscle would not address the problem.

In general, what I have tried to do with my caveat is to stress that research into normal dystrophin function in all the tissues which express it should be actively studied. This does not mean that experiments in myoblast transplantation are unwarranted; only that animal studies need further work before human work should proceed. I would also like to say that I would not be at all surprised if someone in the future will definitively show that lack of dystrophin expression in nerve plays a previously underrated role in the etiology and progression of the disease.

IMMUNOSUPPRESSIVE THERAPY IN DUCHENNE MUSCULAR DYSTROPHY:

CONSIDERATIONS FOR MYOBLAST TRANSFER STUDIES

Jerry R. Mendell

Department of Neurology
Ohio State University
Columbus, Ohio

The issue of immunosuppressive therapy in relation to myoblast transfer is extremely important. Several points must be addressed. The first to consider is whether immuno-suppression is necessary for this treatment to be successful. The second is the choice of drugs to be used and what impact they will have on the natural history of the disease exclusive of the potential benefits of the transplanted cells.

Myoblast transfer studies in mice indicate that hosts may exhibit tolerance to donor cells[1,2]. This may reflect a lack of the expression of Class 1 and Class 2 major histo-compatibility (MHC) antigens on normal muscle fibers[3,4,5]. The potential for problems, nevertheless, definitely exists since these antigens are expressed on muscle under a variety of conditions. Class 1 MHC antigens are observed on nearly all muscle fibers in polymyositis [3,4]. In dermatomyositis, Karpati and colleagues[3], found class 1 receptors confined to perifascicular areas suggesting that regenerating muscle fibers damaged by ischemia were reactive. In relation to the muscular dystrophies, and Duchenne dystrophy in particular, there have been discrepant observations,[3,4]. Karpati et al.[3] found that MHC gene products were expressed only in relation to regenerating fibers. In contrast, Appleyard et al.[4] reported consistently strong expression of Class 1 antigens in Duchenne and Becker dystrophic muscle. In regard to Class 2 MCH gene products, muscle from Duchenne dystrophy patients has shown no reactivity according to some studies[3,4], although HLA-DR was expressed according to a separate report[6]. Clarification of these differences awaits further studies.

Taken together the observations on MCH antigen expression might imply that transplantation of cultured myoblast cells to a foreign host would find an immunologically tolerant environment. This expectation is also supported by findings indicating a low-level expression of Class 1 MHC gene products in aneurally cultured human myoblasts and myotubes[3]. Experimental studies of implanted cultured myoblasts in mdx mouse muscle tend to support this point of view. In work presented by Karpati et al.[1,2], 2-10 day old mdx mice were injected with a suspension of pure, cultured, non-dystrophic myoblasts into multiple sites of the quadriceps muscle. The host cells showed expression of the transplanted nuclei as evidenced by dystrophin staining on the sarcolemma. These studies, done without immunosuppressive therapy, demonstrated that non-dystrophic transplanted myoblasts were able to fuse with host muscle cells and express dystrophin transcribed from the donor gene without signs of rejection for up to 60 days[1,2]. Expression of dystrophin and evidence for donor-host cell fusion was also observed in 5-27 day old mdx mice receiving mononucleated cells derived from neonatal mouse muscle[7]. These studies by Partridge et al.[7] employed athymic mdx/nude double mutants accepting a wide range of allografts and did not specifically address the issue of rejection. Studies by Law and colleagues[8] using the murine dystrophy model (dy^{2J}/dy^{2J}), also provided evidence for fusion following transplantation of donor myoblasts but these studies were also immunocompatible because of the common C57BL/6J background of donor and host.

Available studies to date, therefore, have not provided enough information to establish the unequivocal necessity for immunosuppression in future clinical trials. Nevertheless, a growing consensus supports the use of immunosuppressive therapy for myoblast transfer in human trials. A most important issue would, of course, be the choice of drugs to be used. Extrapolating from the organ transplant experience, cyclosporin emerges as a leading candidate for the choice of drugs. This may be especially true since preliminary experimental myoblast transfer studies demonstrated that cyclosporin treatment does not inhibit muscle regeneration and permits successful fusion of donor myoblasts within host muscle fibers. In experiments by Watt et al.[9], allografts of mononucleated cells became incorporated into host muscle fibers during cyclosporin treatment and were maintained 65 days after the end of immunosuppressive therapy. In studies by Law et al.[10] using donor myoblasts from limb muscle of Swiss Webster mice injected into dystrophic (C57BL/6J-$dy^{2J}dy^{2J}$) mice the efficacy of cyclosporin was also examined. Following myoblast injections into multiple lower extremity muscles and external intercostal muscles, cyclosporin-treated dystrophic muscles showed immature and developing myogenic cells that were of donor origin. Two months, however, did not appear long enough for all of the donor cells to mature but there was significant improvement in muscle structure in 5 of the 14 cyclosporin-treated dystrophic mice. In a control group not receiving cyclosporin, donor myoblasts did not appear to survive. These studies support that cloned myoblasts can survive and develop in immunoincompatible hosts in the presence of cyclosporin.

The final choice for immunosuppression in myoblast transfer studies will be the outgrowth of carefully controlled trials and comparison of various regimens. In organ transplantation, there has been a gradual de-emphasis on the use of cyclosporin as a sole agent because of nephrotoxicity[11,12,13]. Despite the meticulous monitoring of cyclosporin serum levels approximately 15% of patients acquire significant renal dysfunction[14,15]. A series of studies have demonstrated the potential benefits of combined immunosuppressive therapy employing a combination of low doses of cyclosporin, azathioprine and prednisone to avoid cyclosporin-induced nephrotoxicity[11,12,13]. Encouraging results

have emerged and in some studies maintenance doses of
cyclosporin as low as 3.5 mg/kg/day have been used in
combination with azathioprine 1.5 mg/kg/day and prednisone 10
mg/day[12]. These dosage levels are especially appealing since
neuromuscular disease patients with various disorders have
found these drug levels tolerable[16,17].

Decisions regarding choices of immunosuppressive regimens
will be aided by experience in the use of these drugs in
Duchenne muscular dystrophy patients. Not only are the effects
on myoblast fusion and gene expression important but also
interpretation of efficacy of myoblast transfer will require
the knowledge of the effects of any drug on the natural history
of the disease. In this regard only prednisone is well
characterized. Several studies have documented its
benefit[18,19,20] and most recently prednisone was shown to improve
strength and function of patients treated for six months in a
randomized, double-blind six-month trial[21]. This study included
103 boys with Duchenne muscular dystrophy ages 5 to 15 assigned
to one of three regimens: prednisone, 0.75 mg/kg/day (n=33);
prednisone, 1.5 mg/kg/day (n=34); and placebo (n=36). The
change in the score for average muscle strength between
baseline and evaluations at 1,2,3 and 6 months is shown in
Figure 1. Both prednisone groups had significant improvement
when first tested at one month. This improvement continued at
two and three months. The increase in average strength above
baseline at six months was similar to that at three months.
There was no significant difference between the prednisone
groups in the scores for average muscle strength. The placebo
group had a decline in its mean score of 0.21 $\pm$ 0.31, a
decrease identical to that observed in our control data on the
natural history of the disease[22].

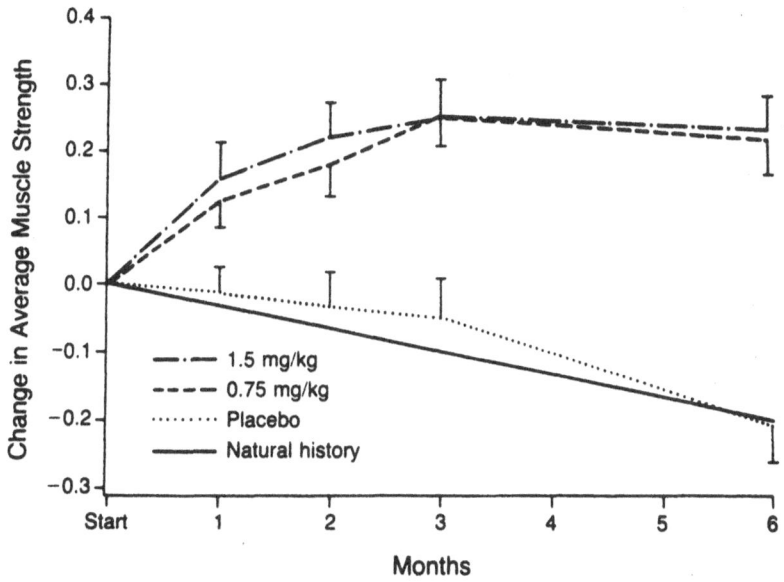

Figure 1. Change (mean $\pm$ SEM) in the score for average muscle
strength in the placebo and prednisone groups after the
initiation of their regimens.

Table 1. Clinical and Laboratory Measurements in the Study Groups after Six Months of Treatment.

Measurement	Placebo	Prednisone		P Values			
		0.75 mg/kg	1.5 mg/kg	Placebo vs. 0.75 and 1.5 mg/kg	Placebo vs. 0.75 mg/kg	Placebo vs. 1.5 mg/kg	0.75 kg vs. 1.5 mg/kg
		mean* (patients completing testing†)					
Muscle-strength score	5.80 (35)	6.23 (30)	6.25 (30)	0.0001	0.0001	0.0001	0.84
Leg-function grade	3.85 (35)	3.25 (31)	3.36 (33)	0.007	0.01	0.03	0.67
Contracture index	-2.36 (35)	-2.37 (31)	-2.25 (33)	0.53	0.89	0.22	0.19
Timed function							
Time to climb stairs (sec)	7.05 (18)	3.87 (23)	4.00 (24)	0.0001	0.0001	0.0001	0.74
Time to travel 9 m (sec)	9.68 (27)	6.81 (25)	7.04 (30)	0.001	0.003	0.005	0.77
Time to stand (sec)	6.17 (16)	4.15 (18)	3.43 (16)	0.0001	0.0002	0.0001	0.055
Lifting weights (kg)	1.17 (28)	1.88 (26)	2.13 (29)	0.0001	0.0001	0.0001	0.06
Pulmonary function							
Forced vital capacity (liters)	1.52 (34)	1.68 (29)	1.66 (28)	0.0001	0.0004	0.002	0.63
Maximal expiratory pressure (mm Hg)	14.74 (30)	17.32 (25)	18.19 (25)	0.001	0.01	0.001	0.43
Maximal voluntary ventilation (mm Hg)	40.64 (34)	45.49 (29)	41.24 (27)	0.17	0.03	0.80	0.09
Creatinine excretion (mg/24 hr)	190 (35)	248 (31)	261 (33)	0.0001	0.0001	0.0001	0.36
Log$_e$ creatine kinase	8.17 (32)	7.91 (29)	7.87 (30)	0.03	0.07	0.04	0.79

*Values represent the least-squares means derived from the average of the first two and last two visits.

†The following patients were dropped from the study: one patient taking placebo, two taking 0.75 mg of prednisone per kilogram, and one taking 1.5 mg per kilogram (never tested).

A compilation of the results for all measurements comparing values at six months is shown in Table 1. The prednisone-treated groups, 0.75 mg/kg/day and 1.5 mg/kg/day, responded nearly identically in each measurement. When the least-square means were used as a basis for comparison, muscle strength was found to be improved in both prednisone groups as compared to the placebo group (p<0.0001). Similar beneficial effects were observed during the timed functional testing. There was a decrease in the time needed to climb stairs (p<0.0001), to stand (p<0.0001), and to travel 9 meters (p<0.001). There was a small but significant improvement in the measures of pulmonary function, including the forced vital capacity and maximal expiratory pressure. The ability to lift a kilogram weight was increased in both prednisone groups. There were no significant changes in any group in the index of joint contracture after treatment.

Of particular interest was that urinary creatinine excretion was significantly increased in the prednisone-treated patients as compared with those given placebo (p<0.0001). Although changes in serum levels of creatine kinase were small, the prednisone-treated patients had values at the conclusion of the study that were significantly lower than those of the placebo-treated patients (p<0.03).

The incidence of side effects in the three study groups is shown in Tables 2 and 3. Weight gain was significantly greater in both prednisone groups than in the placebo group (p<0.0001). Approximately one third of the prednisone treated patients increased their base-line body weight by more than 20%. A cushingoid appearance was present in 55% of the patients who took prednisone 0.75 mg/kg and in 73% of those who took 1.5 mg/kg. Excessive hair growth was significantly more frequent in both prednisone groups (p<0.0005). A variety of other changes tended to occur more often in the prednisone-treated group but did not reach statistical significance (Table 3).

Table 2.   Weight Gain in the Study Groups

PERCENT OF WEIGHT GAINED ABOVE BASE LINE	PLACEBO	PREDNISONE	
		0.75 mg/kg	1.5 mg/kg
	no. of patients (% of group with gain)		
<5	18 (51)	3 (10)	3 (9)
5–10	10 (29)	3 (10)	5 (15)
11–20	5 (14)	13 (43)	14 (44)
>20	2 (6)	11 (37)	10 (32)

*Weight at the first visit was compared with weight at the last visit. In the group taking placebo, 36 patients started the trial and 1 dropped out; in the group taking 0.75 mg of prednisone per kilogram, 33 started, 2 dropped out, and 1 could not be weighed because of his wheelchair; and in the group taking 1.5 mg per kilogram, 34 started, 1 dropped out, and 1 could not be weighed.

The weight gain in the prednisone-treated patients was significant (placebo group vs. both prednisone groups, P<0.0001; placebo vs. 0.75 mg/kg, P<0.001; placebo vs. 1.5 mg/kg, P<0.001). There was no significant difference in weight gain between the prednisone groups (P = 0.93).

Table 3. Side Effects of the Study Groups.*

Side Effect	Placebo (N = 36)	Prednisone		P Values			
		0.75 mg/kg (N = 33)	1.5 mg/kg† (N = 33)	Placebo vs. 0.75 and 1.5 mg/kg	Placebo vs. 0.75 mg/kg	Placebo vs. 1.5 mg/kg	0.75 vs. 1.5 mg/kg
	no. with side effect (%)‡						
Behavioral change§	16 (44)	16 (48)	21 (64)	0.26	0.74	0.11	0.22
Cushingoid appearance	6 (17)	18 (55)	24 (73)	0.0001	0.002	0.0001	0.13
Gastrointestinal symptoms¶	17 (47)	18 (55)	20 (61)	0.32	0.54	0.27	0.62
Excessive hair growth	8 (22)	17 (52)	17 (52)	0.005	0.01	0.01	1.0
Acne	8 (22)	12 (36)	13 (39)	0.11	0.20	0.13	0.80
Easy bruising	5 (14)	1 (3)	2 (6)	0.11	0.14	0.29	0.56
Glycosuria	0	0	0				
Cataracts	0	0	0				

*Includes all side effects reported at visits at one, two, three, and six months.

†Of 34 patients assigned to this group, 1 dropped out before starting medication and thus cannot be included in comparisons of side effects.

‡Percentages do not add to 100 because of rounding.

§Includes hyperactivity, irritability, insomnia, euphoria, and depression.

¶Includes increased appetite, nausea, and stomach discomfort.

There are many potential mechanisms by which prednisone could improve muscle function, which need further study. It is noteworthy that steroid hormones can act as gene regulatory molecules[23,24,25]. One possibility is that a steroid-receptor complex could alter transcription resulting in an increased rate of synthesis of dystrophin or possibly an alternative membrane protein. Another consideration is that the beneficial effects of prednisone are mediated through its anti-inflammatory actions[26,27]. The inflammatory infiltrate seen in muscle tissue from patients with Duchenne dystrophy are composed principally of cytotoxic T cells and macrophages which may have a role in muscle fiber necrosis[28]. Studies attempting to sort out these possibilities are currently underway.

The rather significant effect of prednisone in altering the natural history of Duchenne muscular dystrophy exemplifies the importance of knowledge of the effect of immunosuppressive drugs on the natural history of the disease. In order to interpret myoblast transfer studies, similar data for cyclosporin and azathioprine, used alone or in combination with other drugs, will be essential in order to interpret the therapeutic effects of this exciting approach to gene therapy. At the present time the CIDD study group[20,21,22] is currently comparing the effects of three different drug regimens including: prednisone alone, prednisone and azathioprine and azathioprine alone. Future studies using cyclosporin are in the planning stages. We believe it is most important to have appropriate information available before these drugs are used as immunosuppressive agents for clinical trials in myoblast transfer.

REFERENCES

1. G. Karpati, Y. Pouliot, P. Holland, et al., Dystrophin is detected in mdx skeletal muscle fibers mosaicized by non-dystrophic myonuclei through the implantation of cultured myoblasts, Neurology 39 (Suppl. 1):153-154, (1989).
2. G. Karpati, Y. Pouliot, S. Carpenter, et al., Implantation of non-dystrophic allogenic myoblasts into dystrophic models of mdx mice produces "mosaic" fibers of normal microscopic phenotype, in: "Cell and Molec. Biol. of Muscle Devel." L.H. Kedes, F.E. Stockdale, eds., A.R. Liss, 973-985 (1989).
3. G. Karpati, Y. Pouliot, S, Carpenter, Expression of immunoreactive major histocompatibility complex products in human skeletal muscles. Ann. Neurol. 23:64-72 (1988).
4. S. T. Appleyard, M. J. Dunn, V. Dubowitz, et al., Increased expression of HLA ABL class 1 antigens by muscle fibers in Duchenne muscular dystrophy, inflammatory myopathy and other neuromuscular disorders, Lancet. 2:361-363 (1985).
5. D. A. Isenberg, D. Rowe, M. Shearer, et al., Localization of interferons and interleukin 2 in polymyositis and muscular dystrophy. Clin. Exp. Immunol. 63:450-458 (1986).
6. J. A. Zuk, A. Fletcher, Skeletal muscle expression of class II histocompatibility antigens (HLA-DR) in polymyositis and other muscle disorders with an inflammatory infiltrate, J. Clin. Path. 41:410-414(1988).
7. T. A. Partridge, J. E. Morgan, G. R. Coulton, et al., Conversion of mdx myofibers from dystrophin-negative to

dystrophin-positive by injection of normal myoblasts, Nature 337:176-179 (1989).

8. P. K. Law, T. G. Goodwin, M. G. Wang, Normal myoblast injections provide genetic treatment for murine dystrophy, Muscle and Nerve 11:525-533 (1988).

9. D. J. Watt, J. E. Morgan, T. A. Partridge, Long term survival of allografted muscle precursor cells following a limited period of treatment with cyclosporin A, Clin. Exp. Immunol. 55:419-426 (1984).

10. P. K. Law, T. G. Goodwin, H-J. Lin, Histoincompatible myoblast injection improves muscle structure and function of dystrophic mice, Transplant Proc. Vol. XX:1114-1119 (1988).

11. D. Fries, C. Hiesse, G. Santelli, et al., Triple therapy with low-dose cyclosporin, azathioprine, and steroids: long-term results of a randomized study in cadaver donor renal transplantation, Transplant Proc. Vol. XX:130-135 (1988).

12. S. Caglar, I. Tasdemir, C. Turgan, et al., The effect of triple therapy on graft outcome in renal transplantation, Transplant Proc. Vol. XX: 140-142 (1988).

13. C. Hiesse, P. Prevost, M. Busson, et al., Triple-drug regimen (TDR) of AZA + CsA + P: New definition of factors influencing kidney graft survival at a single center, Transplant Proc. Vol. 21: 1646-1647 (1989).

14. M. I. Lorber, S. M. Flechner, C.T. Van Buren, et al., Cyclosporin, azathioprine, and prednisone as treatment for cyclosporin-induced nephrotoxicity in renal transplant recipients. Transplant Proc. 17:282-284 (1985).

15. S. Kolkin, N. S. Nahman Jr., J. R. Mendell, Chronic nephrotoxicity complicating cyclosporin treatment of chronic inflammatory demyelinating polyradiculoneuropathy, Neurology 37:147-149 (1987).

16. S. A. Richard, M. D. Tindall, J. A. Rollins, et al., Preliminary results of a double-blind, randomized, placebo-controlled trial of cyclosporin in myasthenia gravis, N. Engl. J. Med. 316:719-724 (1987).

17. J. T. Kissel, R. J. Levy, J. R. Mendell, et al., Azathioprine toxicity in neuromuscular disease, Neurology 36:35-39 (1986).

18. D. B. Drachman, R. V. Toyka, E. Myer, Prednisone in Duchenne muscular dystrophy, Lancet. 2:1409-1412 (1974).

19. S. DeSilva, D. B. Drachman, D. Melitis, et al., Prednisone treatment in Duchenne muscular dystrophy: long-term benefit, Arch. Neurol. 44:818-822 (1987).

20. M. H. Brooke, G. M. Fenichel, R. C. Griggs, et al., Clinical investigation of Duchenne dystrophy. Interesting results in a trial of prednisone, Arch. Neurol. 44:812-818 (1987).

21. J. R. Mendell, R. T. Moxley, R. C. Griggs, et al., Randomized, double-blind six-month trial of prednisone in Duchenne's muscular dystrophy. N. Engl. J. Med. 320:1592-1597 (1989).

22. M. H. Brooke, G. M. Fenichel, R. C. Griggs, et al., Clinical investigation in Duchenne dystrophy: 2. Determination of the "power" of therapeutic trials based on the natural history, Muscle and Nerve 6:91-103 (1983).

23. G. M. Ringold, Steroid hormone regulation of gene expression, Ann. Rev. Pharmacol. Toxicol. 25: 529-566 (1985).

24. G. M. Ringold, K. R. Yamamoto, J. M. Bishop, Glucocorticoid-stimulated accumulation of mouse mammary tumor virus RNA: increased rate of synthesis of viral RNA, Proc. Natl. Acad. Sci. 74:2879-2883 (1977).

25. H. A. Young, T. Y. Shih, E. M. Scolnick, et al., Steroid induction of mouse mammary tumor virus: effect upon synthesis and degradation of viral RNA. J. Virol. 21:139-146 (1977).

26. J. E. Parrillo, A. S. Fauci, Mechanisms of glucocorticoid action on immune processes. Ann. Rev. Pharmacol. Toxicol. 19:179-201 (1979).

27. T. C. Cesario, L. Slater, W. J. Poo, et al., The effect of hydrocortisone on the production of gamma-interferon and other lymphokines by human peripheral blood mononuclear cells, J. Interferon Res. 6:337-347 (1986).

28. K. Arahata, A. G. Engel, Monoclonal antibody analysis of mononuclear cells in myopathies. I. Quantitation of subsets according to diagnosis and sites of accumulation and demonstration and counts of muscle fibers invaded by T cells, Ann. Neurol. 16:193-208 (1984).

GENERAL DISCUSSION

MONITORING CLINICAL SUCCESS: PHENOTYPIC TRANSFORMATION

Dr. Engel: Well, I think, with due respect to Lou Kunkel, the findings he showed do not really convince me that the carrier has neurogenic atrophy. We have looked at identical twin sisters who were carriers with antibodies against dystrophin. The more severely affected, manifesting heterozygote, of the twin sisters had much less dystrophin than the more affected one. The muscle of the dystrophic sister is logically like any other Duchenne dystrophy.

Dr. Karpati: A propos to that, could I just say that in every single pair of monozygotic female heterozygotes for the DMD gene that has been studied, the situation is exactly the same. One of the females is severely affected, the other is not affected at all. I think that there are about five or six pairs published, including one of Dr. Andrew Engel's with Dr. Gomez. This phenomenon is probably universal in monozygotic twins, and could be perfectly well explained by a bias of lyonization that might be enhanced by the twinning process itself.

Dr. Kunkel: George, I know of those cases, and actually, I know that this is a phenomenon that one sees. I do not know that you can say that every monozygotic twinning event gives rise to a severely affected twin, and a non-affected twin. Since you do not see twins in whom neither are affected.

Dr. Karpati: Alan Emery actually made a concentrated effort a few years ago to study this issue, and this was also his conclusion. Every monozygotic twin DMD carrier pair showed this pattern of dramatically unequal phenotypic expression.

Dr. Worton: Mr. Chairman, could I just add to that? It seems to me, about two years ago, there was one case published where they checked for exon activation by putting DNA into hybrids. I don't remember how extensive the data was, but it seems to me that it was consistent with a non-random exon-activation of one of the X-chromosomes.

Dr. Bonilla: We also have studied a set of monozygotic twins. At the DNA level, we did not detect any deletion. Dystrophin studies in the clinically affected twin showed about 32% of the fibers were dystrophin-negative. In the unaffected twin, the proportion of dystrophin-deficient fibers was only about 2%.

Dr. Mendell: Thank you.   We have another comment in the back from Dr. Bill Burns.

Dr. Burns: First of all, each organ transplant type is probably going to evolve its own best immunosuppressive therapy, and I don't think one can just look to bone marrow transplant, heart, lung, liver, or kidney, to get some ideas. It may well be, depending on whether or not you can use MHC-matched donors of myoblasts, that you will need less immunosuppression. Certainly, you would not use an agent which makes the natural disease of that patient worse to begin with. If there is some problem with cyclosporin in Duchenne's muscular dystrophy or in the animal model, this would certainly have to be established before trials are begun.   Whether one would go to combinations of immunosuppressive drugs, I think would depend on eventual findings.

My plea would be to keep whatever one does as simple as possible.   You may find that you need less immunosuppression than you do in other kinds of organ transplants, particularly if you are giving transplants in a sort of incremental manner. Myoblast transfer is going to be, I think, a multiple incremental transplant situation.

The plea would be to keep immunosuppressive therapy simple.   It may be that these children are going to be treated for a lengthy period of time, particularly if they are getting multiple transplants, which might be given over a period of years, or perhaps months.   Cyclosporin has a unique property, compared to all other immunosuppressives, and particularly if given prophylactically.   It allows one to build up suppressor cell populations, T-cell suppressor cells populations, that can be specific for the antigens that you are introducing.   That, in the end, is what you are after.   You're trying to allow suppressor cells to be built up over time, which will give you what is a normal kind of immunological tolerance.   Steroids do not do this.

So most people, today, would certainly start out with cyclosporin for these reasons, particularly in a prophylactic mode, and particularly if the graft may not be as immunogenic as some other kinds of transplants.

Dr. Kaufman: This morning, we heard about the potential for rapid expansion of satellite cells for use in transfer therapy.   I would like to make the point that what may also be going on when one is generating 10^{12} cells is a few transformations to tumor genesis.   When one is also immunosuppressing, the incidence of spontaneous tumor genesis is also increasing. It is important to consider these aspects in transfer therapy.

Dr. Cook: I have some comments and questions about strength measurement.   One of the thoughts I had is that sham injections on a control side might induce regeneration of the muscle fibers from the stimulus of the injection.   This muscle regeneration might produce some improvement.   In this regard we have looked at pre- and post-operative Duchenne dystrophy boys using the Cybex isokinetic system.   After thirty days we found no difference between the two sides.   In that case, I might point out that the Cybex system could not detect a difference.

Second of all, I think it is important to realize that if you are going to use any kind of fixed system for measurement you need to measure torque, and not just force-strength. You must take into consideration that the children are growing and will require a different lever arm. Another point with regard to natural history studies is that you need to measure various different muscle groups, to give some kind of idea about distribution of muscle weakness. I think the one thing that Ted Munsat's study and the CIDD study have demonstrated is that you can take away the noise of each measurement and be able to get something statistically meaningful using a small number of children over a short time period. Whereas if you look at the variation for each individual measurement it is more difficult. This is very important.

Lastly, I would ask, why would we need to reinvent the wheel? Why not the CIDD group evaluation system to assess therapy?

Dr. Mendell: Ted, do you want to discuss that issue?

Dr. Munsat: I think one of the problems with measuring strength by manual muscle testing is that yes, you can end up at the same place, but it requires many more patients. An example of this is in a typical manual muscle testing paradigm. To evaluate efficacy outcome, one needs about 100 to 120 patients. If you improve the sensitivity of your measurement instrument, and particularly if you reduce the range of the disability, you can reduce the range of the disability, you can reduce those numbers, literally, down to 20, and even less than that, with the same degree of power.

So, it has, in part, I think, to do with numbers and conservation of resources. I think that if and when transplants get going, there will be many questions that need to be answered. Each time we try to answer one of these questions, we cannot use 120 patients. There has to be a more efficient way to do it.

Dr. Rowland: This has been a wonderful and stimulating discussion, but there sure are a lot of balls in the air right now. We have got dogs and mice and dystrophin and genes and immunosuppression, and twins, and a lot of other things. I would like, if I may, to make one additional comment and ask a question about twins.

The editorial comment is this. Aside from the practicalities of growing myoblasts and inserting them, and making sure, as Bob Brown said, that it is safe, and making sure that patients do not get graft versus host, or drug toxicity, there is another aspect aside from the practicals and ethicals, and that is called money. This concern comes naturally to a departmental chairman. You could buy a lot of experimental dishes, and animal experiments for one clinical trial. Before we go wild with clinical trials, I would suggest that we change the name of tomorrow's session to human experimentation. It would be an entirely different public relations thing, and it would be an entirely different attitude for all of us. And I think that I am just picking up the theme that Steve Hauschka brought up before.

Now to go to the simpler question of twins. I would like to ask Lou Kunkel this question. There are sets of twins, they are all discordant. But they were reported because they were discordant, and that is why people thought they were interesting. We do not know how many twins there are, or whether there are any concordant twins. As a matter of fact, that is sort of a challenge now, if somebody finds a set of concordant twin carriers, they ought to be reported. It is just the reverse of the earlier situation. There seem to be more twin carriers than there are twin boys affected. But the question is, Lou, that you said you expected twins to be discordant. So if you could explain that, I think we would all be in your debt.

Dr. Kunkel: I did not say I expected it; I expected them not to be discordant. If one takes at face value, that the best one would ever expect injecting normal myogenic progenitor cells into a Duchenne muscle is 5-10% replacement, it is important to consider that these twins cannot even at that level compensate for that. What is the reason why? Now, George said, lyonization. Well, that is what everybody says, but that does not make sense. They still have 5% cells in their muscle which are dystrophin competent. They should be able to compensate, by everything we have heard today, everything that has come up and everything I know. Yet, they cannot. The question is why, and I will bet you there is a little more going on, in most twins, as well as potentially in the disease process itself. This is being seen and revealed by the twins, and I think it is worth pursuing. One might caution a little bit about expecting improvement from myoblast transfer based on what these twins represent. They have myoblasts that can work and do fine, and they do not improve.

Dr. Mendell: Earlier in the day the issue of monitoring rejection was omitted from the formal discussion. I have asked Dr. Andrew Engel to comment on this subject.

Dr. Engel: Well, Jerry, you asked me just before lunch and I really have not had too much time to think about it. I would like to say that there are good reasons to think why rejection may not be as much a problem as, for example, the kidney or cardiac transplant. The reasons, I think, were mentioned before: that one is transplanting a relatively homogeneous population of cells which express class 1 antigens, rather than class 2 antigens, and even that may disappear once a myoblast becomes incorporated into the parent fiber. Now, the question that remains is whether or not class 1 expression would reappear on the surface of the muscle fiber after the myoblast has fused or the satellite cell had fused with the recipient fiber. If there is an inflammatory environment, which could exist, or if part of the fiber is damaged, and regeneration takes place, then there may be mechanisms that activate the reexpression of class 1 on the surface of the fiber.

Now, we know from Dr. Blau's work that the nucleus governs a certain, or regulates the protein expression over a certain domain on the muscle fiber. Under these circumstances, a foreign haplotype may be expressed, which in turn could be a sensitization stimulus.

300

Now how to monitor for that? My guess would be by careful study of the biopsy before the transplant is taken, and in particular, looking for evidences of T cell-mediated cytotoxicity, such as invasion by muscle fibers by CD8-positive cells. Secondly, when the transplant is re-evaluated at a given point in time, then repeat the initial measurement.

The second point which might be made is that it is hypothetically possible that sensitization of HLA haplotypes could take place in the form of circulating antibodies. For example, pregnant women became sensitized to the presence of their own fetus, but nothing much happens, because presumably, the pathogenic antibodies do not cross the placental barrier. Nonetheless, pregnant women are an excellent source of HLA antibodies. Now a transplanted patient himself could generate HLA antibodies to the foreign haplotypes, and I think the caution to take here is that before a repeat injection is done, one should look for circulating antibodies in the recipient directed against the leukocyte of the donor.

Dr. Stedman: I recognize that tomorrow's session will have a largely clinical focus, and I just want to make a comment for the basic scientists here that I think will be germane.

It is not generally my style to come forth with such preliminary data on such short notice, but the data looks very exciting.

In looking at a number of Duchenne and Becker patients who are currently dying in respiratory failure, I'd like to very briefly turn some attention toward the diaphragm muscle, which I will follow up on tomorrow.

These observations gave us the idea that the diaphragm would be a very important muscle to look at closely in the Mdx mouse. To our great surprise, there is a substantial amount of fibrosis in Mdx at a stage when limb muscles have, at most, two to three times the normal control level of fibrosis. We decided to see whether we could follow this up by looking for a phenotype. Uncontestably, under the appropriate levels of stress, the ventilatory response of the Mdx mouse is markedly different from that of the normal control. I would be happy to discuss that further at any point today or tomorrow.

Dr. Rieger: I would like to pose a question to Dr. Kunkel. Coming back to these carrier twin girls, which have manifesting symptoms. Is it possible that the normal gene which is present in these girls underwent a mutation during embryogenesis? That is, some precursor cells of these myoblasts have mutated, affecting even the normal gene? Is something like that possible?

Dr. Kunkel: I do not think so. I'm not quite sure I follow you, but basically, the fact that dystrophin is expressed in the myofibers means to me that there is a normal locus. It is functioning normally.

Dr. Rieger: Okay, but the other one that did not have dystrophin, they should also have a copy of the normal gene, which should be operating in 50% of the nuclei.

Dr. Kunkel: That is correct, yes.

Dr. Rieger: What about if that normal copy is also now mutated?

Dr. Kunkel: All right but one would not predict the presence then of normal fibers, positive for dystrophin.

Dr. Fischman: Mr. Chairman, before we leave, I wanted to ask Dr. Karpati and Dr. Griggs if it would be possible before we leave today, to set aside perhaps a 15-minute discussion of what Stephen Hauschka raised. I get a sense that there is going to be a meeting tomorrow largely focused on moving forward on clinical trials. Many in the audience feel strongly that it is premature. I think that it would be time well spent, to at least have an open discussion on this. We have people from all over the world, here. It is going to be hard to reassemble them.

Dr. Fischman: Peter Law has suggested he is moving ahead. High hopes are being raised that this is a valid approach to proceed with. I am not so sure that myoblast transplantation is the proper route right now to proceed. One could make strong arguments that exploring the function of dystrophin and replacing those functions by a molecular biology approach, molecular genetic approach, might be a better one to focus on. I think it might be worth discussing. We have a lot of wise people in the audience, and I would hate to see them leave without having a chance to express their opinions on this.

Dr. Law: I really believe that I should respond to that, because, you know, I am one that is pushing it forward. I must say my attempts will be based on about twenty years of work on my part. In addition, there are other fine workers like Dr. Partridge, Dr. Karpati and others.

I do not believe that the number of cells will be a problem despite what some have suggested. I personally raised the fact that there may be some difficulty with cyclosporin dealing with children. Dr. Mendell earlier mentioned that there may be problems requiring multiple drugs for immuno-suppression. But he was quoting examples from liver and kidney transplants which represent organ transplants, not cell therapy.

I am trying to say, here, that myoblast transfer is ready for clinical trial. There is very little risk of injecting muscle cells into muscle ... a small muscle of the foot. It is not similar to transplantation of nerve cells into brain which carries greater risks. We are talking about muscle cells, even if they do not survive, it is not likely to cause severe damage to the patient. When you start thinking of all of the patients dying, their crying mothers and when you look at the science that we have, I believe we have covered all the necessary grounds including functional, structural, behavioral and genetic considerations.

Dr. Karpati: It is true that tomorrow's session will be dedicated fully to issues that are germane to the human experimentation. I think that Dr. Rowland has a good point that these are not really truly clinical trials, but part of experimentation. Since some of the people will not be there tomorrow, I think that Dr. Fischman's point is well-taken, that a take home message may be appropriate to develop now. I would like to encourage people to speak up, and state their case, not based on emotion, but on hard facts as to what are the areas that they feel that require particular clarification and attention in further animal experimentation or otherwise before a preliminary pilot study in humans can be contemplated. It would be helpful if people would identify particular areas they feel need more work.

I think that we still have time, approximately twenty minutes to do that. Dr. Charash, do you want to say a word about that now, or just tomorrow?

Dr. Charash: Well, I might respond to Dr. Fischman's question.

The Association is not prepared to undertake funding of any human projects yet. I mean, I think Mike Brooke said that the best time to prepare a study protocol is before you have a treatment. I think these preliminary questions should be raised. We are not unappreciative of the fact that the Retriever in North Carolina is called Golden.

Dr. Rowland did raise a question, I guess it is his orientation as chairman of the department to worry about money, but the MDA also worries a little bit about money too. If they can purchase substantially more insight into the physiology/pathology of a good model, they're prepared to do so. No commitment has been made for funding any project, and the papers that have been presented here are all being carefully thought about, and they will be carefully weighed before any decision is made.

Dr. Griggs: George, could I just make a comment reiterating how strongly I feel about these same points. I think Mike Brooke did say it well, and Leon Charash highlighted it. In order to be able to do a clinical trial of myoblast transfer, or other approach analogous to gene therapy, we should have a year or two of natural history myometry for single muscles in Duchenne dystrophy. There is really no data for this type of monitoring. Dr. Munsat is working on this and the CIDD group proposes to work on it. There is also no data on the effects of cyclosporin on the natural history of the disease. The importance of this point is demonstrated by the effects of prednisone. Had myoblast transfer been introduced using prednisone as immuno-suppressive therapy, the results would be confusing. We do not know, and are interested in finding out what the effects of azathioprine, of cyclosporin, and combinations thereof are on muscle strength and function. That is going to take several years to accumulate, even if we start today. We in the CIDD group are just in the process of designing these studies, so that even if we moved forward with all speed to implement a clinical trial, it is going to be several years before we are prepared to do so with the data that has to be attained.

Dr. Partridge: As Peter Law cited, experience is the basis for his decision to go forward with this type of therapy. I would like to present a more cautious view. It seems to me that there are a number of very easily identifiable problems associated with this myoblast implantation that are illustrated by the animal studies. These should be properly addressed. First, not the least of which is the scaling up to provide the number of cells for transfer. It seems to me that you should not hold back human trials necessarily, but there are simultaneous approaches that could go forward. One could identify the very best candidates for myoblast implantation and at the same time do further studies in animals, to establish the actual methods that will eventually be used for therapy.

Unindentified participant: I think we all agree that we as scientists, basic and clinical scientists, also have an obligation to the patients and their families in this issue. There has been a great deal of interest amongst the patient community about myoblast transfer. I think it is incumbent on us to keep them informed of what progress is made, and what our goals are, and also to let them know that we are working hard on the issue. Because there is no denying that we are all excited about the possibility of this. I think it is important to let the patients know, and their families, that we are working hard at it, that we are not treating it casually, and we are not holding back. We are going ahead with due caution, and with everything that needs to be done to make sure that that first transplant is safe.

Dr. Mendell: Well, I think we should turn the session back to the co-chairmen of the entire program, and let them make whatever other comments and announcements are appropriate. Thank you for this session.

Dr. Karpati: Thank you, Jerry, for this very informative and rich session. I think that the many diverse issues that were highlighted here this afternoon nicely set the stage for tomorrow when the same issues will be again dissected in much more detail. Various problems need to be examined in painstaking detail with regard to human experimentation in pilot studies and will be fully discussed. Those of you who will not be there will have to be kept informed through other channels. Hopefully, the meeting tomorrow will identify the problem areas where we need more clarification and more work. Nobody really expected an absolute consensus here. There are so many problems and they are so diverse, and many are so intangible, even after animal experimentation, that it is really difficult to develop an absolute consensus.

SECTION 7

IMPLEMENTATION OF HUMAN TRIALS

SYNOPSIS OF PROCEEDINGS OF THE WORKSHOP ON THE

IMPLEMENTATION OF MYOBLAST TRANSFER IN HUMANS

The third day of the International Conference on Myoblast Transfer Therapy was dedicated to the discussion of the numerous practical issues that required serious attention and solution before myoblast transfer could be contemplated in humans even as an experimental procedure.

Representatives from seven centers (Children's Hospital of Boston, Tufts New England Medical Center, University of Tennessee, Memphis, Montreal Neurological Institute, Stanford University, Duke University and University of Rochester) formed a panel under the chairmanship of Dr. Leon Charash and discussed the following questions with the participation of approximately 150 other delegates:

(1) Definition of the principal aims
(2) Factors that could determine therapeutic efficiency
(3) Outlining the initial goals

    (a) feasibility
    (b) safety
    (c) efficacy

(4) Issues concerning pilot experiments in single muscles

    (a) selection of patients
    (b) selection of donors
    (c) pretransfer evaluation of patients and donors
    (d) selection of injectable muscles
    (e) methods of myoblast production and quality control
    (f) method of myoblast purification (fluorescent-activated cell sorting)
    (g) injection techniques
    (h) immunological aspects, immunosuppression
    (i) control-injected muscles
    (j) end-point measurements (frequency, duration)
       · clinical (force, bulk, etc.)
       · laboratory (dystrophin, imaging)
    (k) ethical, legal and administrative considerations

After animated discussion of these topics, the following suggestions have received support:

(1) Injections of myoblasts into a single muscle of Duchenne patients should only be undertaken by multidisciplinary teams that can provide top-notch expertise

in all the required domains, particularly concerning myoblast generation and quality control, under proper circumstances.

(2) At this time, only the feasibility and safety of the procedure should be determined in phase 1 type of experiments in single muscles with appropriate controls.

(3) Further animal experiments on myoblast transfer, particularly in the dystrophin-deficiency dog model should continue to address many of the unanswered questions.

(4) Myoblast transfer should be viewed in a proper perspective of other possible therapeutic modalities that might emerge in the future for the treatment of DMD/BMD and other inherited muscle diseases.

Editors

PARTICIPANTS

[4]Valerie Askanas, M.D., Ph.D.
University of Southern
    California
Los Angeles

Richard Bartlett, Ph.D.
Duke University Medical Ctr.
Durham

Kurt G. Beam, Ph.D.
Colorado State University
Fort Collins

Richard Bischoff, Ph.D.
Washington University
St. Louis

Helen M. Blau, Ph.D.
Stanford University
Stanford, CA

Michael Brooke, M.D.
University of Alberta
Edmonton

[4]Robert H. Brown, Jr., M.D.
Massachusetts General
    Hospital
Boston

William Burns, M.D.
Johns Hopkins University
Baltimore

Stirling Carpenter, M.D.
Montreal Neurological
    Institute
Montreal

[2]C. Thomas Caskey, M.D.
Baylor College of Medicine
Houston

[1]Leon I. Charash, M.D.
Cornell Medical College
New York City

Nirupa Chaudhari, Ph.D.
Colorado State University
Fort Collins

Barry Cooper, DVM, Ph.D.
Cornell University
Ithaca

Charles P. Emerson, Jr., Ph.D.
University of Virginia
Charlottesville

[2]Andrew G. Engel, M.D.
Mayo Clinic
Rochester, MN

[2]Henry F. Epstein, M.D.
Baylor College of Medicine
Houston

Michel Fardeau, M.D., Ph.D.
INSERM
Paris

[2]Donald A. Fischman, M.D.
Cornell Medical College
New York City

Robert C. Griggs, M.D.
University of Rochester
Rochester, NY

Miranda Grounds, Ph.D.
Queen Elizabeth II
    Medical Center
Nedlands, Western Australia

Richard G. Ham, Ph.D.
University of Colorado
Boulder

Stephen Hauschka, Ph.D.
University of Washington
Seattle

[4]Terry Heiman-Patterson, M.D.
Thomas Jefferson University
Philadelphia

Paul Holland, Ph.D.
Montreal Neurological
    Institute
Montreal

Michael S. Hudecki, Ph.D.
State University of New York
Buffalo

George Karpati, M.D., FRCP
Montreal Neurological
    Institute
Montreal

Stephen Kaufman, Ph.D.
University of Illinois
Urbana

Joe N. Kornegay, DVM, Ph.D.
North Carolina State
    University
Raleigh

[3]Louis M. Kunkel, Ph.D.
Children's Hospital
Boston

Peter Law, Ph.D.
University of Tennessee
Memphis

Alexander Mauro, Ph.D.
Rockefeller Uiversity
New York City

[4]Jerry R. Mendell, M.D.
Ohio State University
    Hospitals
Columbus

Armand Miranda, Ph.D.
Columbia University
New York City

Jennifer Morgan, Ph.D.
Charing Cross and Westminster
    Medical School
London

[1]Theodore L. Munsat, M.D.
Tufts University
Boston

Terry Partridge, Ph.D.
Charing Cross and Westminster
    Medical School
London

Alan Peterson, Ph.D.
Ludwig Center for Cancer
    Research
Montreal

Peter Ray, Ph.D.
Hospital for Sick Children
Toronto

Francois Rieger, Ph.D.
Rockefeller University
New York City

[3]Lee Rubin, Ph.D.
Athena Neurosciences
South San Francisco

Frank Stockdale, Ph.D.
Stanford University
Stanford, CA

Richard Strohman, Ph.D.
University of California
Berkeley

Stephen Tapscott, Ph.D.
Fred Hutchinson Cancer Ctr.
Seattle

Gert van Ommen, Ph.D.
University of Leiden
Netherlands

Frank S. Walsh, Ph.D.
Institute of Neurology
London

Diana Watt, Ph.D.
Charing Cross and Westminster
    Medical School
London

Robert Whalen, Ph.D.
Pasteur Institute
Paris

Jan Witkowski, Ph.D.
Banbury Center
Cold Spring Harbor, NY

Ronald Worton, Ph.D.
Hospital for Sick Children
Toronto

---

[1]Member, MDA Medical Advisory
Committee
[2]Member, MDA Scientific
Advisory Committee
[3]Member, MDA Fellowship Review
Subcommittee
[4]MDA Clinical Director

mitogen, 150,151,153-156,
  195
molecular markers, 273
monoclonal antibody
  (*McAb*), 41-48
    24.1D5, 42
    ERIC-1, 46
    5.1H11, 42,43,46,70,
      167,168
    UJ13A, 46
mosaic muscle fibers, 36,
  81-83,86,90,96,102
    genetic rescue of,
      173-182
mosaicized host fibers,
  70-72
mouse chimeras, 173-181
    dystrophia muscular-
      is (*dy* and *dy*^{2J}),
      173
    mdg, 173
    mdx, 173
mouse fibroblast cell line
  10T½ cells, 3,5,6,50,
    55,206-208
murine dystrophy, 242,260,
  288
muscle *GF*, 103
muscle gene products,
  localization of, 167-172
muscle mass, 235,239
muscle precursor cells
  (*mpc*), 35-39,89-93,101,
    114
muscle regeneration, 103
muscular dysgenesis (*mdg*),
  139, 175,182
    reversion of, 144
    strain of mice, 131
mdg/mdg mice, 140-142,175-
  180
muscular dystrophy (*mdx*),
  180
    mdx/mdx chimeras,
      181,182
myasthenia gravis, 33
myc, 3
myd, 48,205
myf-5, 3,205
myoblast fate, 201
myoblast implantation,
  89-93
myoblast injection, 76-
  78,80-83,86
myoblast lineage, 48-51
myoblast membrane, 47,48,
  51
myoblasts, 3,7-10,14,15,
  26,32,44,47-55,57,59,
  101,131,215,216,260
    chicken, 193

myoblasts *(cont.)*
    cultured, 31
    embryonic, 31
    human, 42,43,97
    proliferation of, 97,
      98,101
    purification of, 70,
      97
myoblast transfer, 7,8,10,
  75,97,180,219,220,
  233,238,246,263,273,
  285,286,304
    allogeneic, 32,33,57,
      127
    histoincompatible,
      31-34
    principles and prac-
      tice, 69-72
myo-D, 3-6,26,27,28,205-
  208
    myoD1, 48,50
myofiber mineralization,
  269
myofibrils, 13
myogenesis, 3,93,102,176,
  189
    terminal, 47-51
myogenic cells, 7,28,35-
  36,49,50,147,159,
  187,190
myogenin, 3,26,48,205,206
myometry, 227
myopathy, 89-93
myosin, 3,6,25,170,172,
  189,206
myosin ATPase, 78
myositis, 33
myotendinous junction, 19
myotonic dystrophy, 237
myotube formation, 51,52
myotubes, 42,44,101,131-
  137,161-165,190
    dysgenic, 139,142-
      145,187,188
    embryonic, 31
    postmitotic, 49
    transplantation, 85

necrosis, 13-15,33,70,86,
  91-93,96,126,154,159,
  246,293
neural cell adhesion
  molecule (*N-CAM*), 14,
  27,70,74,97,127,167-172
    as target cell sur-
      face antigen, 41-46
neuroblastoma line
  B50, 3
neuropathy, 116
neutrophils, 14
nonlethal injury, 13